T0295359

Channel Coding in 5G New Radio

This book provides a comprehensive coverage of major channel codes adopted since the third generation of mobile communication. Modulation schemes suitable for 5G mobile communications are also described based on key new radio application scenarios and performance requirements.

It covers low-density parity check (LDPC) codes, polar codes, tail-biting convolutional codes (TBCCs), and turbo codes. Outer codes and a few advanced coding and modulations are also discussed. In addition, it includes a detailed illustration of each channel coding scheme such as the basic code structure, decoding algorithms, performance evaluation, and complexity analysis. The book offers insights on why and how channel codes are designed and developed in standardization organizations, which significantly facilitates the reading and understanding of 5G channel coding technologies.

Channel Coding in 5G New Radio will be an essential read for researchers and students of digital communications, wireless communications engineers, and those who are interested in mobile communications in general.

Jun Xu is a Senior Expert at the Algorithm Department of ZTE Corporation. With more than 20 years of experience at ZTE Corporation, Xu' s expertise is in channel coding, especially in design and performance evaluation of low-density parity check (LDPC) codes. Xu is also the inventor of compact protomatrices for LDPC of WiMAX and 5G.

Yifei Yuan is the Chief Expert of China Mobile Research Institute. Dr. Yuan graduated from Tsinghua University and Carnegie Mellon University. He specializes in the research and standardization of key air-interface technologies for 3G, 4G, 5G, and 6G mobile networks. He has more than 20 years of experience at Bell Labs, ZTE, and China Mobile.

Channel Coding in 5G New Radio

Jun Xu and Yifei Yuan

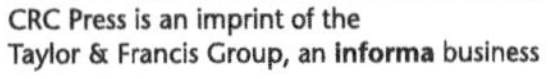

First published 2023
by CRC Press
6000 Broken Sound Parkway NW, Suite 300, Boca Raton, FL 33487-2742

and by CRC Press
4 Park Square, Milton Park, Abingdon, Oxon, OX14 4RN

CRC Press is an imprint of Taylor & Francis Group, LLC

© 2023 Jun Xu and Yifei Yuan

Reasonable efforts have been made to publish reliable data and information, but the author and publisher cannot assume responsibility for the validity of all materials or the consequences of their use. The authors and publishers have attempted to trace the copyright holders of all material reproduced in this publication and apologize to copyright holders if permission to publish in this form has not been obtained. If any copyright material has not been acknowledged please write and let us know so we may rectify in any future reprint.

Except as permitted under U.S. Copyright Law, no part of this book may be reprinted, reproduced, transmitted, or utilized in any form by any electronic, mechanical, or other means, now known or hereafter invented, including photocopying, microfilming, and recording, or in any information storage or retrieval system, without written permission from the publishers.

For permission to photocopy or use material electronically from this work, access www.copyright.com or contact the Copyright Clearance Center, Inc. (CCC), 222 Rosewood Drive, Danvers, MA 01923, 978-750-8400. For works that are not available on CCC please contact mpkbookspermissions@tandf.co.uk

Trademark notice: Product or corporate names may be trademarks or registered trademarks and are used only for identification and explanation without intent to infringe.

Published through arrangement with the original publisher, Posts and Telecom Press Co., Ltd.

Library of Congress Cataloging-in-Publication Data
Names: Xu, Jun (Telecommunications engineer), author. | Yuan, Yifei, author.
Title: Channel coding in 5G new radio / Jun Xu and Yifei Yuan.
Description: First edition. | Boca Raton, FL : CRC Press, 2023. |
Includes bibliographical references. |
Summary: "This book provides a comprehensive coverage of major channel codes adopted since the 3rd generation of mobile communication. Modulation schemes suitable for 5G mobile communications are also described based on key New Radio application scenarios and performance requirements. It covers low density parity check (LDPC) codes, Polar codes, tail-biting convolutional codes (TBCC) and Turbo codes. Outer codes and a few advanced coding and modulations are also discussed. In addition, it includes detailed illustration of each channel coding scheme such as the basic code structure, decoding algorithms, performance evaluation and complexity analysis. The book offers insights on why and how channel codes are designed and developed in standardization organizations, which significantly facilitates the reading and understanding of the of 5G channel coding technologies. Channel Coding in 5G New Radio will be an essential read for researchers and students of digital communications, wireless communications engineers, and those who are interested in mobile communications in general"—Provided by publisher.
Identifiers: LCCN 2022026218 (print) | LCCN 2022026219 (ebook) |
ISBN 9781032372778 (hardback) | ISBN 9781032372785 (paperback) | ISBN 9781003336174 (ebook)
Subjects: LCSH: 5G mobile communication systems. | Radio frequency modulation. | Coding theory.
Classification: LCC TK5103.25 .X83 2023 (print) | LCC TK5103.25 (ebook) |
DDC 621.3845/6–dc23/eng/20220906
LC record available at https://lccn.loc.gov/2022026218
LC ebook record available at https://lccn.loc.gov/2022026219

ISBN: 978-1-032-37277-8 (hbk)
ISBN: 978-1-032-37278-5 (pbk)
ISBN: 978-1-003-33617-4 (ebk)

DOI: 10.1201/9781003336174

Typeset in Minion
by codeMantra

Contents

Foreword

DIGITAL SIGNALS CANNOT BE reliably transmitted without channel coding. Channel coding topics in previous generations of mobile communications were dominated by European and American companies. For instance, the convolutional codes in the second generation were invented by Bell Labs in the USA. Convolutional codes and Reed–Solomon codes, a type of block codes invented by professors at the University of Southern California, formed a golden combination and broke the performance records multiple times. The turbo codes adopted in the third-generation mobile communications were invented by French researchers and refined by companies in Europe, Japan, and the USA. The performance of turbo codes is significantly better than that of convolutional codes. The mainstream standards of the fourth-generation mobile communications continued to use turbo codes as the channel coding scheme, with certain enhancements which were still proposed by European and American companies. This situation has changed significantly in the fifth-generation (5G) mobile communications. Regarding the two most important channel codes for 5G: LDPC codes and polar codes, infrastructure vendors of China have played a significant role and contributed about 50% of key technologies and essential patents. Specification work of 5G channel coding in the first stage was completed in December 2017. The publication of this book is quite in time. The book is not only useful for researchers and developers of 5G systems, but it also serves as a testimony of China's advancement in the state-of-the-art technology.

Channel coding is a highly specialized field which requires a solid background in mathematics and special tools/methods for analysis. This is in contrast to many other fields of digital communications. Books on channel coding would often fall into two extremes. In the first extreme, the contents contain too many obscure mathematic formulae and derivations. Although quite precise and rigorous, it is very difficult to comprehend, except for people who are specialized in channel coding. The other extreme is that the style is too amateurish in level, without enough equations to precisely describe important concepts. In the latter case, while readers may get a glimpse of channel coding, they would hardly grasp the design principles. With these two extremes, engineers working on channel coding often have to look up the prosaic specifications and would not be able to develop the problem-solving skills regarding channel coding, which leads to low output of product development. This book strikes a good balance between these two extremes and begins with the basic concepts, reinforced by reasonable numbers of equations and figures/tables. Plain language is used to describe the main design considerations and optimization directions, in particular for LDPC codes and polar codes. The book can be used by both engineers and university researchers/students.

Zemin Yang

Vice Chair and General Secretary of China Communications Standardization Association

Preface

TELECOMMUNICATION TECHNOLOGIES ARE QUICKLY evolving. In September 2007, 3GPP completed its initial specification work on 4G-LTE which is a fourth-generation mobile communications system based on OFDM and MIMO. In December 2010, LTE was enhanced to LTE-advanced where carrier aggregation and advanced MIMO features were incorporated. In 2012, higher-order modulation was included in LTE-advanced. At the same time, 3GPP started to consider next-generation mobile technologies. After several years of preparation, a study item of new radio (NR) for 5G mobile communications was approved by 3GPP in March 2016 and entered the standardization stage in March 2017. Through the hard work of many companies, the first release of 5G NR specifications (e.g., for eMBB) was completed by 3GPP in December 2017.

Compared to 4G-LTE, many new technologies, such as LDPC codes, polar codes, massive MIMO, and non-orthogonal multiple access (NOMA), have been introduced to 5G NR. To fulfill the demands of the operators, 3GPP is very strict when selecting technologies for standardization. Turbo codes invented in 1993 perform relatively well for low to medium data rates in 3G WCDMA and 4G LTE. However, their performance is significantly inferior to that of LDPC codes in the case of wide bandwidth and high data rate (e.g., 20 Gbit/s, a key performance indicator of 5G NR). Convolutional codes, invented in 1955, were adopted in 2G, 3G, and 4G mobile communications. However, their performance is not competitive and has to relinquish the role to the newcomer polar codes. In the future (e.g., 6G), the further development of mobile communications may spur the introduction of other channel coding schemes. In light of this, design principles, application scenarios, complexities, and performances of those codes are also discussed in this book.

This book features the following: First, the content not only explains the specifications of 5G NR but also has in-depth descriptions of the theories behind the specifications. Hence, it is suitable for both wireless engineers and researchers/students in academia; second, the book contains abundant theoretical performance analysis, as well as computer simulation results; third, the book has a broad coverage, including mainstream channel coding schemes adopted by the industry and the new coding schemes that gain high interest from the academia. Due to the pervasive application of LDPC codes and polar codes in NR, their two codes are described in more detail in this book.

This book was written by Jun Xu, Yifei Yuan, *et al.* of ZTE Corporation. Specifically, Chapter 1 was written mainly by Yifei Yuan; Chapter 2 (e.g., LDPC codes) was written by Jun Xu, Yifei Yuan, Meiying Huang, Liguang Li, and Jin Xu; Chapter 3 (e.g., polar codes) was written by Focai Peng, Mengzhu Chen, and Saijin Xie; Chapter 4 (e.g., convolutional codes) and Chapter 5 (e.g., turbo codes) were written mainly by Jin Xu and Yifei Yuan; Chapter 6 (e.g., outer codes) was written mainly by Liguang Li and Jun Xu; Chapter 7 (advanced codes) was written by Yifei Yuan, Mengzhu Chen, and Focai Peng. The entire book was planned and managed by Yifei Yuan and Jun Xu. In addition, some other colleagues and Professor Kewu Peng have provided constructive suggestions and technical support. Their help is very appreciated.

The authors also thank the support of the research project of the National "863 Plan" of China: "Gbps wireless communications with adaptive and high efficient channel coding technologies" (code number 2006AA01Z71) and the project "Important National Science & Technology Special Project" of China: Key technology study on IMT-Advanced for channel coding with link adaptation (code number: 2009ZX03003-001-04). The authors thank the member companies of the "IMT-2020 (5G) new channel coding & modulation working group" for their technical contributions!

Authors

Authors

Jun Xu is a Senior Expert at the Algorithm Department of ZTE Corporation. With more than 20 years of experience at ZTE Corporation, Xu's expertise is in channel coding, especially in design and performance evaluation of low-density parity check (LDPC) codes. Xu is also the inventor of compact proto-matrices for LDPC of WiMAX and 5G.

Yifei Yuan is the Chief Expert of China Mobile Research Institute. Dr. Yuan graduated from Tsinghua University and Carnegie Mellon University. He specializes in the research and standardization of key air-interface technologies for 3G, 4G, 5G, and 6G mobile networks. He has more than 20 years of experience at Bell Labs, ZTE, and China Mobile.

Abbreviations

Abbreviation	Full name
π/2-BPSK **pi/2-BPSK** **Pi/2-BPSK**	π/2-Binary Phase Shift Keying
128QAM	128-Quadrature Amplitude Modulation
16QAM	16-Quadrature Amplitude Modulation
1G	First-generation mobile communication
256QAM	256-Quadrature Amplitude Modulation
2G	Second-generation mobile communication
32QAM	32-Quadrature Amplitude Modulation
3G	Third-generation mobile communication
3GPP	Third-Generation Partnership Project
4G	Fourth-generation mobile communication
5G **5G NR**	Fifth-generation mobile communication
64QAM	64-Quadrature Amplitude Modulation
APP	A Posteriori Probability
AWGN	Additive White Gaussian Noise
B-DMC	Binary-Discrete Memoryless Channel
BEC	Binary Erasure Channel
BER	Bit Error Rate
BG	Base Graph
BICM	Bit Interleaver, Coding and Modulation
BLER	BLock Error Rate
BP	Belief Propagation
bps	bits per second

bps/Hz	bps per Hz
BPSK	Binary Phase Shift Keying
BSC	Binary-Discrete Symmetric Channel
CA-PC-Polar	CRC-Aided-Parity Check-Polar code
CA-Polar	CRC-Aided Polar code
CA-SCL	Successive Cancelation with List for CRC-aided Polar code
CDMA	Code Division Multiple Access
CNU	Check Node Unit Constraint Node Unit
CQI	Channel Quality Indicator
CRC	Cyclic Redundancy Check
D-CA-Polar **D-CRC-Polar** **Dist-CA-Polar**	Distribution CRC-Aided Polar code
DCI	Downlink Control Information
DE	Density Evolution
eMBB	enhanced Mobile Broadband
EPA	Extended Pedestrian A model
ETU	Extended Typical Urban model
EXIT chart	EXtrinsic Information Transition chart
FAR	False Alarm Rate
FDMA	Frequency Division Multiple Access
FEC	Forward Error Correction
FER	Frame Error Rate
FFT	Fast Fourier Transform
FFT-BP	Belief Propagation with FFT
GF	Galois Field
GSM	Global System of Mobile communications
HARQ	Hybrid Automatic Retransmission reQuest
HSDPA	High-Speed Downlink Packet Access
HSPA	High-Speed Packet Access
HSUPA	High-Speed Uplink Packet Access
IoT	Internet of Things
IR	Incremental Redundancy

KPI	Key Performance Indicator
LDLC	Low-Density Lattice Code
LDPC	Low-Density Parity Check
LLR	Log-Likelihood Ratio
Log-BP	Belief Propagation in Logarithm domain
Log-FFT-BP	Belief Propagation with FFT in Logarithm domain
Log-MAP	Maximum APP in Logarithm domain Maximum A Posteriori Probability in Logarithm domain
LTE	Long-Term Evolution
MAC	Media Access Control
Max-Log-MAP	Maximum–Maximum APP in Logarithm domain Maximum–Maximum A Posteriori Probability in Logarithm domain
ML	Maximum-Likelihood
mMTC	massive Machine-Type Communication
MAP	Maximum APP Maximum A Posteriori Probability
MBB	Mobile Broadband
MCS	Modulation and Coding Scheme
MIMO	Multiple Input Multiple Output
min-sum	minimum-sum-product algorithm
NR	New Radio access technology
OFDM	Orthogonal Frequency Division Multiplexing
OFDMA	Orthogonal Frequency Division Multiple Access
PC-Polar	Parity Check-Polar code
PCCC	Parallel Concatenation Convolutional Code
PDCCH	Physical Downlink Control Channel
PDF	Probability Density Function Probability Distribution Function
PDSCH	Physical Downlink Shared Channel
PER	Package Error Rate
PM	Path Metric
PRB	Physical Resource Block
PUCCH	Physical Uplink Control Channel

PUSCH	Physical Uplink Shared Channel
QAM	Quadrature Amplitude Modulation
QC-LDPC	Quasi-Cyclic LDPC
QLDPC	q-array LDPC Non-binary LDPC
QoS	Quality of Service
QPP	Quadratic Permutation Polynomial
QPSK	Quadrature Phase Shift Keying
RAN	Radio Access Network
RE	Resource Element
RV	Redundancy Version
SC	Successive Cancelation
SCL	Successive Cancelation List
SINR	Signal-to-Interference-plus-Noise Ratio
SNR	Signal-to-Noise Ratio
SOVA	Soft Output Viterbi Algorithm
TB	Transport Block
TBS	Transport Block Size
TBCC	Tail-Biting Convolutional Code
TDMA	Time Division Multiple Access
TD-SCDMA	Time Division-Synchronous Code Division Multiple Access
UCI	Uplink Control Information
UE	User Equipment
UMB	Ultra-Mobile Broadband
UMTS	Universal Mobile Telecommunication System
URLLC	Ultra-Reliable and Low-Latency Communication
VN	Variable Node
VNU	Variable Node Unit
WCDMA	Wideband Code Division Multiple Access
WiMAX	Worldwide Interoperability for Microwave Access

CHAPTER 1

Introduction

Yifei Yuan

FLASHING BACK TO 2012, the "Flower" of 5G mobile communications [1] as shown in Figure 1.1 blossomed around the world. Over the past few years, there have been quite a number of technological achievements in 5G air-interface standards. Among them is channel coding, in particular low-density parity check (LDPC) codes [2] and polar codes [3].

In this book, we try to provide design principles and reveal various key ingredients of channel codes for mobile communications. The first chapter of the book touches upon the evolution of previous generations of mobile communications, followed by the system requirements for 5G new radio (5G NR) [4,5]. Then, channel codes for 5G NR are briefly described. In the end, the aims and the structures of the chapters of this book are presented (Figure 1.3).

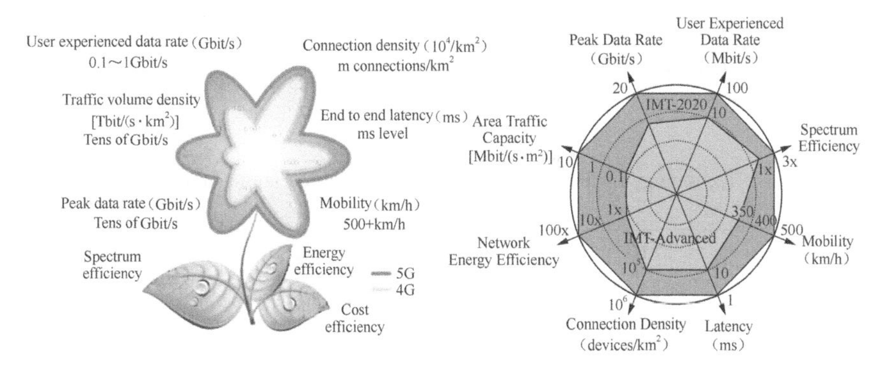

FIGURE 1.1 "Flower" of 5G [1] and spider chart of 5G KPIs as opposed to 4G [5].

DOI: 10.1201/9781003336174-1

1.1 PREVIOUS GENERATIONS OF MOBILE COMMUNICATIONS

Mobile communications (also called cellular communications) trace back to their origin from an invention by AT&T Bell Laboratories in 1968. The layout of cellular communications resembles the cells of natural beehives. These cells are interconnected with each other, forming a large network with continuous coverage. Since the frequency resources can be reused between cells, the capacity of the entire network can be significantly improved. Based on this principle, the first cellular call was made by Motorola Lab in 1973, enshrining the company as one of the pioneers in the cellular era. Over the last 40 years, mobile communications, featuring convenience, mobility, and flexibility, have undergone tremendous growth and drastically changed the lifestyle of human beings, penetrating more than 90% population in many countries. Mobile technologies have evolved from first to fourth generation, as listed in Table 1.1.

First-generation mobile communication (1G) is frequency division multiple access (FDMA) and mostly for voice applications. The physical resource of each user is a fixed allocated frequency, using amplitude modulation (AM) or frequency modulation (FM), similar to the communications in traditional copper wire or AM/FM radio broadcast. Analog voice signals cannot be compressed. Nor can they be protected by error control codes. There is no effective control over the transmit power. The resource utilization is quite low, system capacity is small, and the voice quality is often poor (e.g., with cross-talks). Devices of 1G systems are primarily analog which are difficult to be manufactured in integrated circuits. Hence, the terminals of 1G are quite expensive, bulky and heavy, and out of reach for the general public.

TABLE 1.1 Technology Evolution from First to Fourth Generations of Mobile Communications

Generation	1G	2G	3G	4G
Feature technologies	FDMA Analog modulation	TDMA Digital modulation Convolutional codes	CDMA Digital modulation Turbo codes	Orthogonal division multiple access (OFDMA) Multiple-input-multiple-output (MIMO) TBCCs, QPP turbo codes

Second-generation mobile communication (2G) is primarily time division multiple access (TDMA), and its basic service is still voice. The most widely used air interface of 3G is Global System of Mobile (GSM) communications developed mainly by countries in the European Union. In GSM systems, radio resources are divided into a chunk of frequency blocks, each being a 200 kHz narrow band. Multiple users allocated in each 200 kHz narrow band would reuse the resource in time-shared manner, e.g., occupying a time slot in a round robin. The analog voice signal is first compressed into a digital signal via source coding. The digitized signal can be protected by error correction codes. Digital modulation and power control would be applied before the signal is sent over the air. These technologies greatly improve the transmission efficiency and the system capacity. The channel codes for GSM are block codes and convolutional codes, both having low complexity, with moderate performance.

Third-generation mobile communication (3G) is code division multiple access (CDMA)-based and very robust to interference, which allows full frequency reuse among the neighboring cells and improves the system capacity. There are two major standards of 3G air interface: CDMA2000 and wideband CDMA (UMTS). The standard development body of CDMA2000 is 3GPP2. CDMA2000 is primarily deployed in North America, Korea, China, etc. Each carrier of CDMA2000 has a 1.25 MHz bandwidth. The standard development body of wideband CDMA (WCDMA) is 3GPP where European vendors and operators play important roles. WCDMA is widely deployed globally, including in Europe, Korea, China, etc. The system bandwidth of WCDMA is 5 MHz. To support a higher rate (the original requirement of 3G is 2 Mbps), CDMA2000 and WCDMA each have their own evolutions in 3G: evolution-data-optimized (EV-DO) and high-speed packet access (HSPA), respectively. There is another 3G air-interface standard: time division synchronous CDMA (TD-SCDMA), mostly developed by Chinese and European companies. Its standard development body is also 3GPP. TD-SCDMA is primarily deployed in Mainland China.

The capacity improvement in 3G is largely benefited from the soft frequency reuse, fast power control, and turbo codes. Turbo codes were first proposed in 1993 and are considered as a major breakthrough in telecommunications which can approach the Shannon capacity of a point-to-point channel. Within just a few years, turbo codes quickly found their use in wireless communications and inspired the hot research interest in random codes and iterative decoding.

Fourth-generation mobile communication (4G) is featured by orthogonal frequency division multiplex (OFDM) and orthogonal frequency multiple access (OFDMA) where each user can enjoy a higher data rate. According to Shannon capacity equation $C=B*\log_2(1+SNR)$, using wider bandwidth can directly increase the data rate. The system bandwidth of 4G is much wider than CDMA or WCDMA. OFDM/OFDMA also provides more flexibility in resource allocation when operating in a wide bandwidth.

At the beginning of 4G, there were three standards: ultra-mobile broadband (UMB), worldwide interoperability for microwave access (WiMAX) (based on IEEE 802.16), and long-term evolution (LTE). The specification work of UMB was completed in late 2007. However, due to the lack of interest from the operators, UMB has not been deployed. The specification of WiMAX was released in early 2007. Yet, the industry alliance of WiMAX is rather loose without solid business models, leading to its limited deployment worldwide.

The first release of LTE is Release 8 (Rel-8), completed in September 2007. Propelled by the strong interest from the operators, LTE has become the mainstream air-interface standard of 4G globally. Strictly speaking, LTE Rel-8 is still not 4G since the performance requirement of data rate for 4G is 100 Mbps, yet the maximum rate of Rel-8 LTE is 75 Mbps assuming a 20 MHz carrier bandwidth and single transmit antenna. Beginning in 2009, 3GPP started to specify LTE-advanced. As a major release, LTE-advanced is dubbed as Release 10 whose performances can fully meet what International Telecommunications Union (ITU) IMT-Advanced has required. For instance, assuming aggregation of five carriers (each of 20 MHz system bandwidth), the maximum rate can be 375 Mbps even with a single transmit antenna.

Several air-interface technologies were introduced to LTE-advanced including carrier aggregation, inter-cell interference cancelation (ICIC), wireless relay, enhanced physical downlink control channel (ePDCCH), device-to-device (D2D) communication, etc. These technologies can improve the system performance in several aspects, such as spectral efficiency, peak data rate, system throughput, coverage, etc. Some of these technologies are not only used for macro base station (BS)-based homogeneous networks but also for heterogeneous networks made up of macro BS and low-power nodes.

Channel coding of LTE/LTE-advanced largely reuses the turbo codes for 3G traffic channels and convolutional codes for 3G physical control channels, with a few enhancements in turbo codes to reduce the decoding complexity and improve the performance and the introduction of tail-biting convolutional codes to reduce the overhead.

1.2 SYSTEM REQUIREMENTS OF 5G NR

Different from the previous four generations of mobile communications, the applications of 5G are very diverse where the peak data rate and cell average spectral efficiency are not the only requirements. In addition, the user experience rate, number of connections, low latency, high reliability, and energy efficiency are also very important performance indicators when designing 5G systems. The deployment scenarios would not only be limited to wide area coverage but also include dense hotspots, machine-type communication, vehicle-to-vehicle communication, big outdoor events, subways, etc. All these imply that the technologies of 5G are also richer.

1.2.1 Major Scenarios

For mobile internet users, the goal of 5G is to enjoy an optical fiber level of experience rate. For internet-of-things (IoT), 5G systems should support various applications such as transportation, health care, agriculture, finance, construction, power grid, environmental protection, etc., with the common trait of a massive number of connections. Figure 1.2 shows some notable applications of 5G in mobile internet and IoT.

In IoT, data collection type of services includes low data rate services (e.g., meter reading) and high data rate services (e.g., video monitoring).

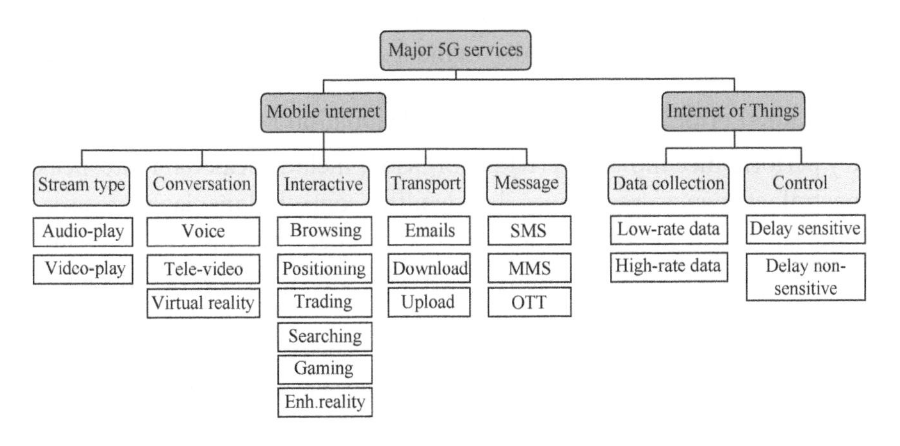

FIGURE 1.2 Major services in 5G.

Meter reading services are characterized by massive connections, very low cost, and low power consumption of terminal devices and small data packets. For video monitoring, not only a high data rate is required, but the deployment density is also high. Control type of services can be grouped into delay-sensitive services, such as vehicle-to-vehicle communications, and delay-insensitive services such as various applications for household activities.

These applications can be mapped to three use scenarios: enhanced mobile broadband (eMBB), ultra-reliable low-latency communication (uRLLC), and massive machine-type communication (mMTC). Among the various services belonging to eMBB, stream-type services require a high data rate and can tolerate a 50~100 ms delay. Interactive services have a more stringent requirement for latency, e.g., 5~10 ms. Virtual reality and online games require both high data rate and low latency, e.g., 10~20 ms. It is expected that by about 2020, ~30% of digital information would be stored in various clouds, meaning that the internet connections between the clouds and the terminals have to be at the optical fiber level. Services belonging to uRLLC can be control-type services that are highly sensitive to delays. Applications and services of mMTC include low-rate data collection, high-rate data collection, and delay-insensitive control-type services.

For eMBB, there are four major deployment scenarios: indoor hotspot, dense urban, rural, and urban macro. In indoor hotspot deployment, the coverage area is rather small where high data rate experience and system capacity are the main goals. Dense urban typical refers to city centers or very densely populated street blocks. The BS layout can be either homogeneous or heterogeneous, and the requirements are high capacity and good coverage both indoors and outdoors. Some parameters for the system performance evaluations of these deployment scenarios are listed in Table 1.2.

1.2.2 Key Performance Indicators and Evaluation Methodology

Key performance indicators (KPIs) for 5G [5] include peak data rate, peak spectral efficiency, maximum bandwidth, control-plane latency, user-plane latency, delay for infrequency small packet transmission, interruption time for mobility, reliability, coverage, battery life, energy efficiency of terminals, spectral efficiency per cell or node, traffic volume per square kilometer, user experience rate, connection density, etc. Channel coding, as a fundamental technology at the physical layer, would have a direct or indirect impact on various KPIs listed above.

TABLE 1.2 Parameters for System Performance Evaluation of Major Deployment Scenarios of eMBB

Deployment Scenario	Indoor Hotspot	Dense Urban	Rural	Urban Macro
Carrier frequency	30, 70, or 4 GHz	4 + 30 GHz (two layers)	700 MHz, 4, or 2 GHz	2, 4, or 30 GHz
Total bandwidth after carrier aggregation	Up to 1 GHz (UL + DL) for 70 GHz carrier Up to 200 MHz (UL + DL) for 4 GHz carrier	Up to 1 GHz (UL + DL) for 30 GHz carrier Up to 200 MHz (UL + DL) for 4 GHz carrier	Up to 20 MHz (UL + DL) for 700 MHz carrier Up to 200 MHz (UL + DL) for 4 GHz carrier	Up to 200 MHz (UL + DL) for 4 GHz carrier Up to 1 GHz (UL + DL) for 30 GHz carrier
Cell layout	Single layer, indoor floor space, open office area	Macro layer (hexagonal) + low-power nodes (randomly distributed)	Single layer, hexagonal cells	Single layer, hexagonal cells
Site-to-site distance	20 m (12 nodes evenly spaced within 120 × 50 m² area)	200 m for the macro layer; Three low-power nodes in each macro cell	1732, or 5000 m	500 m
Number of antenna elements at BS	Up to 256 Tx/Rx	Up to 256 Tx/Rx	Up to 256 (Tx/Rx) for the 4G Hz carrier Up to 64 (Tx/Rx) for the 700 MHz carrier	Up to 256 Tx/Rx
Number of antenna elements at the terminal	Up to 32 (Tx/Rx) for 30 or 70 GHz carrier; Up to 8 (Tx/Rx) for 4 GHz carrier	Up to 32 (Tx/Rx) for 30 GHz carrier; Up to 8 (Tx/Rx) for 4 GHz carrier	Up to 8 (Tx/Rx) for 4 GHz carrier Up to 4 (Tx/Rx) for 700 MHz carrier	Up to 32 (Tx/Rx) for 30 GHz carrier; Up to 8 (Tx/Rx) for 4 GHz carrier
User distribution and mobility	100% indoor, 3 km/h 10 users per node	Macro layer: uniform random distribution, 10 users per cell; 80% indoor @3 km/h; 20% outdoor@30 km/h	50% outdoor vehicle @120 km/h, 50% indoor @3 km/h. Ten users per cell	20% outdoor vehicle @30 km/h, 80% indoor @3 km/h Ten users per cell

- The peak rate refers to the highest theoretically achievable rate when all the radio resources are allocated to a user, discounting the overhead of synchronization signal, reference signal, guard-band, guard period, etc. The downlink peak rate requirement is 20 Gbps, and the uplink peak rate requirement is 10 Gbps. A high peak rate requires that the decoder should finish the decoding in a short time, meaning that the decoding algorithm should not be too complicated, especially for long block length and at a high code rate.

- Peak spectral efficiency is the spectral efficiency when operating at a peak rate. For high-frequency bands, the bandwidth can be very wide; however, the spectral efficiency may not be very high. For low-frequency bands, the bandwidth may be relatively narrow; however, the spectral efficiency can be quite high. Hence, the peak rate may not provide the whole picture about the system bandwidth and spectral efficiency. The KPI for the peak spectral efficiency is 30 bps/Hz for the downlink and 15 bps/Hz for the uplink. A high peak spectral efficiency implies that the code rate should be close to 1, together with high modulation order.

- Control-plane latency refers to the time to switch from the idle state to the start of data transmission in a connected state. The corresponding KPI is 10 ms. User-plane latency refers to the time between the successful reception and the initial transmission of the L2/L3 data packet (e.g., SDU), unrestricted by the discontinuous reception (DRX). For the ultra-reliable low-latency communication (uRLLC) use scenario, the user-plane latency requirement is 0.5 ms for both downlink and uplink. For the eMBB scenario, the user-plane latency requirement is 4 ms for both downlink and uplink. Effective decoding algorithms are key to reducing the user-plane latency.

- Reliability is defined as the probability of error of L2/L3 data packets (SDU) within 1 ms transmission duration. The KPI is that the error rate should be less than 10^{-7}. Reliability is mainly determined by the residual error rate after channel decoding. High reliability requires that the channel code should have a very low error floor and efficient retransmissions to add redundancy.

- Battery life refers to how long a battery can sustain the transceiver operation without recharging. For the mMTC scenario, battery life should consider the extreme coverage, the number of bits to be transmitted in uplink and downlink per day, the capacity of the battery, etc. Low-complexity decoding algorithms can reduce power consumption and extend battery life.

- In the case of the eMBB scenario, under the full-buffer traffic, the requirement of per cell average spectral efficiency of 5G is about three times that of 4G. The requirement for the cell edge spectral efficiency is also about three times that of 4G. By using suitable modulation and channel coding schemes, the spectral efficiency of the system can be further improved.
- The definition of connection density is the total number of terminal devices that can be supported within, for instance per square kilometer, under certain quality of service (QoS) requirements. QoS considers the arrival rate of the service, as well as the transmission time and the bit error rate. For urban deployment, the requirement for connection density is 1 million devices per square kilometer. In order to support a high density of terminal devices, the channel decoder should have relatively low complexity for cost and energy efficiency reasons.

KPIs such as user experienced rate, traffic volume per square area, per cell average spectral efficiency, and cell edge spectral efficiency should be evaluated via system-level simulations. The corresponding performance evaluation is normally carried out in indoor hotspots, dense urban, rural, and urban macro deployment scenarios. For connection density, system-level simulation is often required, under urban macro and rural deployments. In addition, battery life can also be done at the system level.

KPIs such as user-plane latency, control-plane latency, the delay of infrequent small packets, peak data rate, peak spectral efficiency, and battery life can be evaluated analytically.

KPIs such as coverage and reliability can be evaluated either using link-level or system-level simulations.

1.2.3 Simulation Parameters for Performance Evaluation of Modulation and Coding

Performance of modulation and coding is often evaluated via link-level simulations. The results are usually quantified in block error rate (BLER) vs. signal-to-noise ratio (SNR). The simulation settings of eMBB and mMTC/uRLLC are not the same. Some of the parameters are listed in

TABLE 1.3
Basic Simulation Parameters for Channel Coding [6]

Use Scenario	eMBB	mMTC/URLLC
Channel model	First stage: AWGN Second stage: fast fading channels	
Modulation	QPSK, 4QAM	QPSK, 16QAM
Code rate	1/5, 1/3, 2/5, 1/2, 2/3, 3/4, 5/6, 8/9	1/12, 1/6, 1/3
Block length (bits), excluding CRC bits	100, 400, 1000, 2000, 4000, 6000, 8000. Other options include: 12000, 16000, 32000, 64000	20, 40, 200, 600, 1000

Table 1.3. In the first stage of the study, the additive white Gaussian noise (AWGN) channel is assumed. Later on, channels with fast fading would be considered. Note that the focus here is channel coding, whereas traditional modulations such as QPSK, 16QAM, and 64QAM are used to represent low, medium, and high SNR operations, respectively. With respect to the block length, the range is quite wide for eMBB scenarios, starting from 100 bits, reflecting the diverse traffic and usually high data rate characteristics. The range of block length for mMTC and uLRLLC is relatively narrow, starting from 20 bits, reflecting low rate and small packet characteristics. The range of code rate is similar to the situation of the block length. For the uRLLC scenario, the BLER has to be lower than 10^{-5} in order to observe the error floor.

1.3 MAJOR TYPES OF CHANNEL CODES

LDPC codes, polar codes, tail-biting convolutional codes (TBCCs), and turbo codes are the major types of codes that have been studied or specified in 5G.

1.3.1 LDPC Codes

LDPC codes [2] were originally proposed by Robert Gallager in his Ph.D. thesis in 1963. LDPC can be considered as a type of parity check codes based on a bipartite graph whose decoding process is often carried out in an iterative manner. Limited by the processing capability of hardware devices, LDPC codes remained obscure for ~30 years since then. In the 1990s, inspired by the great success of turbo codes, researchers in both academia and the industry drew their attention to LDPC. For long block lengths, LDPC shows excellent performance while having very low decoding complexity. In fact, several performance records were set by LDPC that can most closely approach the Shannon capacity for AWGN channels. Hence, LDPC was first widely used in digital television (e.g., DVB-S2 of satellite TV standards). Later on, LDPC was adopted as an optional feature in WiMAX and WiFi. After many years of research and development, LDPC was finally adopted by 3GPP in October 2016 as the mandatory channel coding scheme for 5G NR eMBB physical traffic channels [7].

Over the last few years, there have been a number of breakthroughs of LDPC in short block length design, the support of flexible block length and code rate, the rate compatible automatic retransmission, etc. In the industry, the effort in LDPC decoding optimization never stops, which keeps improving the maturity of LDPC implementation and facilitates the adoption of LDPC in 5G.

1.3.2 Polar Codes

Polar codes [3] are a new type of block code, proposed by Prof. Erdal Arikan in 2009. Polar codes have a very strict structure, and it can be proved that by using such a structure, polar codes can approach the capacity of the binary discrete symmetric channel (BSC). The principle of polar codes' structure has profound meaning, which provides the direction of code design. Basically, channels become polarized by the polar structure. Information-carrying bits can be transmitted in the "good" channels, whereas the known bits (also called "frozen" bits in the context of polar codes) are transmitted in "bad" channels. Channel polarization is a generic phenomenon, not only for the BSC channel but also for the AWGN channel. The "polar" effect gets more significant as the block length increases.

While the history of polar codes is short, many experiences have been built in the academia and the industry in recent years with regard to the codeword design and decoding algorithms. All these help polar codes to have strong competitiveness in performance, etc., compared to other codes. In November 2016, polar codes were adopted by 3GPP as the mandatory channel coding scheme for 5G NR eMBB physical control channels [8].

1.3.3 Convolutional Codes

Convolutional codes have a long history [9]. It was invented by Peter Elias in the 1950s. In 1967, Andrew Viterbi proposed the maximum-likelihood decoding algorithm – the Viterbi algorithm which uses the time-invariant trellis to efficiently carry out the decoding. Later on, more trellis-based algorithms such as BCJR were proposed. Convolutional codes can be non-recursive or recursive. Many classic convolutional codes are non-recursive. For a very long period of time, roughly from the 1950s to the 1990s, convolutional codes were the best channel coding scheme that can approach (although still with a big gap) the Shannon capacity.

When the constraint length is short, the decoding complexity of convolutional codes is quite low and the performance is fair. Its performance is quite similar to that of turbo when the block length is short. Because of this, convolutional codes have been widely used in various physical control channels, the system information-bearing channels, or for low-cost terminals of 3G and 4G wireless systems.

Normally, a few extra bits are needed for the convolutional encoder to clear up its shift registers, e.g., to return the state of the encoder to zero. In order to reduce this overhead, LTE uses TBCCs where the end state and the initial state of the encoder have to be the same. Because the receiver does not know the initial state of the tail-biting convolutional encoder, the decoding complexity would be increased.

1.3.4 Turbo Codes

The advent of turbo codes in 1993 [10] is considered as a "revolution" in the field of channel coding. For the first time, a practical channel coding scheme can approach so close to the Shannon limit. People began to realize the power of random codes and iterative decoding. Since then, "random coding and iterative decoding" have become the mainstream approaches for channel coding. The basic idea of turbo codes is to introduce a random interleaver to the channel encoder where two recursive convolutional encoders are serially or parallelly concatenated. By doing this, the error correction capability of the codewords is significantly increased. In the decoder, the suboptimal iterative algorithm is used in order to reduce the decoding complexity. The soft information of the bits is passed between the two convolutional decoders where the "belief" continuously improves.

Compared to the traditional LDPC, turbo codes have the advantages of flexibility of block length, code rate, and rate compatibility automatic retransmission. Hence, turbo codes were used in 3G and 4G systems as the mandatory channel coding scheme. However, the decoding complexity of turbo codes is higher than that of LDPC in many cases, especially when the block length is large and the code rate is high.

1.3.5 Outer Code

In order to further improve the error correction/check capability, outer coding [11] can be considered in addition to the physical layer channel coding. In 2G systems, widely used inner codes were block codes or convolutional codes that have limited capability in error correction. In such a case, outer codes were indispensable to maintain the good quality of a link. For 3G, 4G, and 5G systems, some outer codes, for instance, cyclic redundancy check (CRC), are still used for error correction or error checking.

One of the key use scenarios of 5G is ultra-reliability low-latency communications (uRLLC). Whenever uRLLC traffic arrives, resources originally scheduled for other services, such as eMBB or mMTC, would be punctured in order to immediately accommodate uRLLC traffic. Outer codes are expected to improve the robustness of eMBB/mMTC communications under these bursty interference situations when these use scenarios coexist in a system. In addition, outer codes can enhance the link adaptation capability, e.g., HARQ can operate in a more efficient manner.

1.3.6 Other Advanced Coding Schemes

Non-binary LDPC [12,13] was first proposed by Davey and MacKay in 1998, and is different from binary LDPC. Non-binary LDPC works in the GF(q) (where q is usually a power of 2) domain. Its decoding complexity is much higher than that of binary LDPC. Due to its capability of reducing the short loops, especially length-4 loops, non-binary LDPC has better error correction capability and lower error floors.

Repetition accumulation (RA) codes [14] is a coding scheme built on top of turbo codes and LDPC codes. RA codes possess the merit of simple encoding of turbo codes, as well as the advantage of parallel decoding in LDPC. In addition, RA codes have more flexibility in choosing non-zero value elements and thus can potentially avoid the short loops. Compared to turbo codes, binary LDPC, and binary RA codes, non-binary RA codes have better error correction capability, especially for high-order modulation.

The lattice code [15] was originally proposed by David Forney of Codex company in 1988. It strikes a good balance between the coding gain and decoding complexity. In 2007, Naftali Sommer at the University of Tel Aviv proposed a new coding scheme based on LDPC: low-density lattice code (LDLC) which is a practical code to approach the channel capacity of the AWGN channel. Its decoding complexity grows with the block length linearly.

The spinal code [16] is a rate-less code suitable for time-varying channels. The spinal code can also approach the Shannon capacity. Its basic concept is to continuously apply the pseudo-Hash function as well as the Gaussian mapping function to generate pseudo-random Gaussian symbols. Compared with other channel codes, the spinal code can approach the Shannon limit even when the block length is short. When the channel condition is good, the spinal code's performance can be better than the combination of the legacy channel coding scheme and high-order modulation.

1.4 MOTIVATION AND STRUCTURE OF THIS BOOK

During 3GPP RAN#78 meeting held in Lisbon in December 2017, the first release of the 5G NR specification was completed [17], marking the end of the first-stage standardization of 5G (particularly for the eMBB-use scenario). As a key ingredient of 5G air-interface technologies, advanced channel coding schemes play an important role in fulfilling the KPIs. This book is to provide readers with a whole picture of the 5G channel coding study and standard development.

The structure of this book is illustrated in Figure 1.3, starting from Chapter 1 of background introduction, followed by a series of detailed descriptions of major channel coding schemes: Chapter 2 of LDPC codes, Chapter 3 of polar codes, and Chapter 6 of outer codes. In addition, a couple of channel codes, convolutional codes, and turbo codes are also discussed in Chapters 4 and 5, respectively, as the candidate schemes for

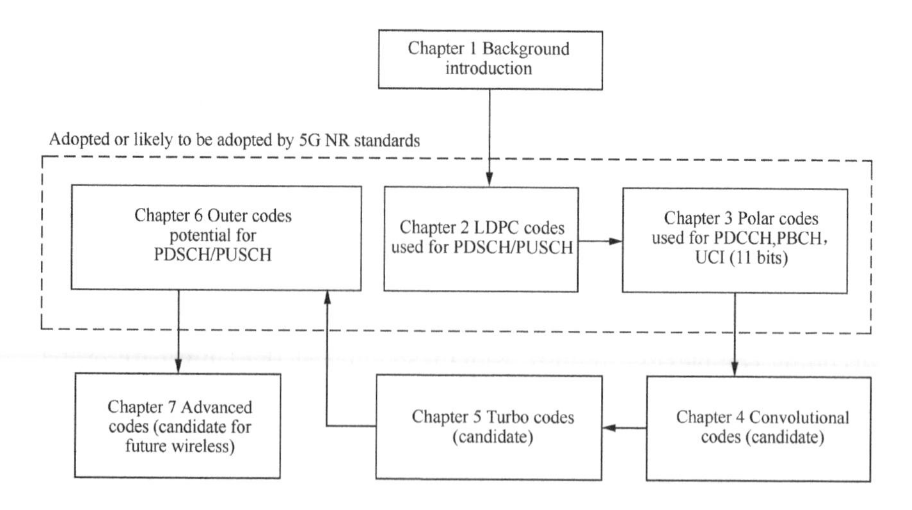

FIGURE 1.3 Structure of this book.

5G. In the end, some more advanced channel coding schemes are overviewed, which may be used in future wireless systems. In each chapter, the discussion of the channel codes includes the code structure, the encoding scheme(s), the decoding scheme(s), the performance, etc. In 5G, repetition and Reed–Muller (for bit length between 3 and 11 bits in 3GPP specification) are also used. However, since they are exactly the same as in 4G LTE, these two coding schemes will not be discussed in this book and interested readers are referred to [18,19] for more information.

REFERENCES

[1] 3GPP, RWS-150089, Vision and Technologies for 5G, CMCC, RAN Workshop on 5G, September, 2015.

[2] R. G. Gallager, "Low-density parity-check codes," *IRE Trans. Inf. Theory*, vol. 8, Jan. 1962, pp. 21–28.

[3] E. Arikan, "Channel polarization: A method for constructing capacity achieving codes for symmetric binary-input memoryless channels," *IEEE Trans. Inf. Theory*, vol. 55, July 2009, pp. 3051–3073.

[4] 3GPP, RWS-150082, Update on ITU-R Work on IMT-2020 for 5G, ITU-R Working Party 5D, AT&T, RAN Workshop on 5G, September, 2015.

[5] 3GPP, TR 38.913 V14.3.0, Study on Scenarios and Requirements for Next Generation Access Technologies (Release 14), 2017.06.

[6] 3GPP, TR 38.802 V14.1.0, Study on New Radio Access Technology, Physical Layer Aspects (Release 14), 2017.06.

[7] 3GPP, R1-1611081, Final Report of 3GPP TSG RAN WG1 #86bis v1.0.0, RAN1#87, November 2016.

[8] 3GPP, Draft_Minutes_report_RAN1#87, Nov. 2016. http://www.3gpp.org/ftp/tsg_ran/WG1_RL1/TSGR1_87/Report/Final_Minutes_report_RAN1%2387_v100.zip

[9] P. Elias, "Coding for noisy channels," *IRE Conv. Rec.*, vol. 3, May 1955, pp. 37–47.

[10] C. Berrou, A. Glavieux, P. Thitimajshima, "Near Shannon limit error-correcting coding and decoding: Turbo Codes," *Proc. IEEE Intl. Conf. Communication (ICC 93)*, May 1993, pp. 1064–1070.

[11] 3GPP, R1-1608976, Consideration on outer codes for NR, ZTE, RAN1#86bis, October 2016.

[12] M.C. Davey, "Low-density parity check codes over GF(q)," *IEEE Commun. Lett.*, vol. 2, no. 6, June 1998, pp. 165–167.

[13] D. J. C. Mackay, *Evaluation of Gallager codes of short block length and high rate applications*, Springer, New York, 2001, pp. 113–130.

[14] G. Tu, *Belief propagation algorithm for RA codes*, Master Thesis, XiDian University, 2014.

[15] N. Sommer, "Low-density lattice codes," *IEEE Trans. Inf. Theory*, vol. 54, no. 4, 2008, pp. 1561–1585.

[16] J. Perry, "Spinal codes," *ACM SIGCOMM Conference on Applications*, 2012, pp. 49–60.

[17] 3GPP, Draft_MeetingReport_RAN_78_171221_eom, 18 Dec. -21 Dec. 2017. http://www.3gpp.org/ftp/tsg_ran/TSG_RAN/TSGR_78/Report/Draft_MeetingReport_RAN_78_171221_eom.zip
[18] 3GPP, TS 36.212, Multiplexing and channel coding, Sept. 2016.
[19] S. Lin, *Error correction codes*, 2nd Edition (translated by Jian Yan), Mechanical Industry Press, Beijing, China, June 2007.

CHAPTER 2

Low-Density Parity Check (LDPC) Codes

Jun Xu, Yifei Yuan, Meiying Huang, Liguang Li and Jin Xu

LDPC CODES WERE PROPOSED BY GALLAGER in 1963 as a type of linear block codes [1,2]. Gallager chose "low density" as the name because the majority of the elements in its parity check matrix **H** are "0", e.g., only a small portion of the elements are "1". After more than 50 years of research, people already have accumulated quite a lot of knowledge of the code design, encoding schemes, and decoding schemes of LDPC. Its application spans data storage, optical communications, wireless communications, etc.

This chapter starts with the description of the inception and development of LDPC, followed by the principle of LDPC, quasi-cyclic LDPC (QC-LDPC), the structure of LDPC, and the standardization of LDPC in 5G NR. Next, decoding complexity, data throughput, decoding latency, and link-level performance of 5G LDPC are presented. In the end, the specification of LDPC in 3GPP is briefly explained, and the future direction of LDPC is provided. The structure of this chapter is illustrated in Figure 2.1.

2.1 INCEPTION AND DEVELOPMENT OF LDPC

The foundation of channel coding was built on the famous paper by C.E. Shannon [3] in 1948, which proves the existence of a channel coding scheme that can ensure the error rate as low as possible when the signal passes through a noisy channel. However, this paper did not provide constructive ways to design such codes. Since then, various coding schemes

DOI: 10.1201/9781003336174-2

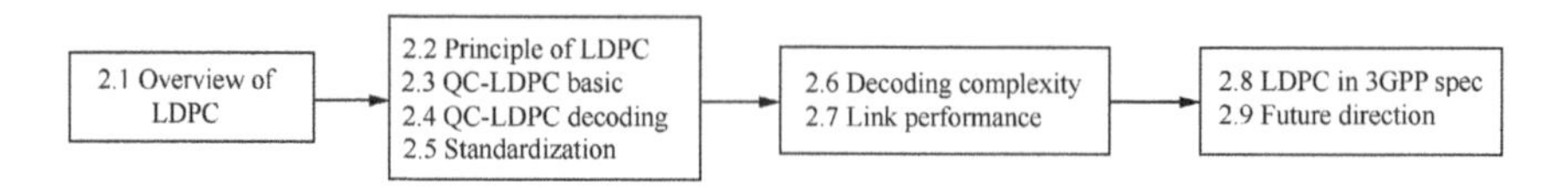

FIGURE 2.1 The structure of Chapter 2.

have been proposed. In 1950, R. W. Hamming invented Hamming codes [4] which is a type of linear block codes. Without losing the generality, the GF(2) linear block code can be represented as Eq. (2.1):

$$x = u \bullet G \tag{2.1}$$

where **u** is the information bit vector before the channel coding and **x** is the bit vector (e.g., codeword) after the coding. **G** is the generator matrix. For the above generator matrix, if it is full-rank, **G** can be converted into the following form (if **G** is not full-rank, the first few rows can also be converted into this form).

$$G = [Q, I] \tag{2.2}$$

where **Q** is the sub-matrix corresponding to the parity check matrix and **I** is the identity matrix corresponding to the information bits. If Q is full-rank, the parity check matrix of this linear block code can be represented by Eq. (2.3).

$$H = [I, Q^T] \tag{2.3}$$

where $\mathbf{Q}^{\mathrm{T}}$ is the conjugate of **Q**. For the above linear block code, we have

$$H \bullet x = 0 \tag{2.4}$$

By computing Eq. (2.4), the receiver would know whether an error occurs during the transmission.

As the above equations have indicated, the generator matrix **G** of a linear block code and its parity check matrix **H** are dual. That is, a linear block code can also be represented by its parity check matrix. In another word, we can first construct a parity check matrix and then convert it to the generator matrix. More specifically, the encoding process can be conducted by certain computations of the parity check matrix.

In 1957, Eugene Prange proposed cyclic codes [5] which is also a type of linear block code. In cyclic codes, a codeword can still be a valid codeword (although different) after a certain amount of cyclic shift. This makes the decoding of cyclic codes much easier. If a codeword becomes invalid after one time of cyclic shift, however, is still valid after multiple times of cyclic shift, this type of codes is called quasi-cyclic code. The quasi-cyclic property can also reduce the decoding complexity.

The LDPC code invented by Gallager in 1963 [1,2] is also a type of linear block code. In general, it is easier to represent LDPC using parity check matrix H which has low density, e.g., more elements of value "0" than "1". The iterative decoding of LDPC was considered too expensive for the hardware back then. Hence, the LDPC did not get much attention for a long period of time.

In 1981, R. M. Tanner studied LDPC using graph theory [6]. However, his work was still not widely noticed. The invention of turbo codes [7] in 1993 inspired researchers to revitalize the belief propagation and iterative decoding proposed decades ago by Gallager [2]. LDPC found its new life and its performance can be comparable to that of turbo [7], ~0.3 dB from the Shannon limit. In 1999, LDPC was adopted by WiMAX (IEEE 802.16e) [8] and WiFi (IEEE 802.11n) [9]. In 2011, LDPC set a performance record, e.g., the LDPC codes developed by S. Y. Chung can approach the Shannon limit as close as 0.0045 dB [10], which has not been broken so far. In 2004, LDPC was adopted by the second-generation digital TV broadcasting standards (DVB-S2) [11] in Europe. In 2016, quasi-cyclic LDPC (QC-LDPC) was adopted by 3GPP as the channel coding scheme for physical traffic channels of 5G NR.

2.2 BASIC PRINCIPLE OF LDPC CODES

2.2.1 Gallager Codes

LDPC is a type of linear block code proposed by Gallager in 1963 [1,2]. In his Ph.D. thesis [2], Gallager provided a detailed performance-bound analysis of the LDPC codes. He also suggested two decoding schemes: (1) simple algebraic method and (2) decoding based on probability theory.

The basic idea of [1,2] is that even though the low-density property of LDPC does not have the advantage in code distances, the low weight of the codewords can reduce the decoding complexity and improve the performance of iterative decoding. In ~40 years after Gallager's thesis, LDPC was rather obscure, partially because the probability theory-based decoding is much more complicated than that of simple algebraic decoding. Even with low density, the computation is still too burdensome for the hardware in those years.

Another reason is the limited understanding of probability theory-based decoding, including its performance potential. Traditional channel decoding relies on the codeword distance criterion, or more specifically the minimum distance and the distribution of the distance, to predict the performance of a coding scheme. By doing so, the performance bounds can have a closed form; thus, the prediction would be more precise. The codeword distance criterion can be quite effective for algebraic block codes of short length. However, for long codes designed using probability theory, the distance criterion would not be accurate enough. LDPC codes' powerfulness lies in the long block length and probability-based decoding. Without a deep understanding of probabilistic decoding, the potential of LDPC would not be fully realized.

The turbo decoding algorithm proposed in [7] brought about a big "fad" of probabilistic decoding which works on soft bits, rather than "hard" bits, e.g., either "true" or "false". The word "soft" refers to a probability distribution of a bit's "belief" which is a real number, representing the "grayness" between "true (white)" and "false (black)". Normally, probabilistic decoding and iterative decoding would go together in order to deliver good performance. By digesting the idea of turbo decoding, people started to find out that the original LDPC proposed by Gallager was indeed using iterative decoding. Both turbo codes and LDPC codes can be represented by a factor graph [12] in a uniform setting. The iterative decoding algorithm is fundamentally similar to the belief propagation algorithm [13] and message passing algorithm [14] widely used in the field of artificial intelligence

A classic LDPC encoder can be described as follows. Let us use $\mathbf{u}$ to represent a binary bit stream of length k. After adding $\mathbf{m}$ parity check bits, a length n code stream is obtained, denoted as $\mathbf{t}$. The corresponding code rate is k/n. Since it is a linear code, the codeword $\mathbf{t}$ can be represented as $\mathbf{u}$ multiplied by a generator matrix $\mathbf{G}^T$:

$$\mathbf{t} = \mathbf{G}^T \mathbf{u} \tag{2.5}$$

where the generator matrix $\mathbf{G}^T$ contains two parts:

$$\mathbf{G}^T = \begin{bmatrix} \mathbf{I}_{k\times k} \\ \mathbf{P}_{m\times k} \end{bmatrix} \tag{2.6}$$

If the hard-decision-based algebraic decoding is used, the corresponding parity check matrix can be written as

$$\mathbf{A} = \left[\mathbf{P} \,|\, \mathbf{I}_{m \times m}\right] \tag{2.7}$$

When there is no error encountered during the transmission, the parity check would pass, e.g., $GA^T = 0$. It should be pointed out that sometimes the parity check matrix of LDPC is not written in the form of a right-sided identity matrix as seen in Eq. (2.7), especially when probabilistic decoding (e.g., belief propagation of soft information) is used. In such cases, certain linear algebra transformation is needed to obtain the generator matrix for the purpose of encoding.

The characteristic of LDPC is that the parity check matrix **A** of m rows and n columns is sparse, e.g., most of the elements are zeros. The matrix **A** can be randomly generated, via structured designs, by exhaustive search, etc. Details can be found in Sections 2.3 and 2.5.3.

Figure 2.2 shows an LDPC code block passing through an AWGN channel and then being decoded.

In Figure 2.2, the information bit stream **u** is first encoded into the coded stream **t** to be passed through an AWGN channel. The output of the channel is the observed sequence **y**. In the LDPC decoder, the soft information is exchanged back and forth between the variable nodes and the check nodes. After a certain number of iterations, the decoder would output the final hard decisions, denoted as $\hat{\mathbf{u}}$.

It is necessary to ensure the sparsity of the parity check matrix **A** for LDPC in the sense that:

- The decoding of LDPC relies on the "sum-product" type of algorithm. Such an algorithm can only perform well when the bipartite

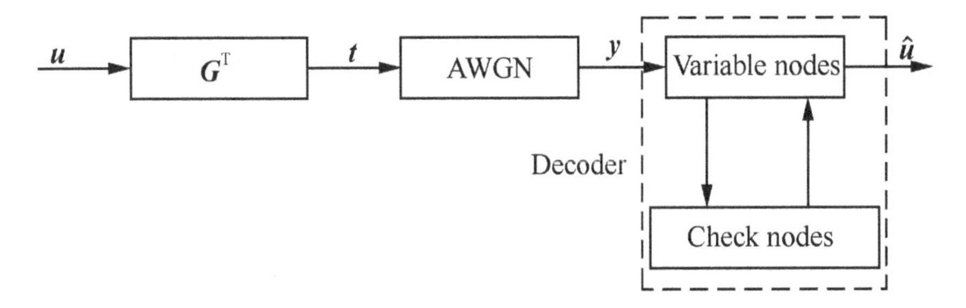

FIGURE 2.2 Block diagram of LDPC code block passing through AWGN channel and decoder.

graph of the parity check is either cycle-free (cycle-free) or without short cycles (cycles). By having the property of sparsity, the chance of having short cycles (cycles) would be low.

- Sparsity means that the connection density between the variable nodes and the check nodes is low. Hence, the number of addition and multiplication operations can be lowered, which is helpful in reducing the decoding complexity.
- When the parity check matrix **A** is sparse, its generator matrix $\mathbf{G}^T$ is often not sparse. This means that it is possible to design the codeword **t** to have good distance properties.

2.2.2 Regular LDPC and Irregular LDPC

LDPC codes can be described using a bipartite graph. The word "bipartite" emphasizes that there are two types of nodes in the graph: variable nodes and check nodes. Nodes of the same type should not be connected directly, e.g., no direct communication. However, it is allowed to have the nodes of the other type pass the information. Each connection is often called an "edge". The entire connections are solely determined by the parity check matrix.

There are two main categories of LDPC codes: regular and irregular [15]. The nodes of the same type (e.g., variable nodes or check nodes) in a regular LDPC have the same degree. Here the degree refers to the number of edges. Figure 2.3 shows an example of a bipartite graph of a regular LDPC which contains nine variable nodes (denoted as $Y_i, i = 1,2,3,......,9$) and six check nodes (denoted as $A_i, i = 1,2,3,......,6$). Each bit node is connected with $q = 2$ check nodes, e.g., the number of "1" elements in a column

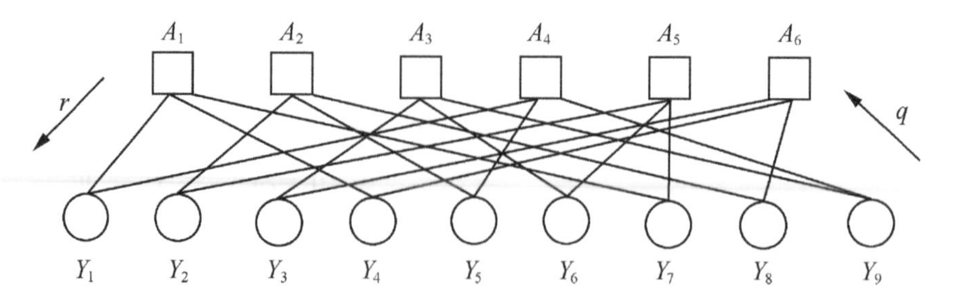

FIGURE 2.3 Bipartite graph of an example of LDPC parity check matrix.

is $d_v = 2$, e.g., the column weight is 2. Each check node is connected to $r=3$ bit nodes, and the number of "1" elements in a row is $d_c=3$, e.g., the row weight is 3. It is noticed that the column weight is the same for all the columns in this parity check matrix. So is the row weight. With respect to the entire 54 elements in this matrix, only 18 elements are non-zero. Hence, it can be considered as sparse. The sparsity would become more pronounced as the size of the parity check matrix increases.

$$\mathbf{A} = \begin{bmatrix} 1 & 0 & 0 & 1 & 0 & 0 & 1 & 0 & 0 \\ 0 & 1 & 0 & 0 & 1 & 0 & 0 & 1 & 0 \\ 0 & 0 & 1 & 0 & 0 & 1 & 0 & 0 & 1 \\ 1 & 0 & 0 & 0 & 1 & 0 & 0 & 0 & 1 \\ 0 & 1 & 0 & 0 & 0 & 1 & 1 & 0 & 0 \\ 0 & 0 & 1 & 1 & 0 & 0 & 0 & 1 & 0 \end{bmatrix} \tag{2.8}$$

In irregular LDPC, the degree of variable/check nodes can be different, following certain distribution. Irregular LDPC codes are often described by a pair of parameters (λ, ρ):

$$\lambda(x) := \sum_{i=2}^{d_v} \lambda_i x^{i-1} \tag{2.9}$$

which defines the distribution of degree of variable nodes, and

$$\rho(x) := \sum_{i=2}^{d_c} \rho_i x^{i-1} \tag{2.10}$$

which defines the distribution of degree of check nodes. More specifically, the coefficients λ_I and ρ_i represent the percentage of outgoing edges with the degree of i, from the variable nodes and the check nodes, respectively. Such representation can also be used to describe a regular LDPC. For instance, the bipartite graph in Figure 2.3 can be written as $\lambda(x) := x^1$ and $\rho(x) := x^2$.

Compared to regular LDPC, irregular LDPC has more design flexibility and room for optimization. Both theoretical analysis and simulation results show that irregular LDPC codes perform better than regular LDPC [16–18]. Therefore, most LDPC codes considered in the industry are irregular.

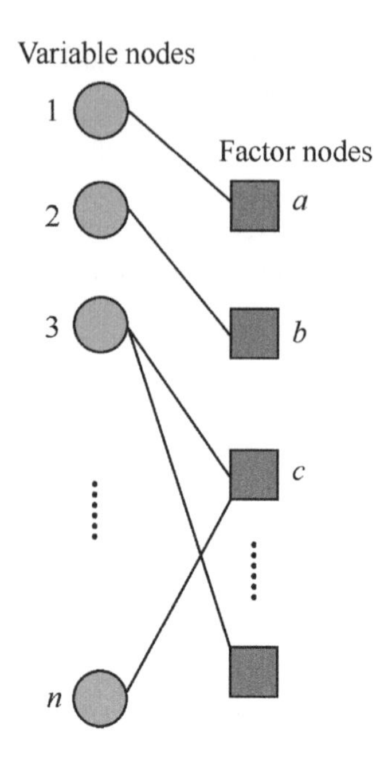

FIGURE 2.4 An example of factor graph.

2.2.3 Principle of Belief Propagation and Its Application

Belief propagation [13], also called message passing [14], is a basic algorithm in probability theory. It has high computation efficiency, which allows its wide use in digital communications, artificial intelligence, computer science, operation research, etc.

The bipartite graph in the context of LDPC can be generalized into a factor graph as shown in Figure 2.4 that consists of variable nodes and factor nodes.

Essentially, belief propagation or message passing can efficiently compute marginal probability distribution, given the connection relationship between variable nodes and factor nodes in a factor graph. Such computation has an important assumption: variable nodes are not related to each other, nor are the factor nodes. In another word, their joint probability distribution can be represented as the product of marginal probability distributions. The computation of marginal probability in a factor graph can be envisioned propagation of belief over a tree structure. That is, a factor graph can be expanded into a tree (assuming that there is no cycle, e.g., acyclic) where the iterative inference progresses as illustrated in Figure 2.5.

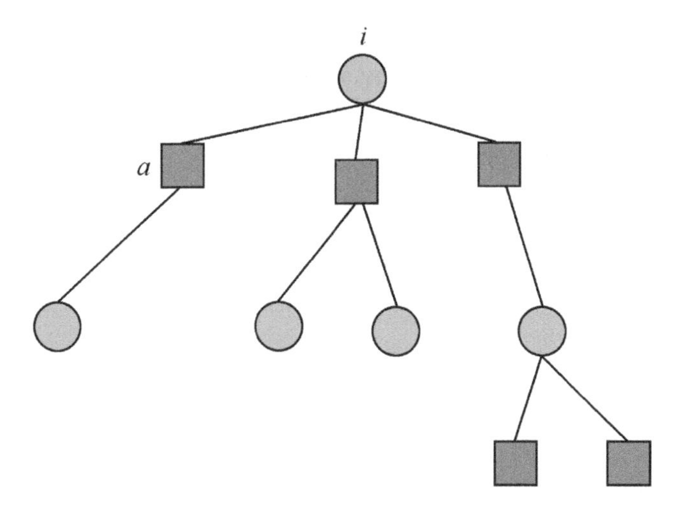

FIGURE 2.5 A factor graph expanded into a tree structure (without cycle, e.g., acyclic).

The information from the factor node a to the variable node i, denoted as $m_{a\to i}(x_i)$, represents the confidence that the factor node a would believe that the variable node i would be in State x_i. The joint "belief" of the variable node i being in State x_i is the product of "belief", e.g., confidence, of all the factor nodes (denoted as $N(i)$) connecting to the variable node i as seen in Eq. (2.11):

$$b_i(x_i) \propto \prod_{a\in N(i)} m_{a\to i}(x_i) \tag{2.11}$$

For an arbitrary factor node, e.g., A, its belief is the joint belief of all the variable nodes connecting to this factor node, represented as

$$b_A(X_A) = b_L(\mathbf{x}_L) \tag{2.12}$$

Note here L refers to a set. Then according to Eq. (2.11), the belief of each variable node in the set L is proportional to the multiplication of the belief by all the factor nodes connecting to this variable node, regarding the probability of this variable node being in State x_k. The joint probability can be represented as follows:

$$b_L(\mathbf{x}_L) \propto \prod_{a\in N(L)} m_{b\to L}(\mathbf{x}_k) \propto f_A(X_A) \prod_{k\in N(A)} \prod_{b\in N(K)/A} m_{b\to k}(x_k) \tag{2.13}$$

Thus, the information (or the message or belief) from a factor node to a variable node can be calculated by summing over the information from all the variable nodes connecting to this factor node, except the message originally from this variable node

$$b_i(x_i) = \sum_{X_A \backslash x_i} b_A(X_A) \tag{2.14}$$

That is,

$$m_{A\to i}(x_i) \propto \sum_{X_A \backslash x_i} f_A(X_A) \sum_{k \in N(A) \backslash i} m_{k\to A}(X_k) \tag{2.15}$$

where the information (message or belief) from a variable node to a factor node can be represented as

$$m_{k\to A}(x_k) \propto \prod_{b \in N(k) \backslash A} m_{b\to k}(x_k) \tag{2.15}$$

which is the product of the information from all the factor nodes, except the recipient factor node.

The above equations can be written in the iterative form. For a tree structure expanded from a factor graph, it is proved that the belief propagation algorithm can converge to the marginal distribution within finite a number of iterations (e.g., twice the depth of the tree structure) if the tree contains no cycles, e.g., acyclic.

Trees expanded by factor graphs in real systems may contain some cyclic sub-structures as illustrated in Figure 2.6. When there are "regional" cycles, the belief propagation algorithm may not converge. Or even when converged, it would not settle to the marginal distribution. There are several ways to solve the cycle structure issue, which has more significance in theoretical studies rather than for practical engineering. Hence, we will not elaborate more in this book.

Belief propagation can be applied in many decoding algorithms for channel coding, for instance, LDPC. Each row in a LDPC parity check matrix represents a parity check. In traditional algebraic decoding, the only thing to check is whether the sum of the values on all variable nodes equals 0. If yes, the parity check passes. Otherwise, the parity check fails. In probabilistic decoding, each parity check node, or the factor node in the factor graph, computes the probability whether the parity check would

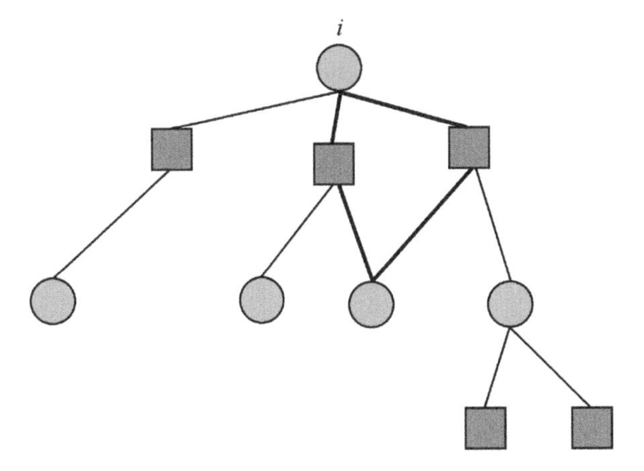

FIGURE 2.6 An example of tree that has cyclic sub-structures (four nodes connected with bold lines).

pass. The computation is based on the edges (e.g., how the check nodes and variable nodes are connected) and certain prior information (e.g., distribution) about the variable nodes. Each variable node computes the probability of the bit taking a value of 1 or 0. This computation is based on the parity check pass probabilities suggested by all the check nodes connecting to this variable node.

Let us use r_{ij} to denote the information probability computed by the parity node. More specifically, r_{ij}^0 represents the probability that the check node j can pass, when the information bit t_i is 0. Similarly, r_{ij}^1 represents the probability that the check node j can pass when the information bit t_i is 1. According to the principle of belief propagation, these two probabilities can be computed by using Eq. (2.17):

$$\begin{aligned} r_{ij}^0 &= \frac{1}{2}\left[1 + \prod_{i' \in row[j] \setminus \{i\}} \left(q_{i'j}^0 - q_{i'j}^1\right)\right] \\ r_{ij}^1 &= \frac{1}{2}\left[1 - \prod_{i' \in row[j] \setminus \{i\}} \left(q_{i'j}^0 - q_{i'j}^1\right)\right] \end{aligned} \tag{2.17}$$

where the notation $i' \in row[j] \setminus \{i\}$ represents the bit indices in the j-th row whose values are 1, except its own bit index i. The probability q_{ij} is computed by the variable node. q_{ij}^0 represents the probability of t_i=0, given the information by all its connected parity check nodes, except the check node j. Similarly, q_{ij}^1 represents the probability of $t_i = 1$, given the information by all its connected parity check nodes, except the check node j. According to the principle of belief propagation, q_{ij} can be calculated as

$$
\begin{aligned}
q_{ij}^0 &= \alpha_{ij} p_i^0 \prod_{j' \in col[i] \setminus \{j\}} r_{ij'}^0 \\
q_{ij}^1 &= \alpha_{ij} p_i^1 \prod_{j' \in col[i] \setminus \{j\}} r_{ij'}^1
\end{aligned}
\tag{2.18}
$$

where the notation $j' \in col[i] \setminus \{j\}$ represents the check node indices in the i-th column whose values are 1, except the check node index j. The coefficient α_{ij} is for normalization to ensure $q_{ij}^0 + q_{ij}^1 = 1$. In Eq. (2.18), p_i^0 and p_i^1 are the prior probabilities of the i-th bit, based on the previous iteration. The extrinsic information e_i^0 and e_i^1 can be calculated as

$$
\begin{aligned}
e_i^0 &= \alpha_i \prod_{j' \in col[i]} r_{ij'}^0 \\
e_i^1 &= \alpha_i \prod_{j' \in col[i]} r_{ij'}^1
\end{aligned}
\tag{2.19}
$$

where the notation $j' \in col[i]$ represents the check node indices in the i-th column whose values are 1. The coefficient α_i is to ensure $e_i^0 + e_i^1 = 1$. In the initial stage, p_i^0 and p_i^1 can be based on whether there is prior information. If not, they are equally probable, e.g., $p_i^0 = p_i^1$. Assuming AWGN channel, the initialization can be carried out column by column. For instance, $q_{ij}^0 = P(t_i = 0 \mid y_i) = \frac{1}{1 + e^{-2y_i/\sigma^2}}$ and $q_{ij}^1 = P(t_i = 1 \mid y_i) = \frac{1}{1 + e^{2y_i/\sigma^2}}$, where $\sigma^2/2$ is the power of AWGN noise. The iteration process can be terminated by a hard decision after each iteration. For example, the information bit is decided to be 0 if $e_i^0 > 0.5$. If the parity check of the hard-decision bits passes, or if the maximum number of iterations is reached, the decoding iteration would be terminated.

2.2.4 Practical Decoding Algorithms

Probability domain belief propagation involves a large number of multiplications which are complicated. It requires a rather wide dynamic range and often suffers numerical instability issues. In practical systems, log-domain belief propagation (log-BP) is often used which works on the log-likelihood ratio (LLR). Let us define

$$LLR(c_n) = \log \frac{\Pr(x_n = +1 \mid y_n)}{\Pr(x_n = -1 \mid y_n)} \tag{2.20}$$

$$LLR(r_{mn}) = \log \frac{r_{mn}^0}{r_{mn}^1} \tag{2.21}$$

$$LLR(q_{mn}) = \log \frac{q_{mn}^0}{q_{mn}^1} \tag{2.22}$$

$$LLR(q_n) = \log \frac{q_n^0}{q_n^1} \tag{2.23}$$

1. The log-BP algorithm includes the following steps: To initialize $LLR(q_{mn})$ according to:

For $n = 0, \ldots, N-1$

For $m = 0, \ldots, M-1$

$$LLR(q_{mn}) = \log(q_{mn}^0 / q_{mn}^1) = LLR(c_n) = 2y_n / \sigma^2$$

End

End

2. Updating the information in check nodes, e.g., to compute $LLR(r_{mn})$ according to:

For $m = 0, \ldots, M-1$

For $n \in N(m)$

$$LLR(r_{mn}) = \left(\prod_{n' \in N(M) \backslash n} \alpha_{mn'} \right) \Phi \left(\sum_{n' \in N(M) \backslash n} \Phi(\beta_{mn'}) \right)$$

End

End

where

$$\alpha_{mn'} = sign(LLR(q_{mn'}))$$

$$\beta_{mn'} = |LLR(q_{mn'})|$$

$$\Phi(x) = -\log(\tanh(x/2)) = \log\frac{e^x+1}{e^x-1}$$

3. Updating the information in variable nodes, e.g., to compute $LLR(q_{mn})$ according to

For $n = 0,\ldots,N-1$

For $m \in M(n)$

$$LLR(q_{mn}) = LLR(c_n) + \sum_{m' \in M(n)\backslash m} LLR(r_{m'n})$$

End

End

4. Updating $LLR(q_n)$:

For $n = 0,\ldots,N-1$

$$LLR(q_n) = LLR(c_n) + \sum_{m \in M(n)} LLR(r_{mn})$$

End

In the above log-BP algorithm, the computation of $\Phi(x)$ is important and it has the following characteristics:

- can be approximated in order to reduce the computation complexity:

$$\Phi\left(\sum_i \Phi(\beta_i)\right) = \min_i(\beta_i),\ \beta_i > 0$$

The above approximation is called the Min-Sum algorithm which does not require knowledge about the noise variance σ^2, e.g., the signal-to-noise ratio (SNR) of the codeword does not need to be known. However, there is a significant error between the result of Min-Sum and the true value of $\Phi\left(\sum_i \Phi(\beta_i)\right)$. To reduce this error, a constant (less than 1) can be used for the adjustment.

$$\Phi\left(\sum_i \Phi(\beta_i)\right) = \mathrm{A}\times\min_i(\beta_i), \quad \beta_i > 0, 1 > \mathrm{A} > 0 \tag{2.24}$$

where the constant A is related to the row weight of the parity check matrix of LDPC, usually ranging between 0.6 and 0.9. The actual value should be determined by simulations. For the AWGN channel, the modified Min-Sum algorithm is worse than the log-BP algorithm by about 0.1 dB. In addition, the result of Min-Sum can be further adjusted by the offset-min-sum algorithm as

$$\Psi = \mathrm{A}\times\min_i(\beta_i) - \rho, \quad \beta_i > 0, 1 > \mathrm{A} > 0, 1 > \rho > 0 \tag{2.25}$$

$$\Phi\left(\sum_i \Phi(\beta_i)\right) = \begin{cases} \Psi & \text{if } \Psi > 0 \\ 0 & \text{else} \end{cases} \tag{2.26}$$

2.2.5 Theoretical Analysis of Performance

Performance of LDPC can be analyzed theoretically. Assuming maximum-likelihood (ML) decoding, the performance of LDPC can be computed via code distance distribution. However, ML decoding of LDPC has very high complexity and is rarely used in practice. Quite often, probabilistic decoding methods such as the BP algorithms discussed in Section 2.2.4 are used. Hence, it is better to analyze the performance of LDPC by considering the BP. In addition, performance analysis based on probabilistic decoding can also reflect the evolution of belief during the course of iterative decoding and provide guidance on the convergence behavior.

It should be pointed out that the performance analysis here is not for a specific parity check matrix. It is rather the average performance over the ensemble of codewords, given the column weight d_v and the row weight d_c. To make the analysis more tractable, it is assumed that within the number of iterations being considered, there is no cyclic sub-structure in the tree expanded from the factor graph.

In his thesis, Gallager provided closed-from formula to compute the performance of regular LDPC under binary symmetric channel (BSC) [1,2]. Let us use $p_1^{(l)}$ and $p_{-1}^{(l)}$ represent in the l-th iteration the probabilities of an information bit taking values 1 and –1, respectively. $q_1^{(l)}$ and $q_{-1}^{(l)}$ represent the probabilities that a check node would believe that in the l-th iteration, the information bit would take a value of 1 and –1, respectively. Ψ_v and Ψ_c represent the mappings (or edges) of variable nodes and check nodes, respectively. Then we have

$$\begin{aligned}\left(q_{-1}^{(l)}, q_1^{(l)}\right) &= \Psi_c\left(\left(p_{-1}^{(l-1)}, p_1^{(l-1)}\right), \ldots, \left(p_{-1}^{(l)}, p_1^{(l)}\right)\right) \\ &= \frac{1}{2}\left(1-\left(1-2p_{-1}^{(l-1)}\right)^{d_c-1}, 1+\left(1-2p_{-1}^{(l-1)}\right)^{d_c-1}\right)\end{aligned} \tag{2.27}$$

$$\begin{aligned}&\left(p_{-1}^{(l)}, p_1^{(l)}\right) = \Psi_v\left(\left(q_{-1}^{(l)}, q_1^{(l)}\right), \ldots, \left(q_{-1}^{(l)}, q_1^{(l)}\right)\right) \\ &= \left(p_1^{(0)}\left(q_{-1}^{(l)}\right)^{d_v-1} + p_{-1}^{(0)}\left(1-\left(q_1^{(l)}\right)^{d_v-1}\right), p_{-1}^{(0)}\left(q_1^{(l)}\right)^{d_v-1} + p_1^{(0)}\left(1-\left(q_{-1}^{(l)}\right)^{d_v-1}\right)\right)\end{aligned} \tag{2.28}$$

Therefore,

$$\begin{aligned}p_{-1}^{(l)} = p_{-1}^{(0)} - p_{-1}^{(0)}&\left[\frac{1+\left(1-2p_{-1}^{(l-1)}\right)^{d_c-1}}{2}\right]^{d_v-1} \\ &+\left(1-p_{-1}^{(0)}\right)\left[\frac{1-\left(1-2p_{-1}^{(l-1)}\right)^{d_c-1}}{2}\right]^{d_v-1}\end{aligned} \tag{2.29}$$

Equation (2.29) clearly shows the probability evolution during the iterative process. It is expected that there exists a threshold ε^*:

$$\lim_{l\to\infty} p_{-1}^{(l)} = 0, \quad \text{when } p_{-1}^{(0)} < \varepsilon^* \tag{2.30}$$

For binary AWGN channel (e.g., the input of AWGN is either 1 or –1), the output probabilities from the nodes in the bipartite graphs are all having continuous values, e.g., the probability is represented by the density functions, rather than discrete functions. In this case, a key procedure is to compute the probability of the sum of random variables.

$$\Psi\left(\mathrm{m}_0,\mathrm{m}_1,\ldots,\mathrm{m}_{d_v-1}\right):=\sum_{i=0}^{d_v-1}\mathrm{m}_i \tag{2.31}$$

which is essentially a convolution operation if the random variables are all independent of each other.

$$^{*}\Psi_v\left(P_0,P_1,\ldots P_{d_v-1}\right)=P_0\otimes P_1\otimes\cdots\otimes P_{d_v-1} \tag{2.32}$$

The analysis can be simplified by using the Laplace transform or Fourier transform. Let us use F to represent the Fourier transform and F^{-1} as inverse Fourier transform.

$$^{*}\Psi_v\left(P_0,P_1,\ldots P_{d_v-1}\right)=F^{-1}\left(F(P_0)F(P_1)\ldots F\left(P_{d_v-1}\right)\right) \tag{2.33}$$

Considering a regular LDPC of column weight and row weight as (d_v, d_c), using $P^{(l)}$ to represent the common probability density from variable nodes to check nodes in the l-th iteration, P_0 to represent the probability density of observation at the receiver, $\tilde{P}^{(l)}$ to represent the density after Fourier transform of $P^{(l)}$, we have

$$\begin{aligned} F\left(\tilde{P}\right)(s,0)&=\hat{\tilde{P}}^0(s)+\hat{\tilde{P}}^1(s)\\ F\left(\tilde{P}\right)(s,1)&=\hat{\tilde{P}}^0(s)+\hat{\tilde{P}}^1(s) \end{aligned} \tag{2.34}$$

The probability density $\tilde{Q}^{(l)}$ can be written as

$$\begin{aligned} \hat{\tilde{Q}}^{(l),0}-\hat{\tilde{Q}}^{(l),1}&=\left(\hat{\tilde{P}}^{(l-1),0}+\hat{\tilde{P}}^{(l-1),1}\right)^{d_c-1}\\ \hat{\tilde{Q}}^{(l),0}+\hat{\tilde{Q}}^{(l),1}&=\left(\hat{\tilde{P}}^{(l-1),0}+\hat{\tilde{P}}^{(l-1),1}\right)^{d_c-1} \end{aligned} \tag{2.35}$$

Combining Eqs. (2.34) and (2.35), a complete iteration can be obtained:

$$F\left(P^{(l+1)}\right)=F\left(P^{(0)}\right)\left(F\left(Q^{(l)}\right)\right)^{d_v-1} \tag{2.36}$$

TABLE 2.1 Maximum Allowed Noise Variance of a Few Regular LDPC for Binary AWGN Channel

d_v	d_c	Code Rate	Threshold of Noise Variance (σ)
3	6	0.5	0.88
4	8	0.5	0.83
5	10	0.5	0.79
3	5	0.4	1.0
4	6	0.333	1.01
3	4	0.25	1.549

Table 2.1 shows a few examples of different (d_v, d_c) and code rates of regular LDPC, and the maximum allowed noise variance in binary AWGN channel.

The performance of irregular LDPC can be analyzed in a similar fashion as of regular LDPC, except that the derivations and computations are more complicated, which is not elaborated in this book. Table 2.2 shows the minimum allowed SNR, e.g., Eb/No for some good choices of (λ, ρ) for irregular LDPC of code rate 1/2. AWGN channel are assumed.

According to Shannon's formula, for a binary AWGN channel, the limit of Eb/No is 0.187 dB for half-rate, e.g., 1 bps/Hz with QPSK modulation. It is observed in Table 2.2 that one of the irregular LDPC codes can achieve Eb/No = 0.2485 dB, which is only 0.06 from the Shannon limit. Using the weight distribution in the last column in Table 2.2, codewords of length of 10^6 can be designed. Simulations show that at a bit error rate of 10^{-6}, the required Eb/No can be as low as 0.31 dB, e.g., only 0.13 dB from the Shannon limit. This verifies that the performance analysis discussed in this section is effective and accurate.

2.3 QUASI-CYCLIC LDPC (QC-LDPC)

There are two main construction methods for LDPC: randomly generated and structure-based. Randomly generated parity check matrices do not have definite structures. Compared to regular LDPC, the irregular LDPC proposed by T. Richardson [19] can noticeably improve the performance. Such a method is suitable for long block lengths. As discussed in Section 2.2, when the block length is at the level of million bits, the performance can approach very closely the Shannon limit, if irregular LDPC is used. However, a randomly generated matrix has the issue of a lack of efficient decoding algorithms. This results in complicated decoding and low data throughput, which hampers its wide use in practical systems.

TABLE 2.2 Suitable Choices of (λ, ρ) for Rate = 1/2 Irregular LDPC and the Lowest Eb/No Allowed for Successful Decoding

d_v	**4**	**5**	**6**	**7**	**8**	**9**	**10**	**20**	**50**
λ_2	0.3836	0.3465	0.3404	0.3157	0.3017	0.2832	0.2717	0.2326	0.1838
λ_3	0.0424	0.1196	0.2463	0.4167	0.2840	0.2834	0.3094	0.2333	0.2105
λ_4	0.5741	0.1839	0.2202				0.0010	0.0206	0.0027
λ_5		0.3699							
λ_6			0.3111					0.0854	
λ_7				0.4381				0.0654	0.0001
λ_8					0.4159			0.0477	0.1527
λ_9						0.4397		0.0191	0.0923
λ_{10}							0.4385		0.0280
λ_{11}									
λ_{12}									
λ_{15}									0.0121
λ_{19}								0.0806	
λ_{20}								0.2280	
λ_{30}									0.0721
λ_{50}									0.2583
ρ_5	0.2412								
ρ_6	0.7588	0.7856	0.7661	0.4381	0.2292	0.0157			
ρ_7		0.2145	0.2339	0.5619	0.7708	0.8524	0.6368		
ρ_8						0.1319	0.3632	0.6485	
ρ_9								0.3475	0.3362
ρ_{10}								0.0040	0.0888
ρ_{11}									0.5750
σ^*	**0.9114**	**0.9194**	**0.9304**	**0.9424**	**0.9497**	**0.9540**	**0.9558**	**0.9649**	**0.9718**
Eb/No (dB)	**0.8085**	**0.7299**	**0.6266**	**0.5153**	**0.4483**	**0.4090**	**0.3927**	**0.3104**	**0.2485**

LDPC can have many structures. Among them, the quasi-cyclic structure is the most widely used. This structure was first used for regular LDPC. Due to its well-formed structure, theoretical analysis can be relatively tractable. Later on, the quasi-cyclic structure was extended to irregular LDPC which provides more design freedom and room for performance improvement. However, the structure-based design has a certain impact on the performance, e.g., not suitable for setting the record to approach the Shannon limit. Yet, the decoding algorithms can be significantly simplified and thus increasing the data throughput and reducing the processing delay.

QC-LDPC has been widely adopted by IEEE 802.11, IEEE 802.16, DVB-S2 standards, etc. where LDPC parity check matrices of several code rates are specified in order to ensure low decoding complexity, good performance, and high data throughput. For instance, in IEEE 802.11n/ac, 12 parity check matrices are specified to support four choices of code rate and three block lengths. In IEEE 802.11ad, four parity check matrices are specified to support four choices of code rate with a fixed block length. In IEEE 802.16e, six parity check matrices are specified to support four choices of code rates and 19 block lengths.

The technical merits of QC-LDPC can be summarized as follows:

- Performance can approach the Shannon limit.
- Relatively low error floor, suitable for high-reliability systems.
- Decoding is highly parallel, thus fast decoding and high data throughput.
- Decoder hardware can be commonly used for multiple code rates for a fixed block length.
- Decoding complexity decreases as the code rate is increased. Suitable for fulfilling the peak data rate requirement.

2.3.1 Matrix Lifting

The concept of quasi-cyclic structure can trace back to the original paper of LDPC [1,2] where Gallager only addressed permutation. The parity check matrix shown below has 20 columns and 15 rows. Its row weight and column weight are 4 and 3, respectively

$$H_g = \left[\begin{array}{cccccccccccccccccccc}
1&1&1&1&0&0&0&0&0&0&0&0&0&0&0&0&0&0&0&0\\
0&0&0&0&1&1&1&1&0&0&0&0&0&0&0&0&0&0&0&0\\
0&0&0&0&0&0&0&0&1&1&1&1&0&0&0&0&0&0&0&0\\
0&0&0&0&0&0&0&0&0&0&0&0&1&1&1&1&0&0&0&0\\
0&0&0&0&0&0&0&0&0&0&0&0&0&0&0&0&1&1&1&1\\
\hline
1&0&0&0&1&0&0&0&1&0&0&0&1&0&0&0&0&0&0&0\\
0&1&0&0&0&1&0&0&0&1&0&0&0&0&0&0&1&0&0&0\\
0&0&1&0&0&0&1&0&0&0&0&0&0&1&0&0&0&1&0&0\\
0&0&0&1&0&0&0&0&0&0&0&0&0&0&1&0&0&0&1&0\\
0&0&0&0&0&0&0&0&1&0&0&0&0&0&0&1&0&0&0&1\\
\hline
1&0&0&0&0&1&0&0&0&0&0&1&0&0&0&0&0&1&0&0\\
0&1&0&0&0&0&1&0&0&0&1&0&0&0&0&1&0&0&0&0\\
0&0&1&0&0&0&0&1&0&0&0&0&1&0&0&0&0&0&1&0\\
0&0&0&1&0&0&0&0&1&0&0&0&0&1&0&0&1&0&0&0\\
0&0&0&0&1&0&0&0&0&1&0&0&0&0&1&0&0&0&0&1
\end{array}\right]$$

Note that this matrix does not restrict circulant permutation. Let us divide the matrix into 12 sub-matrices, each having three rows and four columns. It is observed that the sub-matrices of row weight=2 are in fact the shift of 5 by 5 identity matrix. Here there is no strict limitation on the shift, as long as it satisfies certain requirements for the column weight and the row weight.

If a parity check can be lifted from a base matrix, the representation of parity check matrices can be very concise, regardless of the block length. This is compatible to the modular design concept. Consider the following lifted matrix **H**:

$$H = \begin{bmatrix} P^{h_{00}^b} & P^{h_{01}^b} & P^{h_{02}^b} & \cdots & P^{h_{0nb}^b} \\ P^{h_{10}^b} & P^{h_{11}^b} & P^{h_{12}^b} & \cdots & P^{h_{1nb}^b} \\ \cdots & \cdots & \cdots & \cdots & \cdots \\ P^{h_{mb0}^b} & P^{h_{mb1}^b} & P^{h_{mb2}^b} & \cdots & P^{h_{mbnb}^b} \end{bmatrix} = P^{H_b} \tag{2.37}$$

Here the subscripts i and j represent the row index and column index of the sub-block matrix, respectively. If $h_{ij}^b = -1$, we define $P^{h_{ij}^b} = 0$, e.g., the sub-block matrix is a 0 square matrix. If h_{ij}^b is a non-negative integer, we define $P^{h_{ij}^b} = (P)^{h_{ij}^b}$. Each non-zero **P** is a $z \times z$ permutation matrix, circularly shifted from an identity matrix. The below is a permutation matrix of exponent=1, e.g., cyclic shift=1.

$$P = \begin{bmatrix} 0 & 1 & 0 & \cdots & 0 \\ 0 & 0 & 1 & \cdots & 0 \\ \cdots & \cdots & \cdots & \cdots & \cdots \\ 0 & 0 & 0 & \cdots & 1 \\ 1 & 0 & 0 & \cdots & 0 \end{bmatrix}$$

The exponent (permutation factor) h_{ij}^b can be used to uniquely define each sub-block matrix. The exponent of the identity sub-block matrix is 0. The exponent of the 0 sub-block matrix is –1, or empty. Hence, the parity check matrix **H** can be represented using its exponents, e.g., an exponent matrix $m_b \times n_b$, denoted as H_b. Here, H_b is the base matrix and **H** is the lifted matrix. In practical encoding, z=block length/number of columns in the base matrix, which is often called lifting factor or lifting size.

For instance, the parity matrix **H** can be lifted by the lifting factor $z=3$ from a base matrix $\mathbf{H}_b$ of size $m_b \times n_b = 2\times 4$.

$$H_b = \begin{bmatrix} 0 & 1 & 0 & -1 \\ 2 & 1 & 2 & 1 \end{bmatrix}$$

Tanner (bipartite, or factor) graph of this base matrix is shown in Figure 2.7.

The lines (edges) in Figure 2.7 represent non-zero elements in the base matrix. The lines connecting to check node 1 are highlighted in bold, circled with a solid line, dashed line, and dotted line. When the lifting factor $z=3$, the lifted matrix is

$$H = \left[\begin{array}{c} 1\ 0\ 0 \,\|\, 0\ 1\ 0 \,\|\, 1\ 0\ 0 \,\|\, 0\ 0\ 0 \\ 0\ 1\ 0 \,\|\, 0\ 0\ 1 \,\|\, 0\ 1\ 0 \,\|\, 0\ 0\ 0 \\ 0\ 0\ 1 \,\|\, 1\ 0\ 0 \,\|\, 0\ 0\ 1 \,\|\, 0\ 0\ 0 \\ \hline 0\ 0\ 1 \,\|\, 0\ 1\ 0 \,\|\, 0\ 0\ 1 \,\|\, 0\ 1\ 0 \\ 1\ 0\ 0 \,\|\, 0\ 0\ 1 \,\|\, 1\ 0\ 0 \,\|\, 0\ 0\ 1 \\ 0\ 1\ 0 \,\|\, 1\ 0\ 0 \,\|\, 0\ 1\ 0 \,\|\, 1\ 0\ 0 \end{array}\right]$$

Figure 2.8 is the Tanner graph of the lifted matrix.

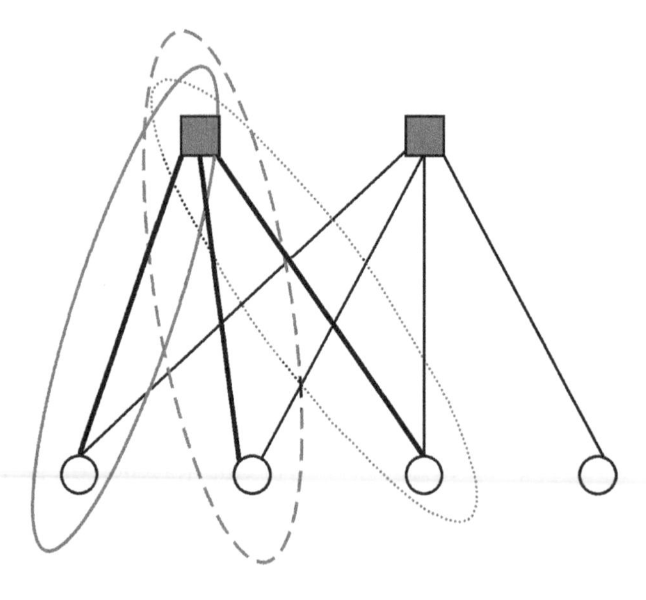

FIGURE 2.7 Tanner (bipartite, or factor) graph of 2 by 4 base matrix.

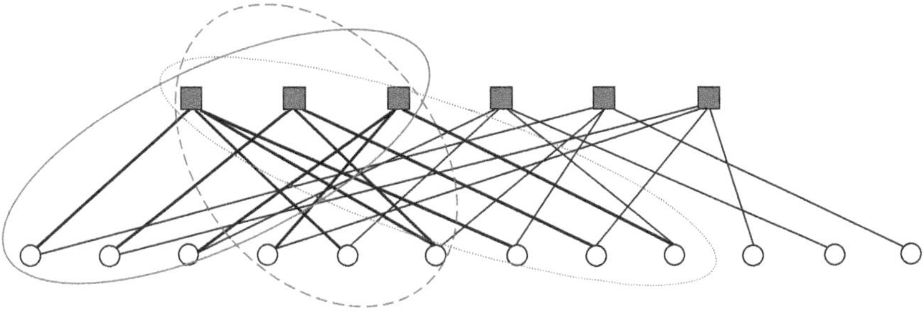

FIGURE 2.8 Tanner graph of lifted matrix.

It can be seen from Figure 2.8 that in the lifted matrix, the nodes (including both variable nodes and check nodes) of the base matrix are copied three times (including the edges). When the element in the base matrix is 0, the permutation matrix is an identity matrix, meaning that the relationship in the edges is kept unchanged, as shown in the solid-line-circled section and in the dotted-line-circled section. When the element in the base matrix is a positive integer, elements in the permutation matrix are circular shifted accordingly, resulting in the cyclic shifting of edges, as seen in the dashed-line-circled section.

It is well known that cyclic shift can be easily implemented in digital circuits. The essence of base matrix lifting-based construction is to convert the design of a multi-block-length parity check matrix into two sub-tasks: the design of the base matrix with a fixed block length and the design of the sub-block permutation matrix to fit different block lengths. The fixed part (base matrix) can be hard coded into the core hardware where basic encoding and decoding algorithms of LDPC can be used to iteratively update the soft information. The adjustable part (quasi-cyclic permutation matrix) only copies the soft information in the core hardware and conducts the cyclic shift within the sub-block matrix, e.g., to adjust the input and output nodes for the soft information. Hence, during each iteration, this processing can be simultaneously carried out in different sub-block matrices. If there are enough hardware elements, parallel processing can be fully achieved. Based on the above discussion, it is clear that a QC-LDPC can be uniquely defined by the base matrix, the lifting factor, and the permutation matrix.

For QC-LDPC, a lifted matrix can be obtained by one cyclic permutation matrix as seen in the above examples so far. This means that the row weight and the column weight of each sub-block matrix are 1. This is often called a single-edge matrix. By contrast, if a sub-block matrix has

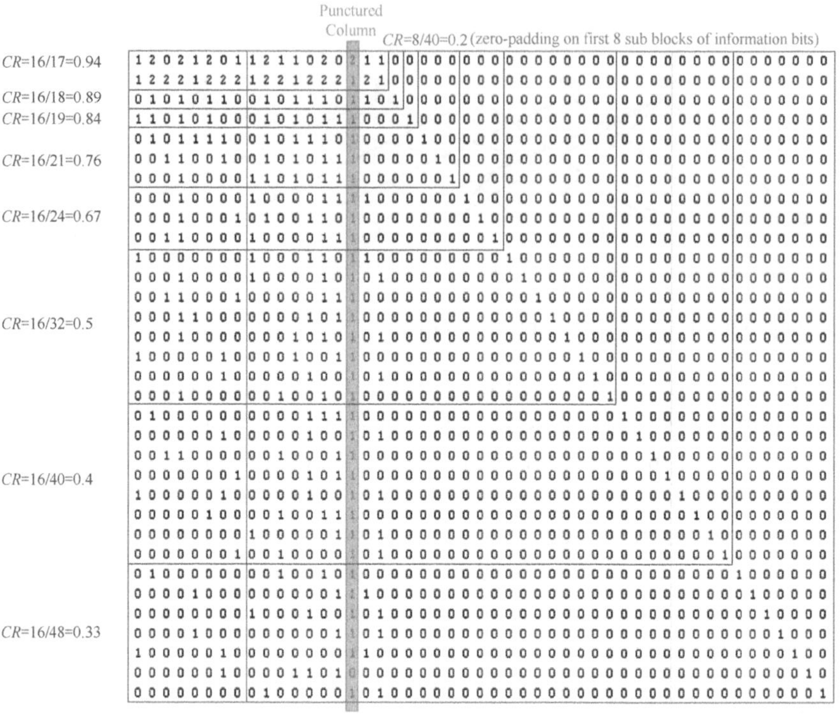

Punctured Column

CR=8/40=0.2 (zero-padding on first 8 sub blocks of information bits)

Code rate	Matrix row
CR=16/17=0.94	12021201 1211020 2 110000000000000000000000000000000
	12221222 1221222 1 210000000000000000000000000000000
CR=16/18=0.89	01010110 0101110 1 101000000000000000000000000000000
CR=16/19=0.84	11010100 0101011 1 000100000000000000000000000000000
	01011110 0101110 1 000010000000000000000000000000000
CR=16/21=0.76	00110010 0101011 1 000001000000000000000000000000000
	00010000 1101011 1 000000100000000000000000000000000
	00010000 1000011 1 100000010000000000000000000000000
CR=16/24=0.67	00010001 0100110 1 000000001000000000000000000000000
	00110000 1000011 1 000000000100000000000000000000000
	10000000 1000110 1 100000000010000000000000000000000
	00010000 1000010 1 010000000001000000000000000000000
	00110001 0000011 1 000000000000100000000000000000000
CR=16/32=0.5	00011000 0000101 1 000000000000010000000000000000000
	00010000 0001010 1 010000000000001000000000000000000
	10000010 0001001 1 000000000000000100000000000000000
	00000010 0000100 1 010000000000000010000000000000000
	00010000 0010010 1 000000000000000001000000000000000
	01000000 0000111 1 000000000000000000100000000000000
	00000010 0000100 1 010000000000000000010000000000000
	00110000 0010001 1 000000000000000000001000000000000
CR=16/40=0.4	00000001 0000101 1 000000000000000000000100000000000
	10000010 0000100 1 010000000000000000000010000000000
	00000100 0010011 1 000000000000000000000001000000000
	00000000 1000001 1 010000000000000000000000100000000
	00000001 0010000 1 010000000000000000000000010000000
	01000000 0010010 1 000000000000000000000000001000000
	00001000 0000001 1 100000000000000000000000000100000
	00000000 1000100 1 010000000000000000000000000010000
CR=16/48=0.33	00001000 0000001 1 010000000000000000000000000001000
	10000010 0000000 1 100000000000000000000000000000100
	00000010 0001101 0 000000000000000000000000000000010
	00000000 0100000 1 010000000000000000000000000000001

FIGURE 2.9 An example of multi-edge base matrix of LDPC [20].

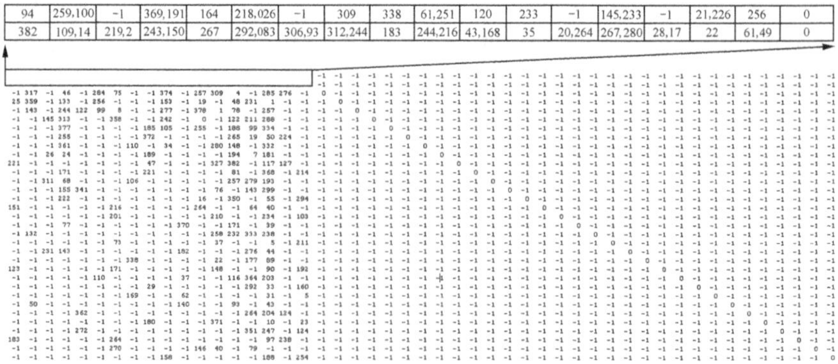

94	259,100	−1	369,191	164	218,026	−1	309	338	61,251	120	233	−1	145,233	−1	21,226	256	0
382	109,14	219,2	243,150	267	292,083	306,93	312,244	183	244,216	43,168	35	20,264	267,280	28,17	22	61,49	0

[illegible]

FIGURE 2.10 Exponents of permutation matrices of a multi-edge LDPC [20].

a column weight or row weight more than 1, it is often called a multi-edge matrix. A multi-edge matrix can be understood as the summation of multiple different sub-block matrices, each being a cyclic shifted matrix from an identity matrix. Figures 2.9 and 2.10 [20] show some examples of

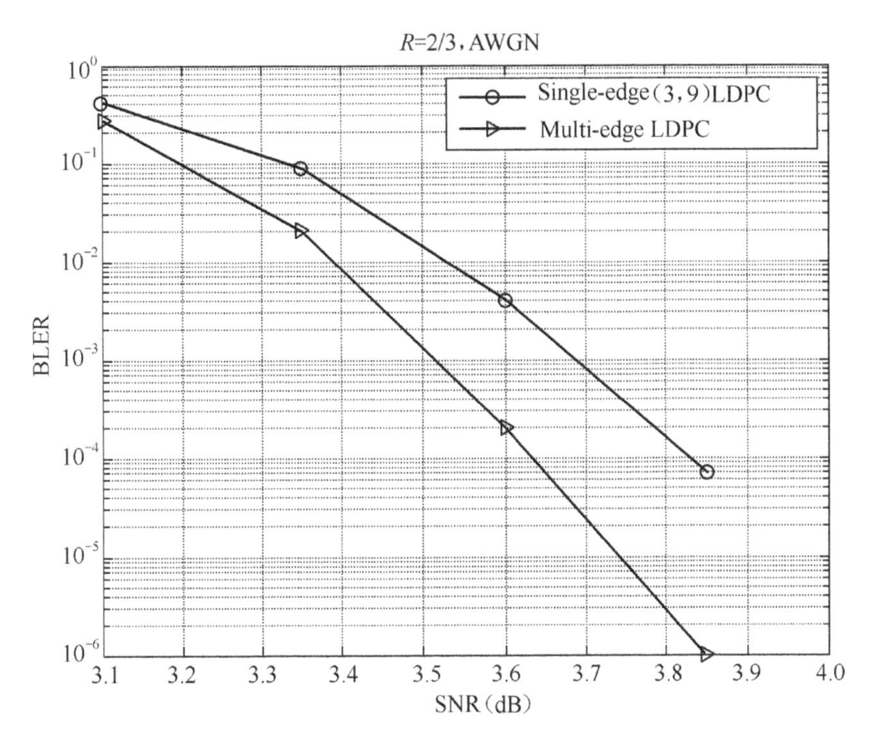

FIGURE 2.11 Performance comparison between single-edge LDPC and multi-edge LDPC [21].

multi-edge base matrices. In particular, some of the cells in Figure 2.10 have two different shifting values.

The merit of multi-edge construction is to bring more design flexibility for QC-LDPC. For example, when the code rate is high, the number of columns and rows is typically small. With the multi-edge approach, the column weight can be increased in certain columns, thus improving the performance. As shown in Figure 2.11 [21], the link performance of multi-edge LDPC is better than that of single-edge LDPC by about 0.15 at block error rate (BLER) = 1% for AWGN channel.

Even though multi-edge LDPC can improve the performance, it brings some issues in hardware implementation of the decoding. When reading data from the memory because of the conflict of the address, multi-edge matrices do not allow outputting the data and writing on the memory to be finished all at once. More control logics need to be added, which would increase the decoder complexity and latency. With these considerations, multi-edge LDPC was not adopted in 5G NR channel coding.

2.3.2 Basic Structure of Base Matrix

A base matrix of mother-rate QC-LDPC has two parts $\mathbf{H}_b$: systematic bit part H_b^{system} and parity bit part H_b^{parity}.

$$\mathbf{H}_b = \left[\mathbf{H}_b^{\text{system}} \mid \mathbf{H}_b^{\text{parity}}\right] \tag{2.38}$$

LDPC codes in 5G NR have a "Raptor-like" structure. Their parity check matrix can be gradually extended to the lower code rate from a high-code-rate "kernel matrix", so that flexible code rates can be supported. More specifically, the parity check matrix has the following structure as shown in Figure 2.12.

Matrices **A** and **B** in Figure 2.12 form the high-code-rate kernel part. More specifically, matrix **A** corresponds to the information bits, and matrix **B** is a square matrix having a dual-diagonal structure, corresponding to the parity bits for high code rate. Matrix **C** is an all-zero matrix. Matrix **E** is an identity matrix, corresponding to the parity bits of a low code rate. Matrices **D** and **E** form a single-parity check relationship. Two sizes of base matrices are specified: 46*68 and 42*52, to support long block length of a high code rate and short block length of a low code rate, respectively.

Part of the base matrix of LDPC for 5G NR is shown in Figure. 2.13 where k_{bmax} is the maximum number of columns of systematic bits. m_b is the number of columns (or rows) in the parity check part. n_b is the total number of columns of the base matrix. The gray-filled elements in the systematic bit part have non-negative values. 0 elements correspond to the identity permutation matrix.

The upper left part circled by the solid lines in Figure 2.13 is the kernel matrix, corresponding to the highest code rate designed for 5G LDPC codes. In the parity check part of the kernel matrix, in addition to the dual-diagonal structure, a weight-3 column is added to the left in order to reduce the encoder's complexity, e.g., to avoid matrix inversion computation during the encoding. For low to medium code rate part, a single-parity check is used to reduce the code rate. That is, once the codeword of

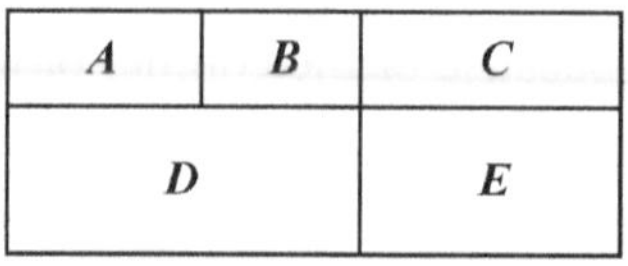

A	*B*	*C*
D		*E*

FIGURE 2.12 Structure of parity check matrices of LDPC codes for 5G NR.

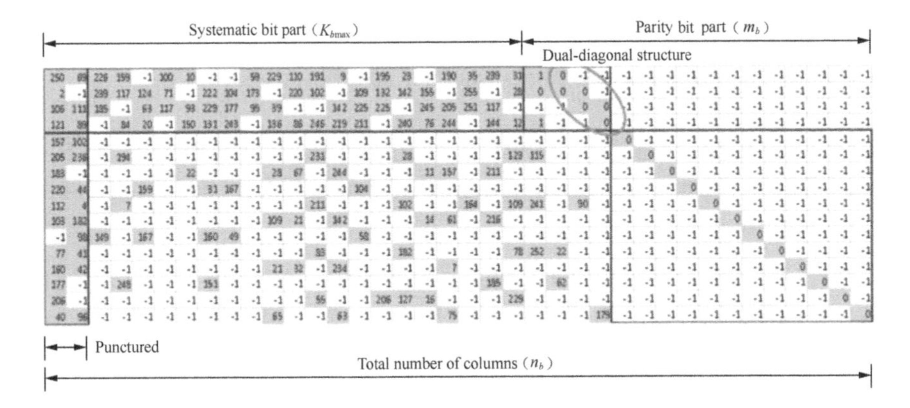

FIGURE 2.13 Part of a base matrix of LDPC for 5G NR.

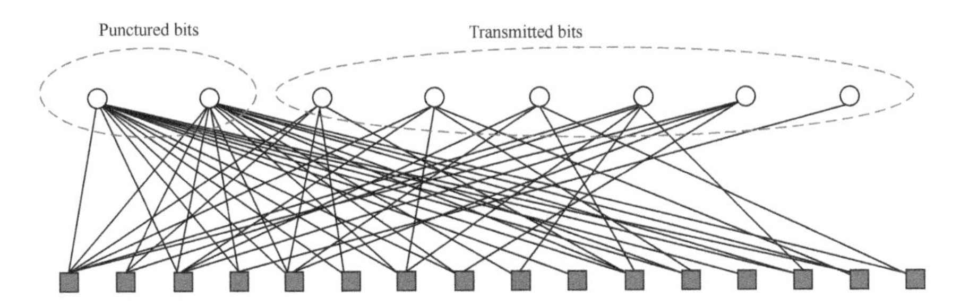

FIGURE 2.14 Tanner graph of lifted matrix (partial).

a high code rate is generated, a simple parity relation can be used to get the parity bits for the low code rate.

It can be seen that the two left-most columns have a very high column weight. The purpose is to ensure full connection between the parity check nodes and the first few variable nodes so that the soft information exchange or BP can be made more fluent and smoother. Figure 2.14 shows the Tanner (bipartite or factor) graph of the information bit part in Figure 2.13. Obviously, the first two variable nodes are connected to most of the check nodes. A large amount of analysis and simulations have shown that if the information bits corresponding to the left-most columns can be punctured, the performance of QC-LDPC can be further improved. It should be noted that even though the first two system bits are punctured, their connections to the check nodes still exist, which means that these two variable nodes can still be used for BP (message passing) in the decoder.

2.3.3 Encoding Algorithms

The basic encoding process of 5G LDPC is as follows

1. Encode the information bits according to the kernel matrix (**A**; **B**). Due to the dual-diagonal structure in matrix **B**, fast encoding can be achieved.
2. If the target code rate is higher than the code rate of the kernel matrix, puncture the parity bits. If the target code rate is lower than the code rate of the kernel matrix, use matrix (**D**; **E**) which contains single-parity relations to generate more parity bits.

In addition, to ensure the performance of the initial transmission, the information bits corresponding to the two left-most columns in matrix **A** are punctured.

- **Encoding for the Kernel Matrix.** Since LDPC codes are defined by parity check matrices and most LDPC codes are systematic codes, it is possible to encode the information bits according to the parity check matrices, without calculating the generation matrices of LDPC. Also, a QC-LDPC can be uniquely defined by a base matrix, a lifting factor, and a permutation matrix. Hence, the encoding of QC-LDPC can be based on these three key ingredients. The principle in the computation of QC-LDPC encoding is still that the product of the parity check matrix and the codeword is 0.

$$\mathbf{P}^{\mathbf{H}_b} \times \mathbf{C}^T = 0 \tag{2.39}$$

where $\mathbf{P}^{\mathbf{H}_b}$ is the parity check matrix of QC-LDPC. **P** is the standard permutation matrix (circularly right-shifted identity matrix) whose size is the lifting factor. $\mathbf{H}_b$ is the base matrix and **C** is the codeword of QC-LDPC. **0** is an all-0 vector whose size equals the number of rows in the parity check matrix. The codeword **C** can be written as $\mathbf{C} = [\mathbf{u}, \mathbf{v}]$ where **u** is the systematic bits of the codeword and **v** is the parity bits of the codeword. **u** and **v** can be partitioned into a number of groups, each of size z (lifting factor) where **u**(i) or **v**(i) represent the i-th group of length z. Similarly, the parity check matrix can also be written into two parts: $\left[\mathbf{H}_{\text{system}}, \mathbf{H}_{\text{parity}}\right] = \left[\mathbf{P}^{\mathbf{H}_b^{\text{system}}}, \mathbf{P}^{\mathbf{H}_b^{\text{parity}}}\right]$ where $\mathbf{H}_{\text{system}}$ is the systematic bit part in the parity check matrix and $\mathbf{H}_{\text{parity}}$ is the parity bits part of the parity check matrix. Since LDPC is a type of systematic codes, the encoding process of LDPC is essential to compute the parity bits. Hence, Eq. (2.39) can be rewritten as

$$[\mathbf{H}_{\text{system}},\mathbf{H}_{\text{parity}}]\times[\mathbf{u},\mathbf{v}]^T=0 \tag{2.40}$$

Due to the binary operation, we have

$$\mathbf{P}^{\mathbf{H}_b^{\text{system}}}\times\mathbf{u}^T=\mathbf{P}^{\mathbf{H}_b^{\text{parity}}}\times\mathbf{v}^T \tag{2.41}$$

Compute the adjugate matrix λ, e.g., the systematic bits part (left part in Eq. (2.41)), written in the form of matrices and vectors:

$$\lambda(i)=\sum_{j=0}^{k_b-1}\mathbf{P}^{\mathbf{H}_b^{(i,j)}}\cdot\mathbf{u}(j)\qquad i=0,1,\cdots,m_b-1 \tag{2.42}$$

It can be seen that $\lambda(0)$ is the dot product of the systematic bits in the codeword and the first-row elements in matrix H_b^{system}. In the binary domain, all "+" and "−" operations are XOR. The notation $\mathbf{P}^{\mathbf{H}_b^{(i,j)}}\cdot\mathbf{u}(j)$ means left cyclic shift of the vector $\mathbf{u}(j)$ by the value of $\mathbf{H}_b^{(i,j)}$. Similarly, $\lambda(1)$ is the dot product of the systematic bits in the codeword and the second-row elements in matrix H_b^{system}. $\lambda(2)$ is the dot product of the systematic bits in the codeword and the third-row elements in matrix H_b^{system}, and so on.

It is noticed that has $\mathbf{H}_b^{\text{parity}}$ dual-diagonal structure. Extending each row of Eq. (2.42), we have

$$\begin{aligned}
&\lambda(0)=\mathbf{v}(0)+\mathbf{v}(1);\\
&\lambda(0)=\mathbf{v}(1)+\mathbf{v}(2);\\
&\vdots\\
&\lambda(x)=\mathbf{P}^{\mathbf{H}_b^{(x,kb_{\max})}}\mathbf{v}(0)+\mathbf{v}(x)+\mathbf{v}(x+1);\\
&\lambda(m_b'-1)=\mathbf{v}(0)+\mathbf{v}(m_b'-1).
\end{aligned} \tag{2.43}$$

where m_b' is the number of parity check columns in the kernel matrix. x is the row index of the second non-negative element in the column whose weight is 3. Adding all the equations in Eq. (2.43) together, we get

$$\mathbf{P}^{\mathbf{H}_b^{(x,kb\max)}} \cdot \mathbf{v}(0) = \sum_{i=0}^{m_b'-1} \lambda(i) \tag{2.44}$$

Then $\mathbf{v}(0)$ can be calculated:

$$\mathbf{v}(0) = \mathbf{P}^{(z-\mathbf{H}_b^{(x,kb\max)}) \bmod z} \sum_{i=0}^{m_b'-1} \lambda(i) \tag{2.45}$$

The rest of parity bits $\mathbf{v}(i)$ can be obtained in a recursive manner. This is because that

$$\lambda(0) = \mathbf{v}(0) + \mathbf{v}(1); \tag{2.46}$$

Then we get $\mathbf{v}(1)$, and so on

$$\mathbf{v}(i+1) = \mathbf{v}(i) + \lambda(i) \tag{2.47}$$

Note that both forward and backward recursion can be carried out in dual-directional recursion. The forward process is from v(0) to v(1) and then to v(2), and so on. The backward process is from $\mathbf{v}(m_b'-1)$ to $\mathbf{v}(m_b'-2)$ and then to $\mathbf{v}(m_b'-3)$, and so on. This is because

$$\lambda(m_b'-1) = \mathbf{v}(0) + \mathbf{v}(m_b'-1) \tag{2.48}$$

We get $\mathbf{v}(m_b'-1)$. Similarly,

$$\lambda(m_b'-2) = \mathbf{v}(m_b'-2) + \mathbf{v}(m_b'-1) \tag{2.49}$$

$\mathbf{v}(m_b'-2)$ can be obtained. Dual-directional recursion can improve the speed of encoding. By following the above procedure, the codeword $\mathbf{c}' = [\mathbf{u}, \mathbf{v}]$ can be generated from the kernel matrix.

- **Encoding Process for a Low Code Rate.** Since the extended part of the base matrices of 5G LDPC has a single-parity check structure, it is quite easy to compute the parity bits when the code rate is low. The computation is as follows.

$$\mathbf{c}(i)=\sum_{j=0}^{i-1}\mathbf{P}^{\mathbf{H}_b^{(i,j)}}\times\mathbf{c}(j) \tag{2.50}$$

where $i=m_b',m_b'+1,m_b$. The first m_b' vectors in the codeword $\mathbf{c}$ denoted as $\mathbf{c}'$, are calculated from the kernel matrix.

2.3.4 QC-LDPC Design for Flexible Block Length

According to the principle of QC-LDPC, a QC-LDPC can be uniquely defined by its base matrix, lifting factor and permutation matrix. The block length of a QC-LDPC can be determined by the number of columns in the systematic bit part of the base matrix, denoted as kb, and the lifting factor z, e.g., $K=kb\times z$, If kb is kept the same and the lifting factor z can be arbitrary, the granularity of the block length of QC-LDPC is kb. If the methods of filler bits [22–25] are used, the granularity of the block length can be 1 bit.

In 3GPP 5G NR, two protomatrices (or base graph) are defined for eMBB. Each base graph has eight base parity check matrices. The first base graph can support up to 8448 information bits (larger than 6144 in LTE) and its maximum number of columns in the systematic bit part $kb_{\text{max}}=22$. The second base graph can support up to 3840 information bits and its maximum number of columns in the systematic bit part $kb_{\text{max}}=10$. Hence, the maximum lifting factor is 8448/22=384. In theory, the lifting factor z can be any positive integer smaller than 384. However, a too flexible lifting factor would increase the burden of hardware implementation, especially the circuit and pin layout. The lifting factor has a relationship with the parallel level of LDPC decoder. When the parallel level is a power of 2, Banyan switch can be used for the shifting network of LDPC decoder, which reduces the implementation complexity.

The shape of a Banyan switch is like a butterfly and it is very widely used in the shifting network. A Banyan switch consists of $J=\log_2(\text{PM})$ levels. In each level, there are K=PM/2 number of "2-2" switches. Each: "2-2" switch is made up of two "2-1 MUX" elements (values). Hence, there are total PM* $\log_2(\text{PM})$ of "2-2" elements. Here PM is called the parallel level, e.g.,

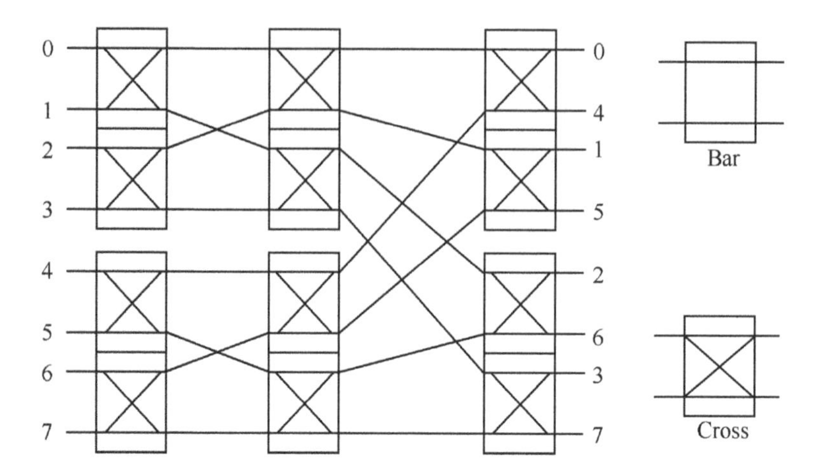

FIGURE 2.15 An example of Banyan switches.

the number of input ports of a Banyan switch. As shown in Figure 2.15, when the switching control of the value is 0 (the state of BAR), the upper output pin of the "2-1 MUX" element is connected to the upper input pin. Otherwise (e.g., the state of CROSS), the upper input pin of the "2-1 MUX" element is connected to the lower output pin, and the upper lower input pin is connected to the upper output pin.

Figure 2.15 shows a Banyan switch of the parallel level up to 8. Each switch "2-2" $S_{k,j}$, (k=1,2,..., K, j=1,2,..., J) can operate in "CROSS" or "BAR" state, via controlling signal. Considering an 8 by 8 permutation matrix (quasi-cyclic shift matrix), when the corresponding element is 1, the identity matrix is shifted to the right by one step. Then the relationship between the input and the output becomes 01234567 → 12345670. It means that switches $S_{2,1}$, $S_{4,1}$, $S_{1,2}$, $S_{3,2}$, and $S_{1,4}$ should be set to the "CROSS" state, whereas the rest of the "2-2" switches should be set to the " BAR" state.

Banyan switch is very efficient if the parallel level is a power of 2. When the parallel level is not a power of 2, other shifting networks would be needed, for instance, the QSN (QC-LDPC shift network) switch which can support arbitrary lifting factors and shifts. However, the complexity of the QSN switch is higher than that of the Banyan switch. A QSN switch has $\log_2$(PM)+1 levels and total number of PM*(2*$\log_2$(PM)-1)+1≈2*PM*$\log_2$(PM) "2-1 MUX" elements. Figure 2.16 shows the overall structure of a QSN switch consisting of a left-shifting network, a right-shifting network, and a combining network.

Since the shifting network constitutes a large portion of the decoding complexity of LDPC, the design of lifting factors should strive to use

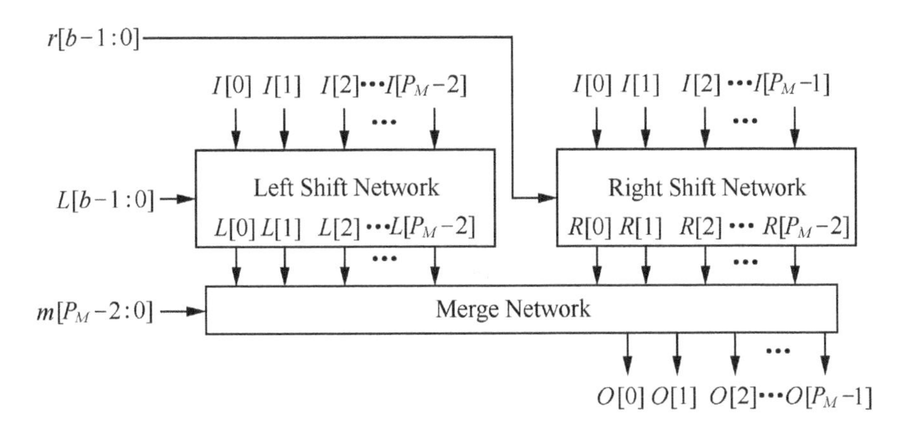

FIGURE 2.16 Overall structure of a QSN switch.

TABLE 2.3 $V_{i,j}$ Values Corresponding to Base Matrix (Base Graph) BG1

Row Index i	Column Index j	Set Index i_{LS}							
		0	1	2	3	4	5	6	7
0	0	250	307	73	223	211	294	0	135
	⋮	⋮	⋮	⋮	⋮	⋮	⋮	⋮	⋮
	23	0	0	0	0	0	0	0	0
1	0	2	76	303	141	179	77	22	96
	⋮	⋮	⋮	⋮	⋮	⋮	⋮	⋮	⋮
	24	0	0	0	0	0	0	0	
				⋮					
45	1	149	135	101	184	168	82	181	177
	6	151	149	228	121	0	67	45	114
	10	167	15	126	29	144	235	153	93
	67	0	0	0	0	0	0	0	0

the Banyan network for LDPC decoding in order to reduce the decoding complexity.

The lifting factor Z_c can be determined according to the block length of information bits. Then based on the lifting factor Z_c and the set index i_{LS} in Table 2.3 to find the shifting value. The element in the i-th row and j-th column of the base matrix can be obtained with the following equation:

$$P_{i,j} = \begin{cases} -1, & V_{i,j} = -1 \\ \mathrm{mod}(V_{i,j}, Z_c), & \text{otherwise} \end{cases} \tag{2.51}$$

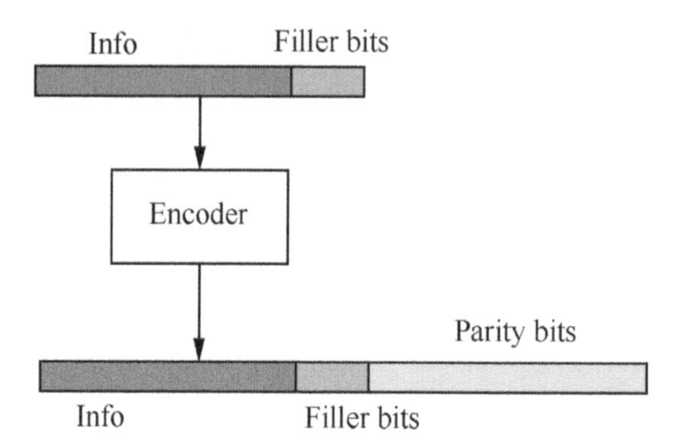

FIGURE 2.17 An illustration of filler bits appended to the end of a block of information bits.

where $V_{i,j}$ and $P_{i,j}$ are the shift values before and after the adjustment, respectively. $V_{i,j}$ values are listed in Table 2.3.

The calculated number of information bits $K=kb*z$ may not be equal to the code block size (CBS), meaning that (K – CBS) filler bits need to be added. As illustrated in Figure 2.17, filler bits are all appended to the end of the information bit block. By adjusting the lifting factor and filler bits, QC-LDPC can support the arbitrary length of block size.

2.3.5 Multi-Code Rate Design for QC-LDPC

In theory, the performance of LDPC can be fully optimized by designing a base matrix specifically for each code rate. However, this would significantly increase the complexity of hardware implementation and standards specification, especially when these base matrices do not have any nested relationship. For a LDPC base matrix, the following two methods [22,24–26] can be used to adjust the code rate. The multi-code rate considered in this section is mostly for high code rates. The case of a low code rate will be discussed in the subsequent section.

As shown in Figure 2.18, there are k_b columns in the systematic part and m_b columns in the parity check part, with a total of $n_b=(k_b+m_b)$ columns in the base matrix. Hence, the mother code rate is $k_b/n_b=k_b/(k_b+m_b)$. If the first few systematic columns are punctured, the mother code rate becomes $k_b/(k_b-p_b+m_b)$. The code rate can be increased by selecting the first m_b'' rows, that is, the first m_b' columns in the parity check part. The corresponding code rate is $k_b/\left(k_b-p_b+m_b''\right)$. As the code rate keeps increasing, when it exceeds $k_b/\left(k_b-p_b+m_b'\right)$, an even higher code rate can be achieved by

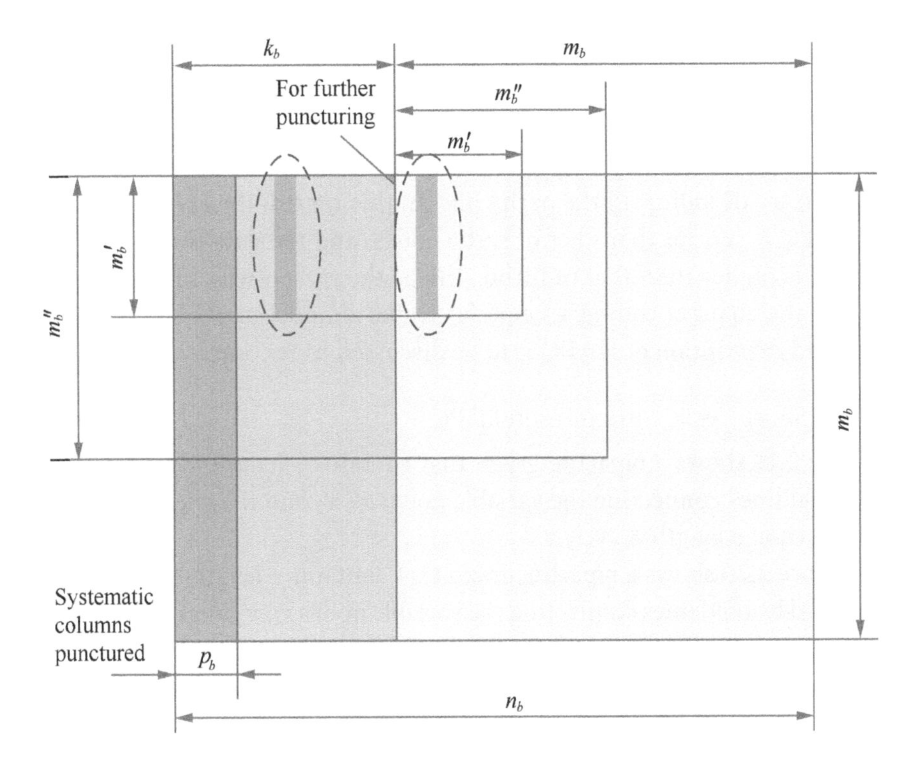

FIGURE 2.18 Illustration of multi-code rate QC-LDPC.

further puncturing the bits to be transmitted, till the highest code rate is supported by the specification. For these high code rates, both the encoding and the decoding operate in the kernel matrix.

In addition to the scaling and puncturing (essentially to puncture with the granularity of lifting factor) of base matrices, the code rate can be more finely adjusted by bit puncturing after getting the lifted matrices.

2.3.6 Fine Adjustment of Code Rate for QC-LDPC

As discussed in Section 2.3.5, different code rates can be achieved by selecting different subsets of a base matrix. As the code rate decreases, more parity bits, or redundant information, would be generated by the encoder. Hence, QC-LDPC inherently supports the incremental redundancy (IR) type of HARQ retransmission, till the code rate is reduced to the mother code rate.

Different from turbo codes, the redundancy reduction of LDPC is not simply to puncture the parity bits. Instead, the punctured parity bits in LDPC's parity check matrix would no longer participate the parity checking operation. Hence, the extrinsic information of these punctured bits will not be updated anymore, in order to reduce the decoding latency and improve the data throughput. During the retransmission, the LDPC decoder only needs to process the parity bits already transmitted, e.g., no need to try decoding all the parity bits. In this regard, the HARQ latency of LDPC is shorter than that of turbo codes, and the data throughput of LDPC is higher than that of turbo. This is the reason why LDPC is chosen to the channel coding scheme for traffic channel in 5G NR. A more detailed description of HARQ is to be discussed in Section 2.5.4.

2.3.7 Short Cyclic Structures in LDPC

Figure 2.19 shows a bipartite graph that contains a length-4 girth where the bold lines connecting the variable nodes x_1, x_2 and the check nodes c_1, c_2 construct a length-4 cycle.

Figure 2.20 shows a bipartite graph that contains a length-4 cycle constructed by bold lines connecting the variable nodes x_1, x_2, x_3 and the check nodes c_1, c_2, c_4.

A girth is often used to describe the short cycles in a bipartite graph. More precisely, in graph theory, the girth refers to the length of the shortest cycle in a bipartite graph. For instance, if a bipartite graph has cycles of length-6, length-8, length-10, and length-12, the girth of this bipartite

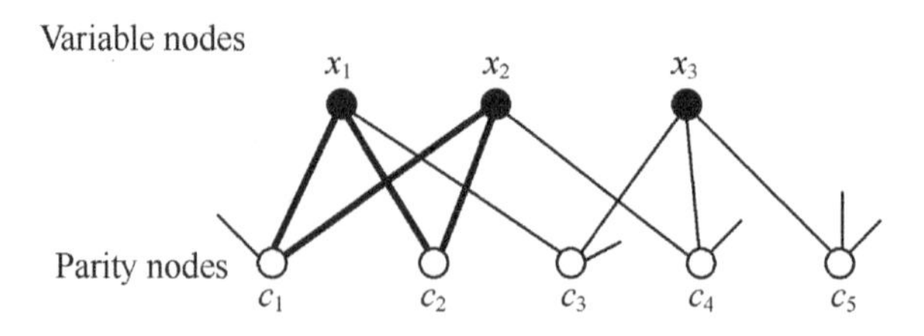

FIGURE 2.19 An illustration of short cycle of length-4 in a bipartite graph.

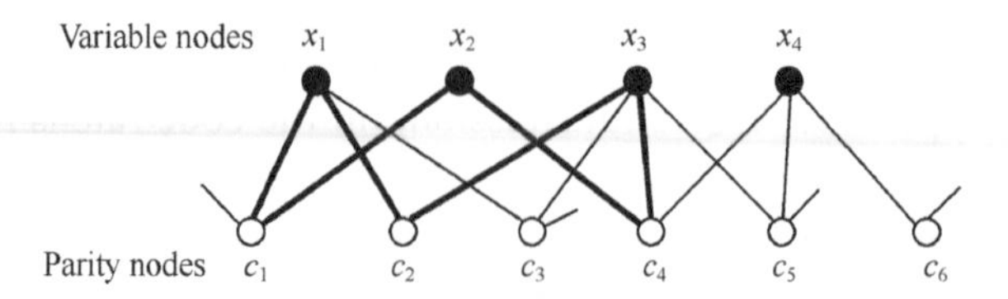

FIGURE 2.20 An illustration of short cycle of length-6 in a bipartite graph.

graph is 6. A girth can be defined respective to a node or to an edge. The girth at node *u* refers to the length of the shortest cycle connecting the node *u*. The girth at edge *e* refers to the length of the shortest cycle that includes the edge *e*.

The girth at a variable node equals the minimum number of iterations between the output message from this node and the input message to this node. In well-designed LDPC codes, the message from this node has already been thoroughly passed to the rest of the nodes in the bipartite graph, well before reaching the minimum number of iterations determined by the girth. In another word, the longer the girth of a variable node, the lower the chance of self-feedback, thus the better the decoder's performance.

The parity check matrix of LDPC can be represented as bipartite graphs in a one-to-one mapping. A parity check matrix **H** of size M*N defines a set of constraints: each codeword of length N satisfies M parity checks. The bipartite graph contains N variable nodes, each corresponding to a bit in **H**. The bipartite graph contains also M parity bits, each corresponding to a parity check in **H**. Each parity bit is connected to all the variable bits to be parity checked. The total number of edges in a bipartite graph is equal to the number of non-zero elements in a parity check matrix.

Message passing for LDPC codes assumes that variable nodes are independent of each other. The existence of short cycles violates the assumption of independence and thus degrades the performance of the iterative decoder due to the positive feedback. Even though turbo codes also rely on iterative decoding, the turbo interleaver can reduce the chance of positive feedback. For the acyclic Tanner graph (cycle-free), message passing algorithm is the optimal. When short cycle exists, message passing is a suboptimal algorithm.

Figure 2.21 shows a length-4 cycle in a parity check matrix. Figure 2.22 shows a length-6 cycle in a parity check matrix.

It is seen in Figures 2.21 and 2.22 that regardless of whether regular or irregular LDPC, it is important to reduce or eliminate short cycles when designing a parity check matrix. More specifically, to maximize the girth or reduce the number of girths.

2.3.8 Short Cycle Characteristics of QC-LDPC

Figure 2.23 shows a base matrix length-4 cycle made up of four identity matrices with certain exponents, which may form a length-4 cycle or cycles of other lengths.

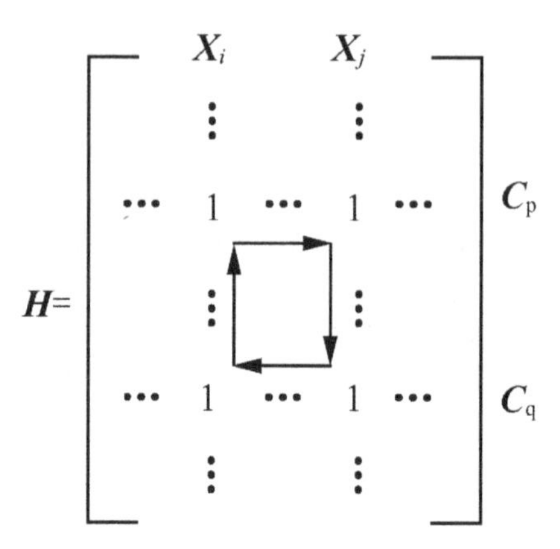

FIGURE 2.21 A length-4 cycle in a parity check matrix.

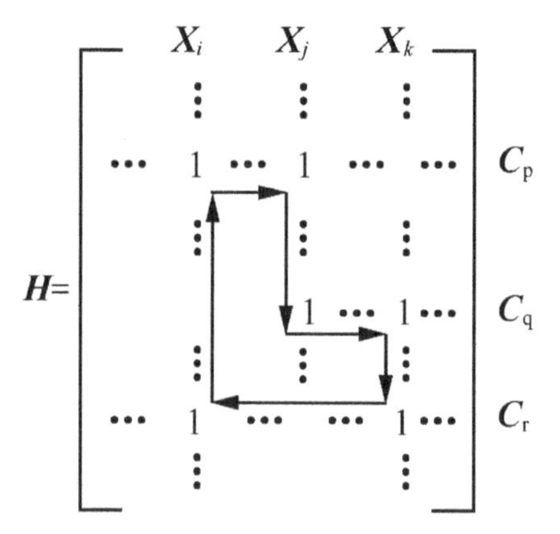

FIGURE 2.22 A length-6 cycle in a parity check matrix.

According to the definition of a base matrix of QC-LDPC, a base matrix and its corresponding parity check matrix are essentially the same, except that the base matrix is a condensed form of the parity check matrix. Given a lifting factor, the girth of a base matrix is the girth of the lifted parity check matrix. As shown in Figure 2.23, comparing the parity check matrix and the bipartite graph, when there is a length-4 cycle in the base matrix $\mathbf{H}_b$, there would be length-4 or longer cycles in the parity check matrix H lifted from $\mathbf{H}_b$. More specifically, the four $z \times z$ matrices P^i, P^j, P^k, and P^l are not all-zero matrices. The four exponents i, j, k, and l may form cycles of different lengths. When $\mathrm{mod}(i - j + k - l, z) = 0$, P^i, P^j, P^k, and P^l form length-4 cycle in the parity check matrix H; when z is an even number, and

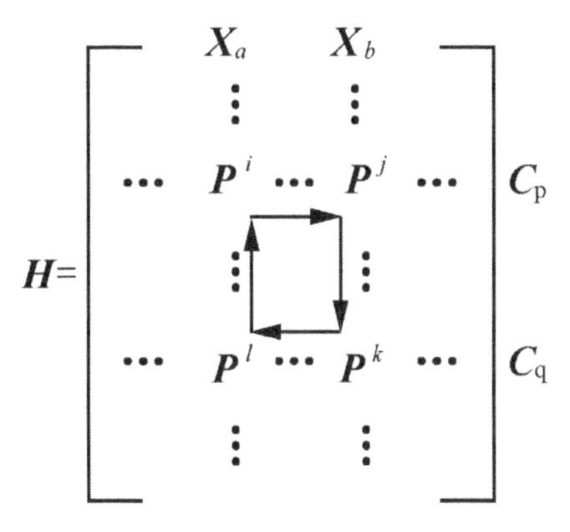

FIGURE 2.23 A lifted parity check matrix that may contain length-4 or longer cycles.

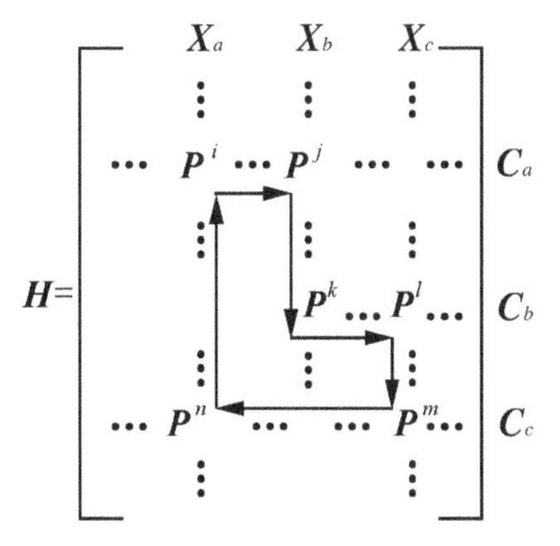

FIGURE 2.24 A lifted parity check matrix that may contain length-6 or longer cycles.

$\text{mod}(i-j+k-l, z/2)=0$, P^i, P^j, P^k, and P^l form length-8 cycles in the parity check matrix H. In all other cases, P_i, P_j, P_k, and P_l form length-12 cycles or do not form cycles in H.

Similarly, Figure 2.24 shows a length-6 cycle in the base matrix $\mathbf{H}_b$, where there would be length-6 or longer cycles in the matrix H lifted from $\mathbf{H}_b$. More specifically, the six $z \times z$ matrices P^i, P^j, P^k, P^l, P^m, and P^n are determined by the column indices $\{C_a, C_b, C_c\}$ and row indices $\{R_a, R_b, R_c\}$. They are not all-zero matrices. The six exponents i, j, k, l, m, and n may form cycles of different lengths. When $\text{mod}(i-j+k-l+m-n, z)=0$, P^i, P^j, P^k, P^l, P^m, and P^n form a length-6 cycle in the parity check matrix H;

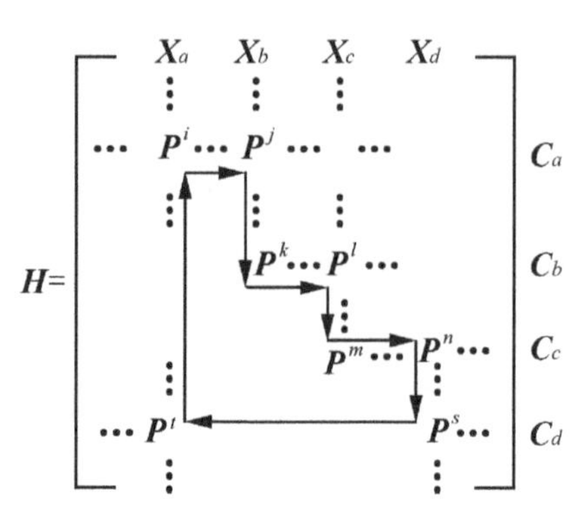

FIGURE 2.25 A lifted parity check matrix that may contain length-8 or longer cycles.

when z is an even number, and mod$(i - j + k - l + m - n, z/2)$=0, P^i, P^j, P^k, P^l, P^m, and P^n form length-12 cycles in the parity check matrix H.

Figure 2.25 shows another example of a length-8 cycle in the base matrix $\mathbf{H}_b$, where there would be length-8 or longer cycles in the matrix H lifted from $\mathbf{H}_b$. More specifically, the eight $z \times z$ matrices P^i, P^j, P^k, P^l, P^m, P^n, P^s, P^t are determined by the column indices $\{C_a, C_b, C_c, C_d\}$ and row indices $\{R_a, R_b, R_c, R_d\}$. They are not all-zero matrices. The eight exponents i, j, k, l, m, n, s and t may form cycles of different lengths. When mod$(i\text{-}j\text{+}k\text{-}l\text{+}m\text{-}n\text{+}s\text{-}t, z)$, P^i, P^j, P^k, P^l, P^m, P^n, P^s, and P^t form length-8 cycle in the parity check matrix H; when z is an even number, and mod $(i - j + k - l + m - n + s - t, z/2) = 0$, P^i, P^j, P^k, P^l, P^m, P^n, P^s, and P^t form length-16 cycles in the parity check matrix H.

In real applications, a parity check matrix is often obtained by lifting the base matrix. Quite often, the size of the permutation matrix is an even number. If some elements of a base matrix do not form short cycles, the block matrices in the lifted matrix that correspond to those elements in the parity check matrix would not form short cycles either.

In summary, if the lifting factor is an even number, the following can be concluded:

- The sufficient and necessary condition of having girth ≥6 in LDPC parity check matrix is for any four elements i, j, k, l that form a length-4 cycle by counter-clockwise (or clockwise) rotation, mod$(i - j + k - l, z) \neq 0$ should be satisfied.
- The sufficient and necessary condition of having girth ≥8 in LDPC parity check matrix is for any four elements i, j, k, l that form a

length-4 cycle by counter-clockwise (or clockwise) rotation, $mod(i - j + k - l, z) \neq 0$ should be satisfied, and for any six elements i, j, k, l, m, n that form a length-6 cycle by counter-clockwise (or clockwise) rotation, $mod(i - j + k - l + m - n, z) \neq 0$ should be satisfied.

- The sufficient and necessary condition of having girth ≥10 in LDPC parity check matrix is for any four elements i, j, k, l that form a length-4 cycle by counter-clockwise (or clockwise) rotation, $mod(i - j + k - l, z) \neq 0$ should be satisfied, and for any six elements i, j, k, l, m, n that form a length-6 cycle by counter-clockwise (or clockwise) rotation, $mod(i - j + k - l + \underline{\mathbf{m}} - n, z) \neq 0$ should be satisfied, and for any eight elements i, j, k, l, m, n, s, t that form a length-8 cycle by counter-clockwise (or clockwise) rotation, $mod(i - j + k - l + m - n + s - t, z) \neq 0$ should be satisfied.
- When girth ≥ 10, further increasing the girth would not significantly improve the decoder's performance. Hence, it is more important to reduce the chance of length-4, length-6, and length-8 cycles.

LDPC codes that satisfy the above properties are called high-girth LDPC. Rate = 3/4 code of [27] and rate = 5/6 code of [26] are all high-girth, which are widely used in LDPC and WiMAX.

2.4 DECODER STRUCTURES OF QC-LDPC

There are roughly three types of decoder architecture of QC-LDPC [28], as illustrated in Figures 2.26, 2.27, and 2.28, respectively. The core processing units are check node unit (CNU) and variable node unit (VNU). The primary function of CNU is to update the check node. Algorithms such as Min-Sum are all carried out inside the logic gates in CNU. The primary function of VNU is to update the variable node. Note that VNU can be achieved by updating the soft information inside the memory. Hence, in some decoders' architecture, VNU is not shown as an explicit module. During the decoding process, soft information propagates between CNU and VNU at parallel level of P.

LDPC decoding can have two scheduling methods: flooding and layered. In flooding type of scheduling, during each decoding iteration, first, calculate all the soft information from the variable nodes to the check nodes. Then calculate all the soft information from the check nodes to the variable nodes. The flooding method is more suitable for the full-parallel structure and computer simulations.

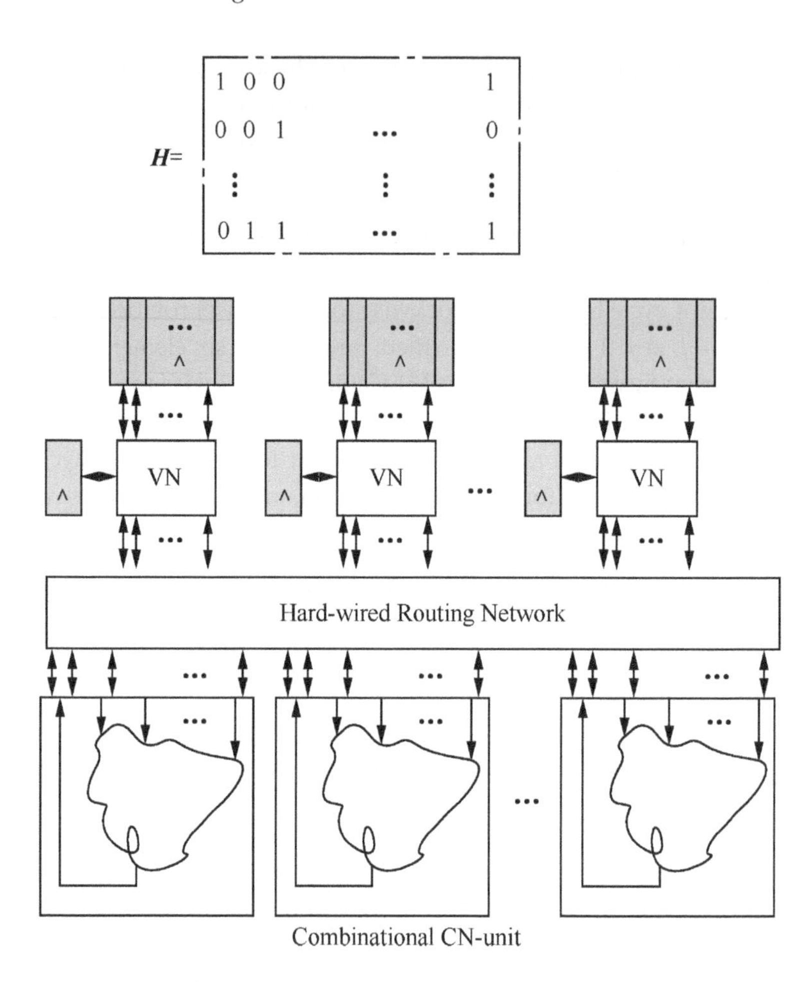

FIGURE 2.26 An illustration of full-parallel LDPC decoder.

In the layered type of scheduling, for each layer, only update the related nodes which are to be used for the soft information in the next layer. The layered method is more suitable for the row-parallel and block-parallel structures. It can also reduce the number of iterations. Compared to the flooding method, layered method can save 50% of the total number of iterations.

2.4.1 Full-Parallel Decoding

Each basic processing unit inside a LDPC decoder carries out various operations related to the parity check equation of LDPC, including the reading of data, the update of variable node, the update of check node, and the writing of data. The BP algorithm of LDPC decoding is essentially

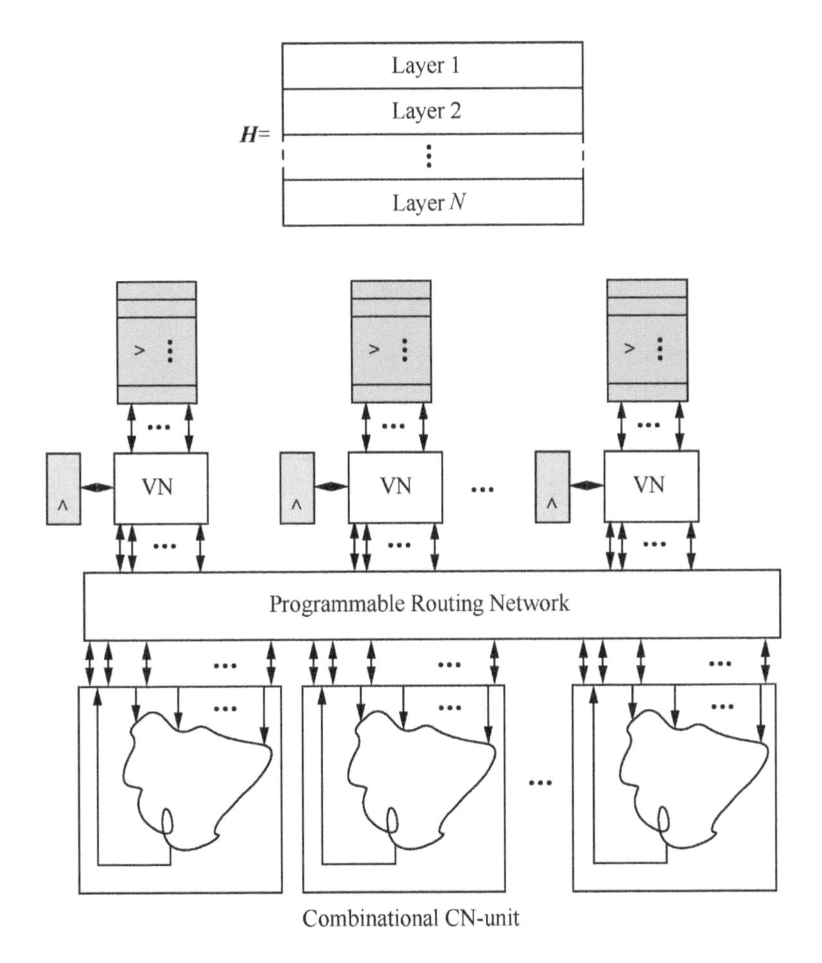

FIGURE 2.27 An illustration of row-parallel LDPC decoder.

parallel processing, which puts LDPC codes in advantage over turbo codes when the code rate is high and the block length is long. Parallel processing helps to improve the data throughput, and is suitable for large bandwidth and high SNR deployments. Hence, if the bottleneck is not the hardware complexity, that is, the number of basic processing units is not strictly restricted, and a full-parallel structure would be preferable. Figure 2.29 shows the soft information flowing inside the Tanner graph. In the upper part of Figure 2.29, all the soft information from the variable nodes is input to all the check nodes for updating. In the lower part of Figure 2.29, all the soft information from the check nodes are input to all the variable nodes for updating. Figure 2.26 is the hardware architecture of the full-parallel decoder which requires quite a number of hardware resources.

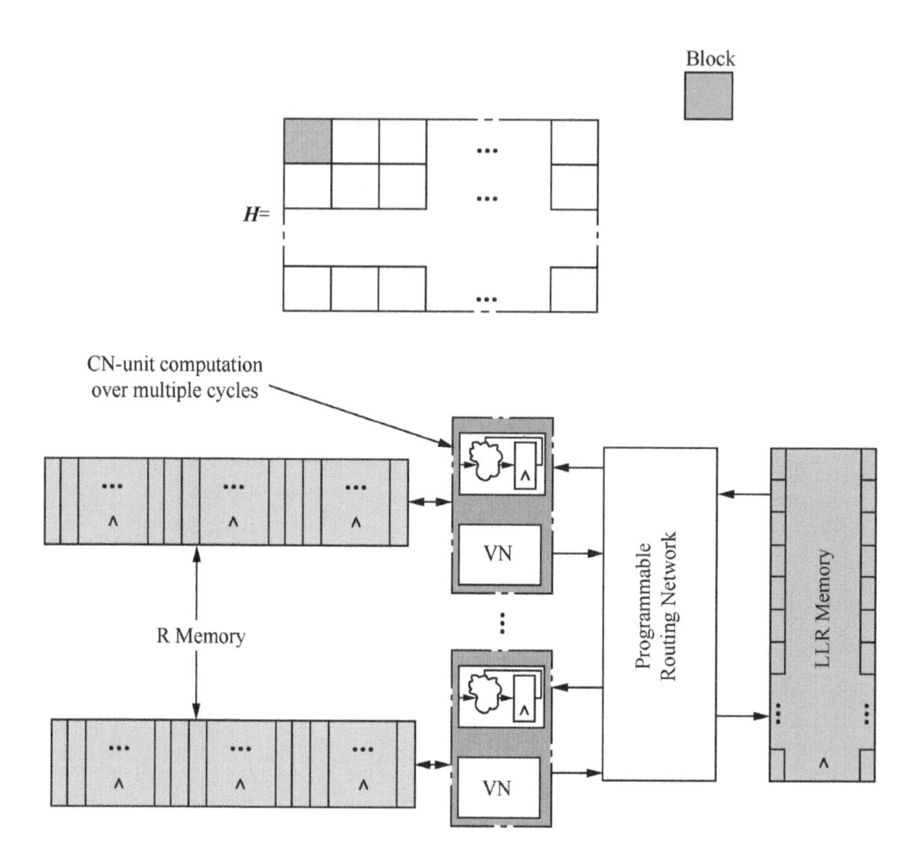

FIGURE 2.28 An illustration of block-parallel LDPC decoder.

In full-parallel decoding, all the parity check equations (each corresponding to a basic processing unit) are calculated simultaneously. Considering that the mother code rate of a base matrix is 1/3, then the base matrix has mb=16 rows, nb=24 columns, and lifting factor z=500. Hence there are totally 16*500=8000 parity check equations, e.g., 8000 basic processing units running at the same time, for each decoding iteration. Assuming the maximum number of iterations is ITER, the total number of clock cycles is Clk=a*ITER, where a is the number of clock cycles for each iteration. Then the throughput can be calculated as follows:

$$\text{Throughput} = f * K / \text{Clk} = f * K / (\text{a} * \text{ITER}) \tag{2.52}$$

where f is the operating frequency and K is the length of code block. Assuming that f=200 MHz, ITER=20, the basic processing unit takes 8 clock cycles to finish, and the throughput is 5 Gbps. Note that here 8000

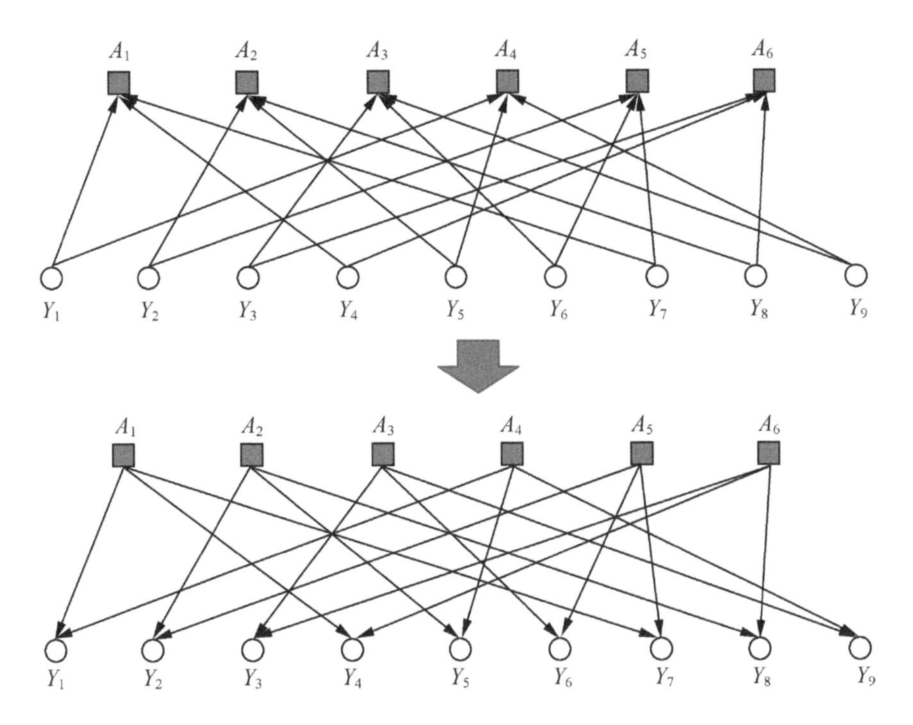

FIGURE 2.29 An illustration of Tanner graph for full-parallel LDPC decoding.

basic processing units are needed, which imposes heavy burden on hardware implementation. Hence, in practical systems, full-parallel architecture is rarely used.

In the above full-parallel architecture, all the check nodes need to be provided with hardware resources. This would cause big waste of hardware, especially for an arbitrary length of LDPC and variable code rate.

2.4.2 Row-Parallel Decoding

Row-parallel decoding is often associated with layered scheduling. Its main difference from full-parallel decoding is that in each decoding iteration, check nodes are divided into several groups. Check nodes belonging to the first group would be updated. Then, the variable nodes connecting to the first group of check nodes are updated. Then, the check nodes connect to those variable nodes and so on till the last group of nodes, as illustrated in Figures 2.30 and 2.31. During the decoding, the soft information from the previously updated variable nodes can be used for the information updating of the check nodes in the current iterations that are connected to the variable nodes of the previous iteration. Because of this,

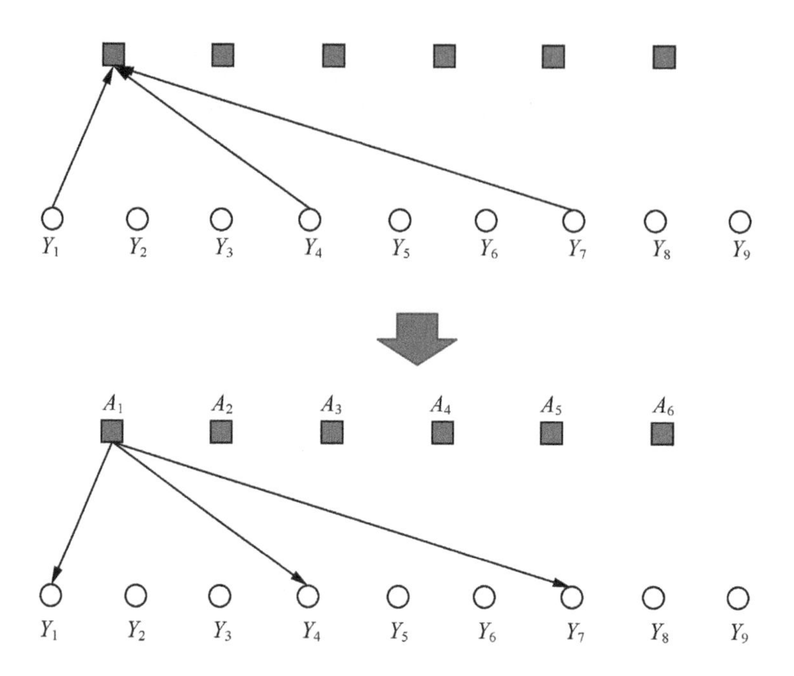

FIGURE 2.30 Row-parallel (layered structure) of LDPC decoding (the first layer).

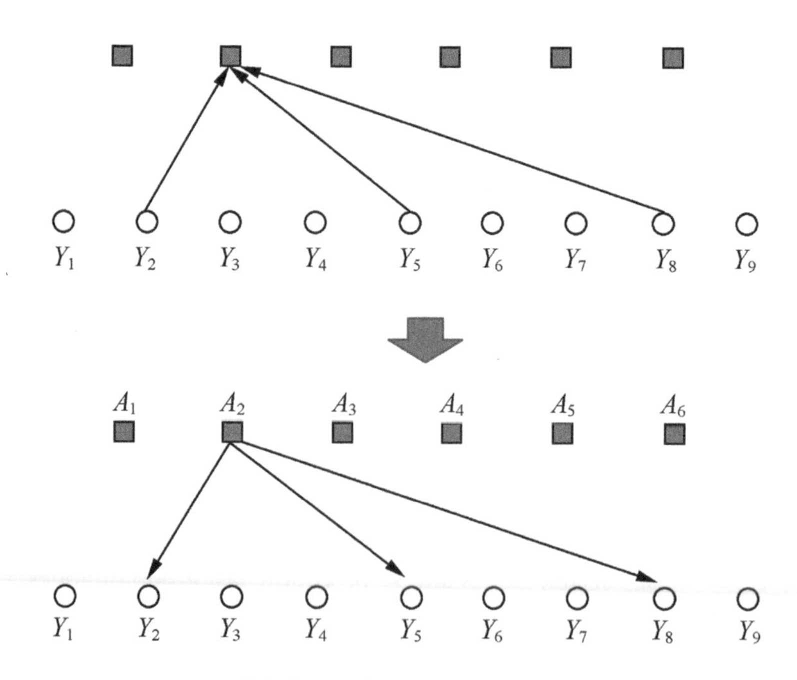

FIGURE 2.31 Row-parallel (layered structure) of LDPC decoding (the second layer).

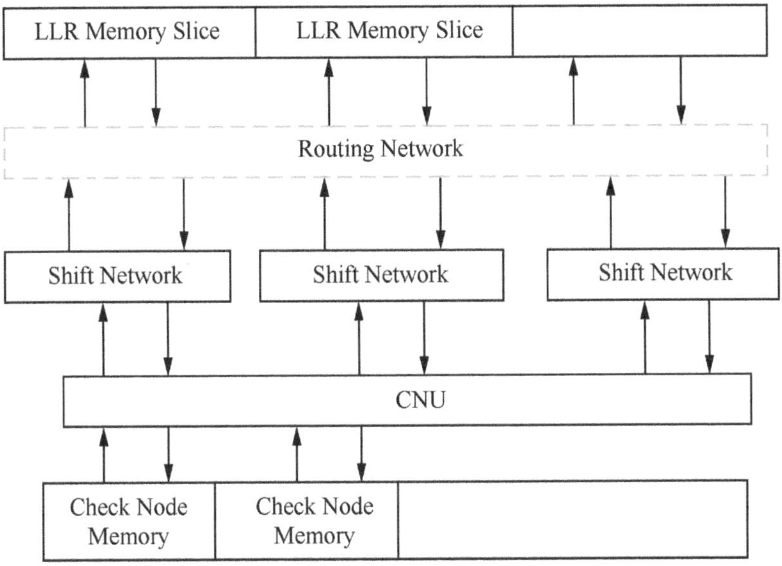

FIGURE 2.32 Structure of a row-parallel decoder.

the decoding convergence is faster, e.g., roughly half the number of total iterations required for full-parallel decoding.

The structure of a row-parallel decoder is illustrated in Figure 2.32, including memory, route network, shifting network, CNU, controller, etc. The number of input pins of CNU is equal to the maximum row weight of the base matrix. The number of memory slices is roughly equal to the number of columns in the kernel matrix part of the base matrix.

Take the scaled Min-Sum algorithm as an example. Each CNU contains a comparison unit, selection unit, addition unit, scaling unit, etc. Figure 2.33 shows the inner structure of the CNU where the number of CNU input pins should be at least equal to the maximum row weight of the base matrix. The more number of input pins is, the more complex is the CNU.

As Figure 2.34 shows, the function of the routing network is to move the cyclic shifted data from the memory slices to the pins of CNU. The layered decoding is accomplished by reading the data from the pins belonging to different groups of nodes. The routing network can reduce the number of input pins required for the CNU. However, if the base matrix is very big, the complexity of the routing network would increase drastically, especially for a decoder to support multiple base matrices, e.g., base graphs (BGs). The routing network in Figure 2.34 supports two BGs with kb=32 and 10, respectively.

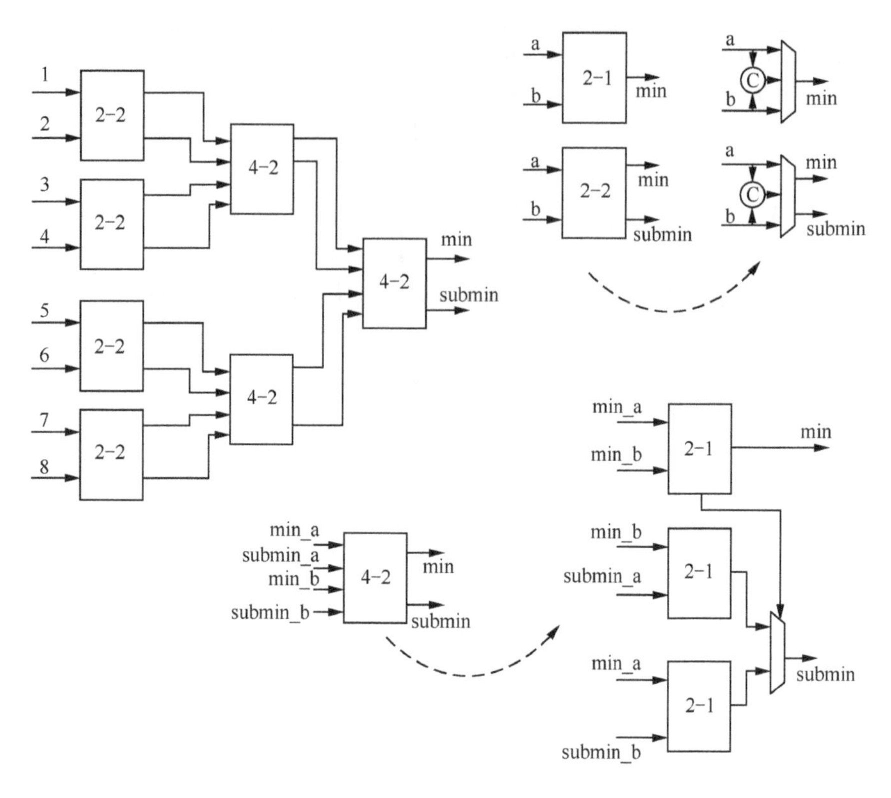

FIGURE 2.33 Inner structure of CNU.

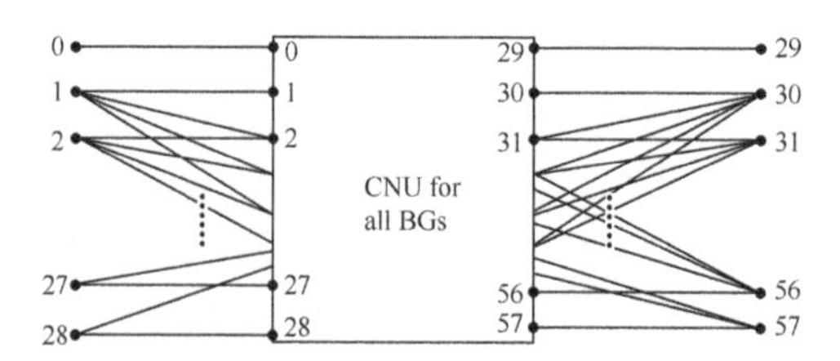

FIGURE 2.34 An illustration of routing network.

When a base matrix (e.g., base graph) is small, the complexity of routing network can be significantly reduced. Sometimes, the one-to-one mapping may be possible to connect each memory slice and each CNU pin, as shown in Figure 2.35 where kb of the base graph is 16.

Shifting networks have been described in Section 2.3.4, for instance, Banyan networks. The primary function of shifting networks is to support parallel processing under different lifting factors. The complexity of shifting networks is related to the base matrix. The number of shifting

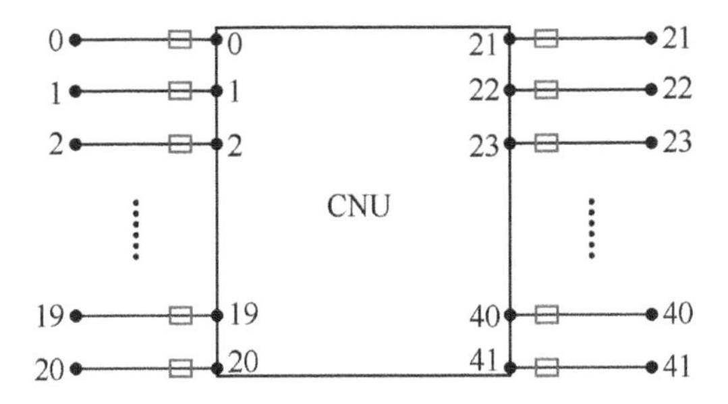

FIGURE 2.35 An illustration of simplified routing network.

networks required is proportional to the number of columns in the systematic bit part of the base matrix.

The major function of the memory is to store the LLR and the check nodes. The memory size is primarily determined by the maximum length of the encoded block. Its complexity is also related to the number of memory slices. Thus, the longer the block of information bits, the lower the code rate, and the more memory is required. Given the same number of information bits, the more number of memory slices, the more complicated is the memory.

2.4.3 Block-Parallel Decoding

For each row/layer of base matrix, the processing can be further divided into multiple cycles or blocks, as seen in Figure 2.36 which comprises memory, shifting network, CNU, controller, and connections. The level of parallel processing of block-parallel decoding is lower than that of row-parallel decoding. Hence, its throughput is lower than that of row-parallel. Since there is no conflict in the memory of each basic processing unit, parallel processing can be carried out for each basic processing unit.

Compared to row-parallel decoding, the number of shifting networks required for block-parallel decoding is much less, e.g., two would often be enough, one for reading from the memory of LLR, the other for writing into the memory of LLR. Hence, the complexity of shifting network and the number of memory slices for block-parallel are almost independent of the row weight of the base matrix. It should be pointed out that if the aim is to increase the throughput of the block-parallel decoder to approach that of the row-parallel decoder, the level of parallel processing of the block-parallel decoder has to be drastically increased, thus increasing the complexity of shifting networks.

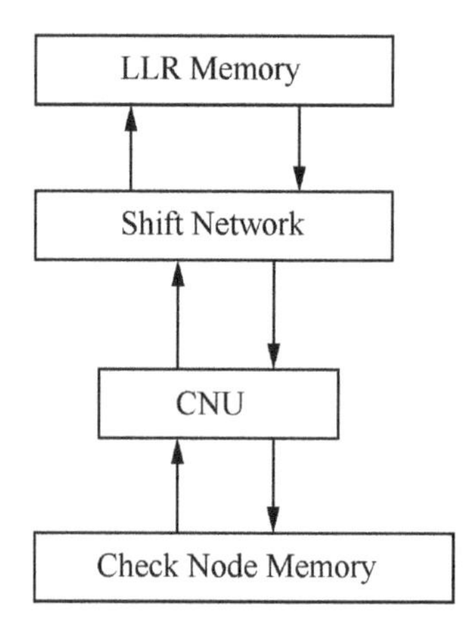

FIGURE 2.36 An illustration of block-parallel decoding.

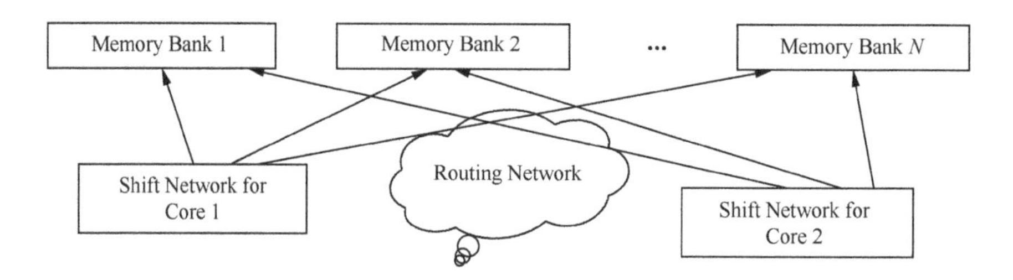

FIGURE 2.37 Routing network and shifting networks for two simultaneously processed blocks.

To improve the processing speed of block-parallel, multiple blocks can be processed simultaneously, which requires adding routing networks and shifting networks. When the number of simultaneous blocks is small, the routing networks and shifting networks can be a little simpler than those of row-parallel. Figure 2.37 shows an example of the routing network and the shifting networks for two simultaneously processed blocks.

The inner structure of the CNU of the block-parallel decoder is shown in Figure 2.38 which is composed of three 2-to-1 operating circuits and two comparators. Compared to row-parallel, the complexity of CNU for block-parallel is lower. The read/write operation for memory slices is more like a serial operation. In this case, the complexity of CNU has nothing to do with the number of memory slices, nor to the row weight of the base matrix. It is only related to the maximum level of parallel processing supported by the CNU.

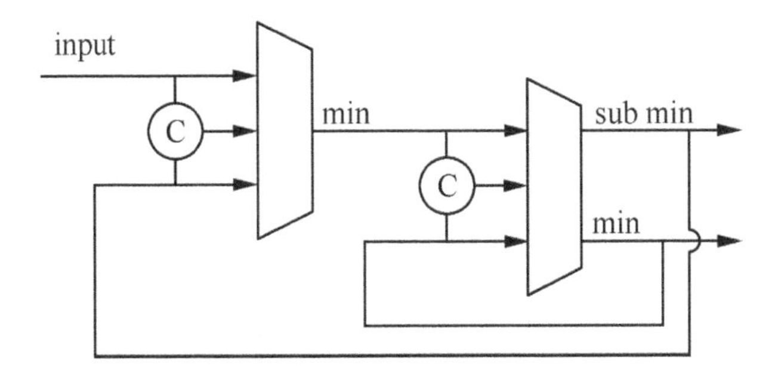

FIGURE 2.38 Inner structure of a CNU.

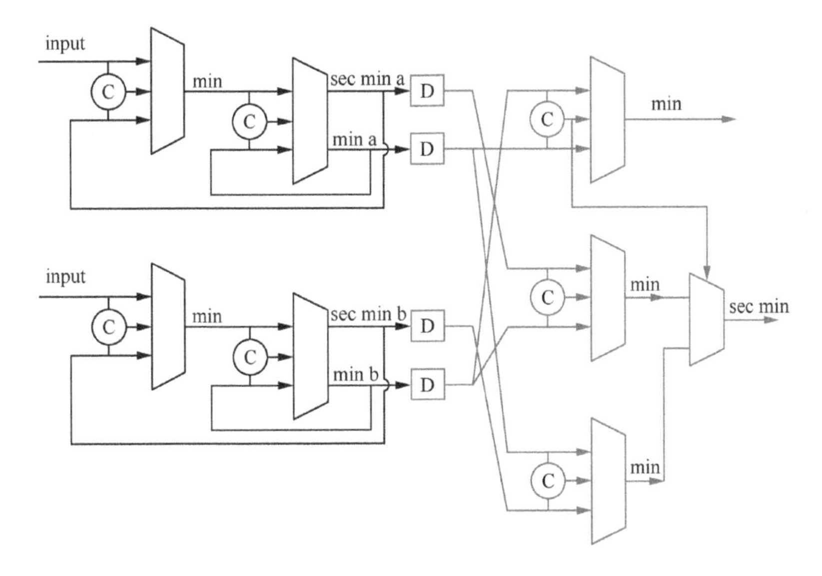

FIGURE 2.39 Inner structure of CNU when supporting two blocks simultaneously processed.

Figure 2.39 shows the inner structure of CNU when two blocks are simultaneously processed.

Similar to the case of row-parallel, the memory size required for block-parallel is determined by the maximum length of the encoded block.

2.5 STANDARDIZATION OF LDPC CODES IN 5G NR

2.5.1 Design of Lifting Factors

As mentioned in Section 2.3.4, Banyan networks have advantages over QSN networks due to the lower complexity. In order to maximize the use

TABLE 2.4 Values of Shifting Factor (Z)

Set Index (i_{LS})	α	Set of Lifting Sizes (Z)
0	2	{2, 4, 8, 16, 32, 64, 128, 256}
1	3	{3, 6, 12, 24, 48, 96, 192, 384}
2	5	{5, 10, 20, 40, 80, 160, 320}
3	7	{7, 14, 28, 56, 112, 224}
4	9	{9, 18, 36, 72, 144, 288}
5	11	{11, 22, 44, 88, 176, 352}
6	13	{13, 26, 52, 104, 208}
7	15	{15, 30, 60, 120, 240}

of Banyan networks for QC-LDPC decoders, the lifting factors (or lifting sizes) are chosen to be either the powers of 2, or the power of 2 multiplied by a positive integer, whose design principle is elaborated in [29,30]. As shown in Table 2.4, all the lifting factors of 5G NR LDPC can be written in the form of $a \times 2^j$, where a is an element in the set {2, 3, 5, 7, 9, 11, 13, 15}. j is an element in the set {0, 1, 2, 3, 4, 5, 6, 7}. For instance, when $i_{LS}=0$, then $a=2$; when $i_{LS}=1$, then $a=3$; when $i_{LS}=2$, then $a=5$, when $i_{LS}=3$, then $a=7$; when $i_{LS}=4$, then $a=9$; when $i_{LS}=5$时, then $a=11$; when $i_{LS}=6$, then $a=13$; when $i_{LS}=7$, then $a=15$. For each i_{LS}, a base matrix is defined for QC-LDPC of 5G NR.

According to Table 2.4, the maximum lifting factor of 5G NR LDPC is 384. When the block length of the information bits is 8448, the decoder can perform decoding with parallel level=384, e.g., with 384 CNUs, in order to get very high throughput. High level of parallel processing can increase the decoding speed, but at the same time increase the hardware cost and complexity. The maximum level of parallel processing can be considered as a capability of the LDPC decoder. However, this capability does not need to be indicated to the encoder. That is, the encoder does not need to select the lifting factor based on the maximum level of parallel processing. For the above-mentioned information block size of 8448, decoders of either parallel level of 384, or 128 or 64 or even 8 can be used. This is called the "friendly decoding" property of the LDPC encoder.

The lifting factors in Table 2.4 obey the form $a \times 2^j$ which is decoding-friendly. Hence, for decoders with the maximum level of parallel processing PM=2^i, no change is needed in the circuit to decode the LDPC with the lifting factor whose exponent i is less than j. This is because that when the lifting factor z is used in the encoding, the shifting network in the decoder can read z soft information from the memory and write back to the memory after the information is processed by the CNU. However, how to achieve the cyclic shift of size *z* with a shifting network whose level of parallel processing PM is less than *z*?

Denote $\mathbf{X}=\{x_i\}_{z\times 1}=[x_1,x_2,\cdots,x_z]$ as a vector containing z variables. z can be evenly divided by q. **X** can be divided into *q* sub-vectors, denoted as $\mathbf{X}=[\mathbf{u}(0),\mathbf{u}(1),\mathbf{u}(2),\cdots,\mathbf{u}(q-1)]$, where each sub-vector contains $l=z/q$ elements (*l* can be considered as the level of parallel processing or parallel metric PM), where $\mathbf{u}(k)=[x_{k,}x_{q+k,}x_{2q+k,}\cdots,x_{(l-1)q+k}]^T$ $k=0,1,\cdots,q-1$.

Denote $\mathbf{Y}=[y_1,y_2,\cdots,y_z]$ as the vector obtained by left shifting **X** by *s*, which can be represented as $\mathbf{Y}=\mathbf{P}^s\cdot\mathbf{X}$, where **P** is the $z*z$ permutation matrix, z is the lifting factor, and s is the shifting value. Y can also be divided into q sub-vectors $\mathbf{Y}=[\mathbf{v}(0),\mathbf{v}(1),\mathbf{v}(2),\cdots,\mathbf{v}(q-1)]$, where $\mathbf{v}(k)=[y_{k,}y_{q+k,}y_{2q+k,}\cdots,y_{(l-1)q+k}]^T$ $k=0,1,\cdots,q-1$.

Denote $s'=\lfloor s/q\rfloor$ and $n_0=\text{mod}(s,q)$, since z can be evenly divided by *q*, $\mathbf{v}(k)$ is in fact the left-shifted version of $\mathbf{u}(k)$ by s' or $s'+1$, which is

$$\begin{cases} v(k)=Q^{s}u(k+n_0) & k=0,1,\ldots,q-n_0-1 \\ v(k)=Q^{s+1}u\big(k+(q-n_0)\big) & k=q-n_0,q-n_0+1,\ldots,q-1 \end{cases}$$

where Q is a $q*q$ permutation matrix.

The following is an example of how to divide the cyclic shift of a lifting factor $z=42$ into three cyclic shifts of lifting factor $z'=14$. Denote $\mathbf{Y}=[x_{29},x_{30},\ldots x_{41},x_0,x_1,\ldots x_{26},x_{27},x_{28}]$ as a result of left shifting of the vector $\mathbf{X}=[x_0,x_1,x_2,\ldots,x_{41}]$ by 29. First divide the vector X into three sub-vectors u(0), u(1), and u(2) each of length $z'=14$, by uniform sampling, as illustrated in Figure 2.40.

Then, the left cyclic shift is carried out in each sub-vector, with the values of 10, 10, and 9, respectively. After that, three sub-vectors are left-shifted by 2. Then, these three vectors are interlaced and return to **Y**, as illustrated in Figure 2.41.

It is seen from the above example, the process of dividing a large cyclic shift into smaller cyclic shifts contains cyclic shifts within each sub-vector and the cyclic shift across multiple sub-vectors. For the design of lifting factor $z=a*2^j$ in 5G NR LDPC, the length of each sub-vector can be a power of 2 so that the cyclic shift within each sub-sector can be achieved by low-complexity Banyan shifting networks. The cyclic shift across multiple sub-vectors can be achieved by QSN shifting networks. Since the

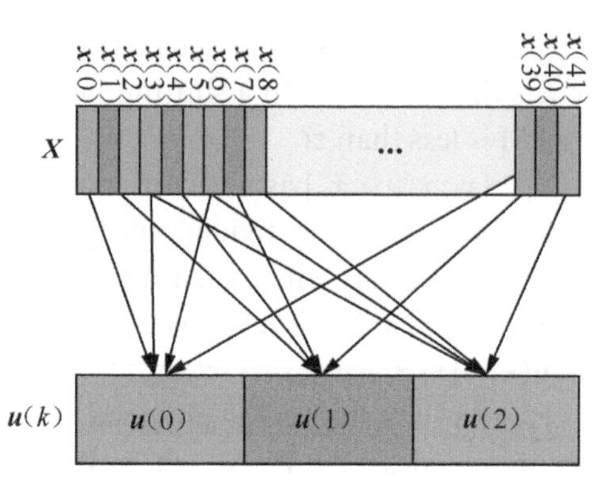

FIGURE 2.40 Vector X divided into three sub-vectors $\mathbf{u}(k)$, $k=0, 1, 2$.

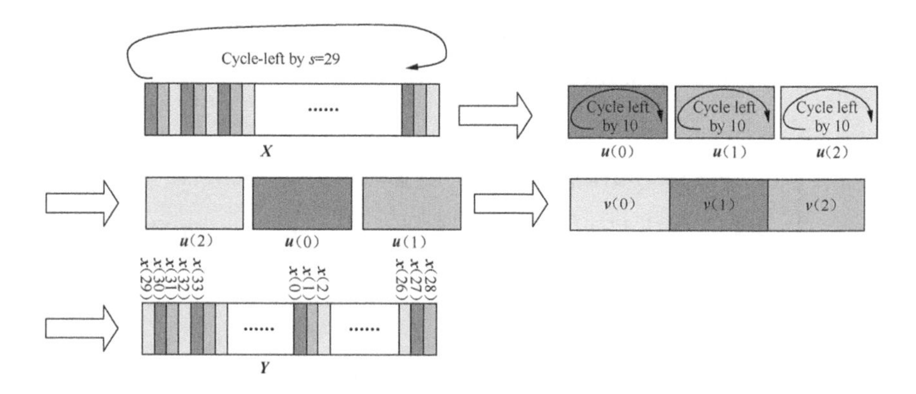

FIGURE 2.41 An example of large cyclic shift divided into smaller cyclic shifts.

number of sub-vectors is rather limited, the complexity of QSN shifting networks can be kept low. This hybrid (Banyan+QSN) shifting networks allows using relatively less complicated hardware to support flexible block lengths and be friendly to various levels of parallel processing

Table 2.5 compares the numbers of 2-1 MUX elements for QSN shifting networks and (Banyan+QSN) hybrid networks, under different levels of parallel processing. Apparently, (Banyan+QSN) hybrid networks are generally less complicated than pure QSN networks.

2.5.2 Design of Compact Base Matrix (Base Graph)

The encoding complexity and the throughput of decoding of QC-LDPC are tightly related to the size of the base matrix. A bigger base matrix would increase the complexity in both memory procedure and computation, as

TABLE 2.5 Complexity Comparison between Pure QSN Shifting Networks and (Banyan+QSN) Hybrid Networks

Level of Parallel Processing	Pure QSN	(QSN+Banyan) Hybrid Networks	Ratio of Complexity Reduction
8	48	48	0
16	128	2*[2*8*log2(8)]+8*[2*log2(2)]=112	**12.5%**
32	320	4*[2*8*log2(8)]+8*[4*log2(4)]=256	**20%**
64	768	8*[2*8*log2(8)]+8*[8*log2(8)]=576	**25%**
128	1792	16*[2*8*log2(8)]+8*[16*log2(16)]=1280	**28.6%**
256	4096	32*[2*8*log2(8)]+8*[32*log2(32)]=2816	**31.25%**
512	9216	64*[2*8*log2(8)]+8*[64*log2(64)]=6144	**33.33%**

TABLE 2.6 Sizes of Base Matrices (BGs) of LDPC in IEEE 802.16e and 802.11ad

Standards	Code Rate	Size of Base Matrix (Base Graph)
IEEE 802.16e/11n/11ac	1/2	kb=12, mb=12, nb=24
	2/3	kb=16, mb=8, nb=24
	3/4	kb=18, mb=6, nb=24
	5/6	kb=20, mb=4, nb=24
IEEE 802.11ad	1/2	kb=8, mb=8, nb=16
	5/8	kb=10, mb=6, nb=16
	3/4	kb=12, mb=4, nb=16
	13/16	kb=13, mb=3, nb=16

well as the longer delay for decoding and reduced throughput. For those mature standards (as listed in Table 2.6), the target throughputs are all at the level of Gbps, and the sizes of the base matrices are relatively small, e.g., the number of systematic columns is no greater than 20. In order to achieve a high throughput and reduce the complexity, smaller base matrix (base graph) is preferable (as elaborated in [31] for compact base matrix design), for instance, to limit the number of systematic columns to ≤26. In general, such base matrices are considered as compact where the number of systematic columns can take one of the values of {6, 8, 10, 12, 16, 20, 22, 24, 26}.

Compared with non-compact base matrices, compact base matrices have the following advantages:

1. Since compact base matrices have less rows and smaller row weights, the number of non-empty permutation matrices in the base graph is smaller. Thus, the complexity of CNU, shifting networks, and routing networks is lower.

2. Given the same length of the information bits, larger lifting factors can be designed, which supports a higher level of parallel decoding in order to achieve higher throughput. When operating near the peak data rate, normally the maximum block length is needed. In this case, the maximum lifting factor of the compact base matrices is greater than that of non-compact base matrices, thus allowing a higher level of parallel processing. This is crucial to fulfill the 20 Gbps peak rate requirement for 5G mobile networks.
3. Many simulation results show that the performance of compact base matrices is comparable to that of non-compact base matrices.
4. As indicated in Table 2.6, the compact base matrix design has been used in many wireless systems where the peak data rate requirement is at the Gbps level. The technology itself is quite mature.
5. For QC-LDPC, the length of information bits is equal to the number of systematic columns multiplied by the lifting factor. Since the number of systematic columns in the compact base matrices is smaller, the granularity of supportable block length is finer, thus allowing more flexible code length.
6. For non-compact base matrices, a 20 Gbps peak rate can only be achieved when the code rate is very high, e.g., 8/9. By contrast, a relatively lower code rate can be used for compact protomatrices to achieve a 20 Gbps peak rate. In another word, more modulation and coding scheme (MCS) levels can be used for the operation near the peak data rate.
7. In the memory blocks for storing the parameters of base matrices, since the number of non-empty permutation matrices is smaller for compact base matrices, the required memory would be much smaller.

Due to the above merits of compact base matrices, compact design is reflected in the NR LDPC, e.g., two sizes of base matrices (BGs) are specified in the standards: the number of systematic columns is 22 and 10, respectively.

2.5.3 Protomatrices (BGs)

During the design of protomatrices (BGs) of 5G QC-LDPC, the following criteria are considered:

- The row weight of the kernel matrix is close to the number of systematic columns (K_b). For instance, the row weights of the first four rows in BG1 are all equal to $K_b - 3$, and the row weights of the first four rows in BG2 are all equal to $K_b - 2$.
- The parity check part of the kernel matrix is similar to that of IEEE 802.16 and IEEE 802.11.
- The column weights of the first two columns are significantly greater than the column weights of the rest of the columns.
- The general trend is as the row index is increasing, the row weight gradually increases. As the column index is increasing, the column weight gradually increases.
- The column weights are all 1 for the column (K_b+5) to the last column.

The above design criteria for LDPC are elaborated in Ref. [26,27,32]. Matrices bearing these characteristics are called raptor-like structures. A more detailed design of the base matrices (BGs) can rely on the bit filling method [32]. In addition, some performance estimation-based tools can be applied, for instance, the probability density evolution algorithm and EXIT chart-based [33]. The minimum Hamming distance can be another performance indicator. Apart from the above-mentioned tools, the design of BGs can also be based on reducing the chance of short cycles and being stuck in the local minimum during iterative decoding.

The design of LDPC base matrices should balance both performance and throughput. Smaller base matrices are beneficial to supporting higher level of parallel processing and higher throughput to meet the peak data rate requirement. However, the number of elements in smaller matrices is smaller, which may limit the design freedom to some extent and pose some difficulty in optimization and enhancing the performance. By contrast, larger base matrices have more elements and more design freedom with more potential for performance optimization at the cost of a lower level of parallel processing and more challenges to fulfill the peak rate requirements, given the same hardware complexity as smaller base matrices.

Another factor for consideration is the number of base matrices (BGs) to be introduced. Using more base matrices can definitely improve the performance. However, the hardware cost is also increased, as well as the effort for standardization. By contrast, using fewer base matrices can reduce the hardware cost and simplifies the specification, albeit increasing the difficulty of performance optimization.

The third design factor is that for long code blocks, the mother code rate should not be too low. There are three reasons for this. Firstly, when scheduling a transmission, long code blocks are usually associated with good channel conditions where the high code rate corresponds to high data rate transmission. Secondly, for long blocks of information bits, if the mother code rate is low, the memory requirement would be high and the hardware implementation would be challenging. Lastly, a very low mother code rate design requires big base matrices which increases the complexity of LDPC decoding.

By jointly considering the above factors, the following principles are reached in 3GPP [34]. The design of base matrices (BGs) is based on a compact base graph design and would support two BGs. The first base graph of the base matrix (BG1) is larger and the number of its systematic columns K_b can be up to 22. The minimum mother code rate is 1/3. The code rate of the kernel matrix is equal to 22/24. It can support the maximum block size (information bits) of 8448. The second base graph of the base matrix (BG2) is smaller and the number of its systematic columns K_b can be up to 10. The minimum mother code rate is 1/5. The code rate of the kernel matrix is 5/6. It can support the maximum block size (information bits) of 3840. While the Kb of BG1 is more than twice the Kb of BG2, due to the lower mother code supported by BG2, the size of overall BG2 is comparable to that of overall BG1. The applicable ranges of use for BG1 and BG2 are illustrated in Figure 2.42 for the initial HARQ transmission.

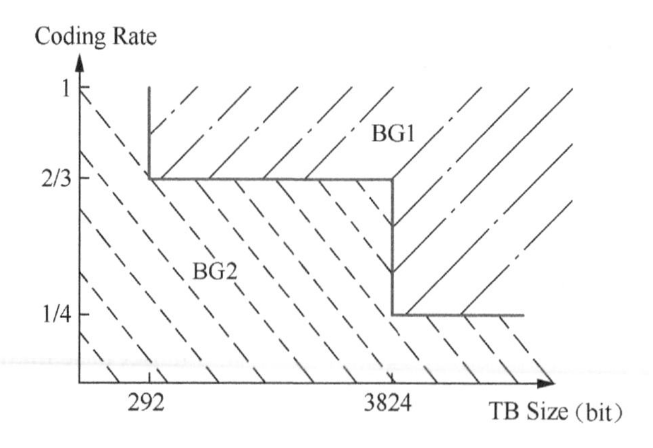

FIGURE 2.42 Applicable ranges of coding rate and TB size for BG1 and BG2 in the initial HARQ transmission.

More comprehensively, if any one of the following three conditions is met, BG2 should be used.

1. The size of the transport block (information bits) is not greater than 292.
2. The transport block size is between 292 and 3824, and the code rate of the initial HARQ transmission is not greater than 2/3.
3. The code rate of the initial HARQ transmission is not greater than 1/4.

In all other cases, BG1 should be used.

2.5.3.1 Base Graph 1 (BG1)

Figure 2.43 shows the basic structure of BG1 which consists of 46 rows and 68 columns. The top left part consisting of four rows and 22 columns are called systematic columns of the kernel matrix (upper left 4×26 matrix in BG1) which has high row weights (many non-zero elements) and can have a big impact on LDPC performance at a high code rate. The middle 16 rows are designed with quasi-orthogonal characteristics, e.g., there is a certain orthogonal relationship between rows but not very strict. The bottom 26 rows are strictly row-orthogonal.

The above design provides a good trade-off between the performance and the complexity. Using large row weight in the kernel matrix can ensure good performance at a high code rate. Although the density of non-zero

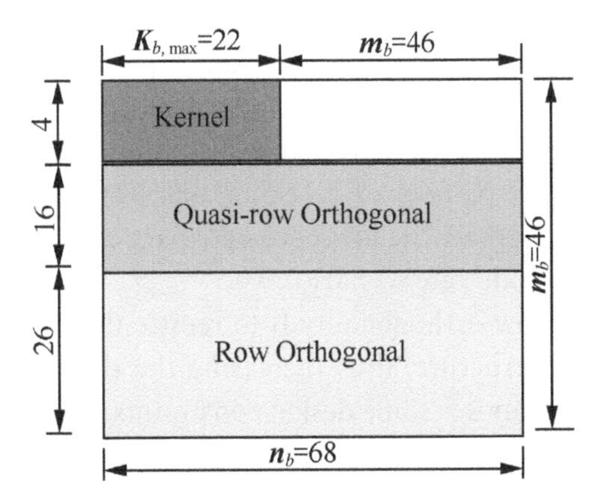

FIGURE 2.43 Basic structure of BG1.

elements in the kernel matrix is quite high, due to the small number of rows and columns in the kernel matrix, the number of non-zero elements in the entire BG1 is not very high. Hence, its impact on the overall decoding latency is not big. Also, the LDPC parity check matrix has the property of expanding to the lower code rate. That is, the kernel matrix would be included in any code rate of LDPC. The impact of the kernel matrix on the LDPC performance at a low code rate is rather small.

Quasi-orthogonal design refers to that except for the elements in the first two columns, all other elements are group-wise orthogonal. Group-wise orthogonality means that some rows in a matrix are orthogonal to each other. For instance, the fifth row to the eighth row form a group. Within this group, there are three rows that are quasi-orthogonal, e.g., except for the elements of the first two columns, other elements of these three rows are row-orthogonal. Another example is the group in BG1: from the 9th row to the 11th row. Note that any two groups of rows may not be quasi-orthogonal. Quasi-orthogonal design balances performance and throughput. In the rows within each group, only the first two columns would have the issue of addressing conflicts during the data reading/writing from or to CNU, due to the non-orthogonality between rows. The difficulty of parallel decoding (thus extra waiting time) is only limited to these two columns, e.g., no other elements would have the address conflict issue, which can significantly increase the decoding speed and throughput. On the other hand, the non-orthogonality in the first two columns provides certain design freedom to further improve the performance with respect to strict row-orthogonal design.

Orthogonal design refers to group-wise row-orthogonal. From the 21st row to the last row in BG1, any adjacent two rows are orthogonal. Such row-orthogonal would increase the decoding speed, because when each CNU of the two orthogonal rows is conducting data reading/writing to/from the memory slices, no address conflict would occur, regardless of whether using a row-parallel or block-parallel decoding structure. The row-orthogonal design for the low-code-rate part of BG can increase the throughput in low-code-rate scenarios.

The purpose of row orthogonality is to reduce the decoding complexity of row-parallel structure, thus increasing the throughput. Certainly, row orthogonality imposes some design constraints such as the number of non-zero elements and their distributions, etc.

Based on the basic structure shown in Figure 2.43, after a large amount of simulations and computer search, as well as the performance optimization

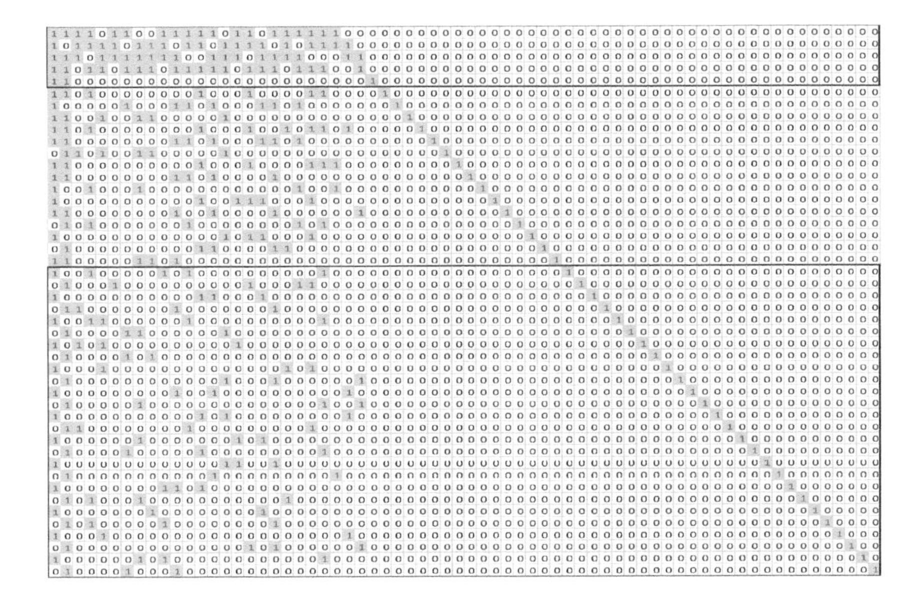

FIGURE 2.44 The positions of non-zero elements in BG1.

under various block lengths and code rates, the actual positions of non-zero elements are specified by 3GPP, as illustrated in Figure 2.44. The number of non-zero elements is 316. Compared to the size of the entire base graph, e.g., 46*68=3128, the base matrix bears the characteristics of low density.

Note that a base graph of the base matrix only provides the positions of non-zero elements, e.g., it does not yet specify the lifting factor (lifting size) for each non-zero element. In another word, a base matrix is the combination of the base graph and the lifting factors so that the base matrix can be lifted to the actual parity check matrix for LDPC encoding and decoding. The design of base matrices can be via computer searching. For instance, to select a smaller set of matrices with less chance of short girth and trapping set, out of the randomly generated set of matrices by the computers. Simulations would be carried out to further verify the performance. Those matrices that result in a lower error floor and have consistent performance across different block lengths or code rates would be finally chosen for the base matrices. In 3GPP, after intensive simulation effort and discussions between companies, eight base matrices each of BG1 and BG2 are specified. Table 2.7 shows part of the eight base matrices for BG1.

TABLE 2.7 Base Matrices ($V_{i,j}$) of BG1

Row Index i	Column Index j	Set Index i_{LS} 0	1	2	3	4	5	6	7
0	0	250	307	73	223	211	294	0	135
	:	:	:	:	:	:	:	:	:
	23	0	0	0	0	0	0	0	0
		:		:		:			
45	1	149	135	101	184	168	82	181	177
	6	151	149	228	121	0	67	45	114
	10	167	15	126	29	144	235	153	93
	67	0	0	0	0	0	0	0	0

2.5.3.2 Base Graph 2 (BG2)

Figure 2.45 shows the basic structure of BG2 which consists of 42 rows and 52 columns. The left part on the top, consisting of four rows and ten columns, is called the systematic columns of kernel matrix (upper left 4×14 matrix in BG2) which has higher row weights (many non-zero elements) and bigger impact on the performance. The 22 rows at the bottom are row-orthogonal. Different from BG1, the 16 rows in the middle of BG2 are not quasi-row-orthogonal. That is, except the elements in the first two columns of the 16 rows, all other elements are not group-wise row-orthogonal. Such design is to improve the LDPC performance at a mid- or low code rate. This is important to extend the use scenarios of BG2. For instance, BG2 can be applied to high-reliability scenarios in the future.

Based on the basic structure shown in Figure 2.45, after a large amount of simulation work and computer search, the performance is optimized over various block lengths and code rates. The positions of non-zero elements for BG2 are decided, which is illustrated in Figure 2.46.

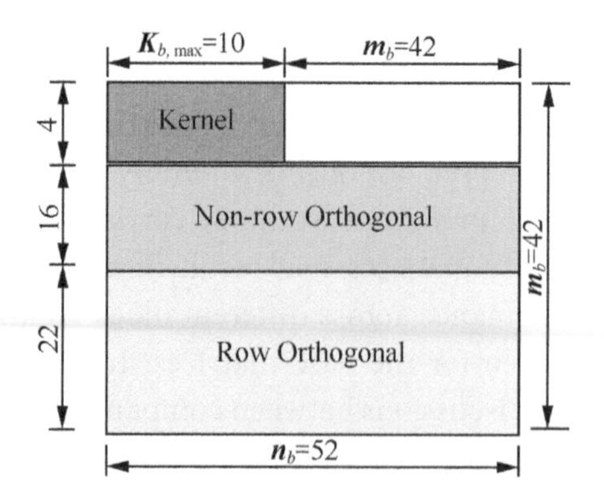

FIGURE 2.45 The basic structure of BG2.

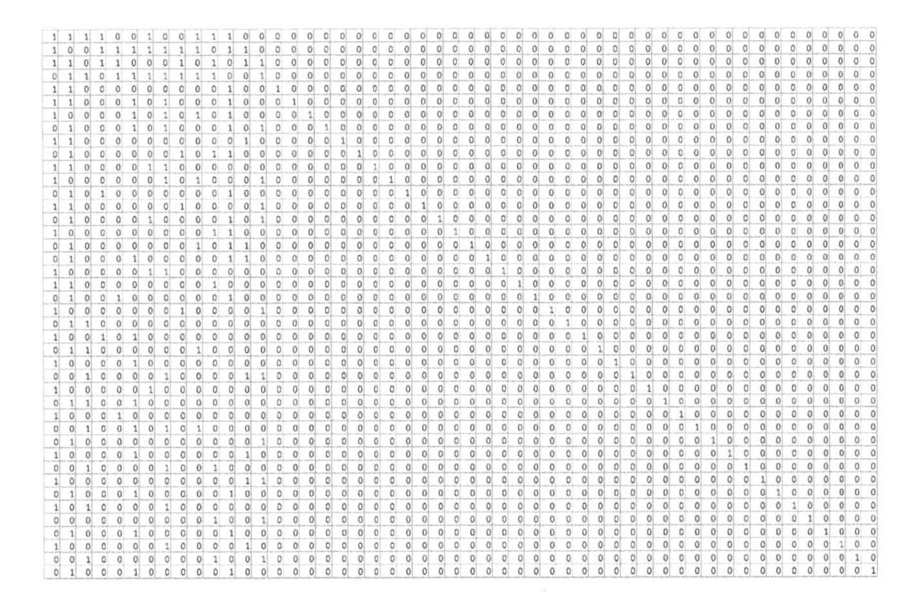

FIGURE 2.46 The positions of non-zero elements of BG2.

TABLE 2.8 Part of Eight Base Matrices of BG2

Row Index i	Column Index j	Set Index i_{LS} 0	1	2	3	4	5	6	7
0	0	9	174	0	72	3	156	143	145
	:	:	:	:	:	:	:	:	:
	11	0	0	0	0	0	0	0	0
41	1	129	147	0	120	48	132	191	53
	:	:	:	:	:	:	:	:	
	51	0	0	0	0	0	0	0	0

Part of the eight base matrices of BG2 is illustrated in Table 2.8.

2.5.4 Rate Matching

As discussed in Section 2.1, information bit stream **u** is encoded by linear block code generator **G** and becomes x=u*G. If the generator matrix G is already converted into the form **G**=[**Q**, *I*] or **G**=[**I**, **Q**], the encoded codeword **x** would include two parts: systematic bit stream **s** and parity bit stream **p**. That is **x**=[**s**, **p**], as shown in Figure 2.47.

LDPC is a type of linear block codes. Without losing the generality, a codeword of LDPC after encoding can also be represented as **x**=[**s**, **p**]. Rate matching is used in the case when the number of encoded bits is not

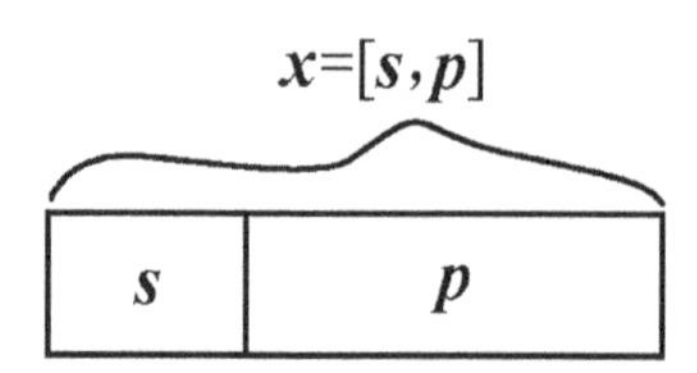

FIGURE 2.47 Systematic bits and parity check bits in a linear code block.

equal to the number of bits supported by the available radio resources. For instance, when the available radio resources are relatively abundant, which of the bits would be transmitted? If the available radio resources are limited, which of the bits are to be punctured? A related technique is HARQ. For instance, a self-decodable version of the codeword would be sent in the initial HARQ transmission. If the receiver cannot decode successfully, the transmitter would send a self-decodable or non-self-decodable version in the second HARQ transmission. The actual bits may be the same or different between the first and the second transmissions (the exact bits can be different even for self-decodable versions). Usually, a version that contains a large portion of systematic bits would be self-decodable. Here, self-decodability refers more specifically to the ability of successful decoding if the version is designed for the initial HARQ transmission.

A circular buffer is used for rate matching and HARQ for QC-LDPC, as illustrated in Figure 2.50 [35]. The systematic bits and parity bits at the output of the encoder are first stored in a circular buffer. For each HARQ transmission, the data is to be sequentially read out from the buffer according to the redundant version (RV). For instance, in the order of RV0 → RV2 → RV3 → RV1. Each RV essentially defines the starting position of each HARQ data block in the circular buffer.

It should be emphasized that the initial HARQ transmission should be self-decodable (consisting of systematic bits). The spacing between the starting positions of different RVs is not uniform. Such non-uniform spacing can improve the performance [36]. According to [37], if the receiver does not use limited buffer rate matching (LBRM) [38], for BG1 (the length of encoded bits is 66*z where z is the lifting factor [39]), the starting positions of RV0, RV1, RV2, and RV3 are 0, 17*z, 33*z, and 56*z, respectively. For BG2 (the length of encoded bits is 50*z), the starting positions of RV0, RV1, RV2, and RV3 are 0, 13*z, 25*z, and 43*z, respectively. It can be found that RV0 and RV3 are self-decodable. In the design of QC-LDPC, the focus is the performance of initial HARQ transmission. Hence, RV0 is preferably used for the initial HARQ transmission. When the transmitter is not

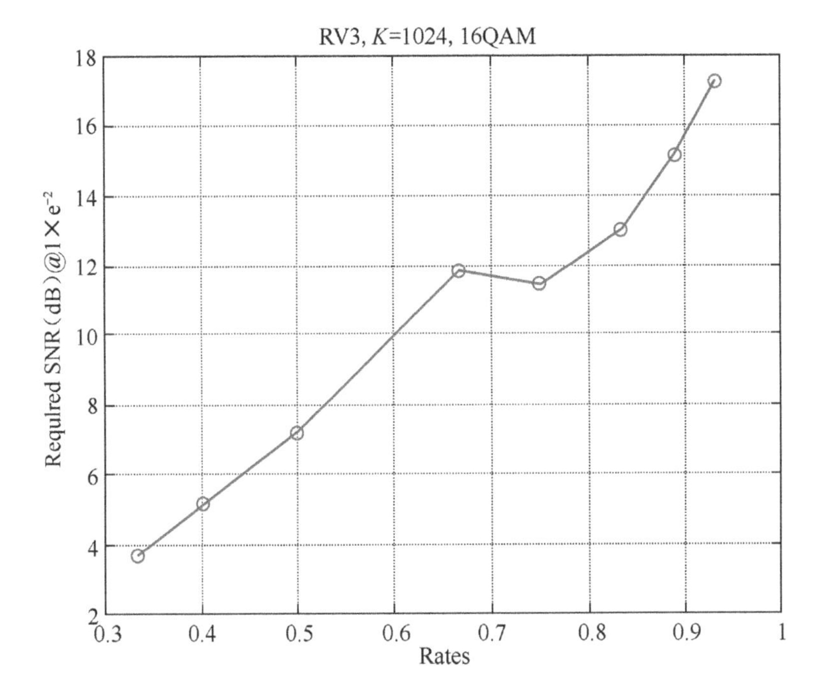

FIGURE 2.48 Self-decodability of RV3.

sure whether the receiver can successfully decode the initial HARQ transmission, RV3 should be used for the initial transmission since RV3 is not only self-decodable but also contains significant portion of redundant bits to exploit the channel coding gain. RV2 which has the highest portion of redundant bits can also be used. The self-decodable design for RV3 was proposed in [40] which can effectively solve the issue of indistinguishable NACK and DTX in carrier aggregation scenarios. Such a design was adopted in 3GPP NR standards. Figure 2.48 shows the self-decodability of RV3, where the length of the redundancy version is 1024 and BLER = 1% is considered. It is observed that within the range of practical code rates, RV3 is self-decodable. In Figure 2.49, the performances of initial RV0 + 2nd RV0, and initial RV0 + 2nd RV3 are compared. It is seen that the soft combining between RV0 and RV3 leads to better performance compared with the soft combining of simple repetition, e.g., RV0 + RV0. More detailed performance can be found in [40].

2.5.5 Interleaving

Interleaving refers to shuffling or scrambling the bits after rate matching. The purpose of interleaving is to protect from the bursty interference. After

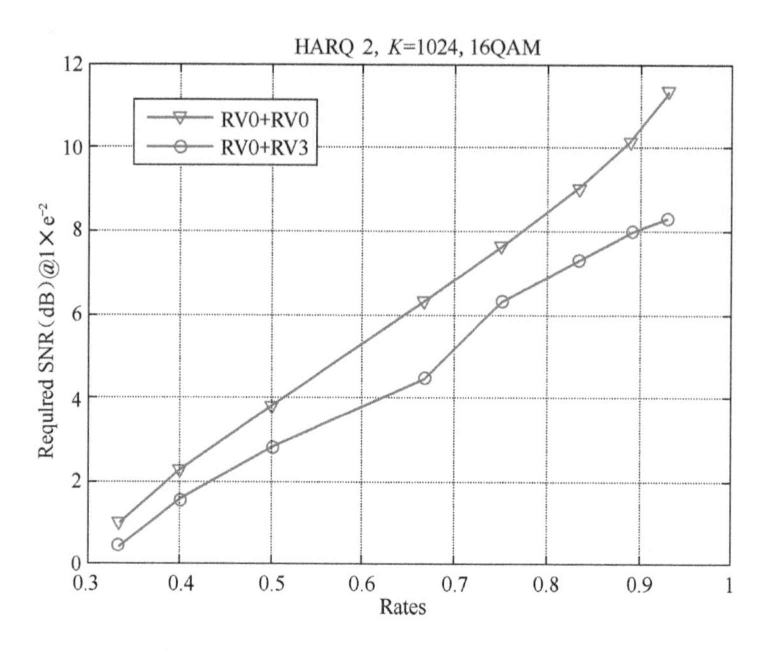

FIGURE 2.49 Performance comparison of soft combining between RV0 (initial) + RV0 (2nd), and RV0 (initial) + RV3 (2nd).

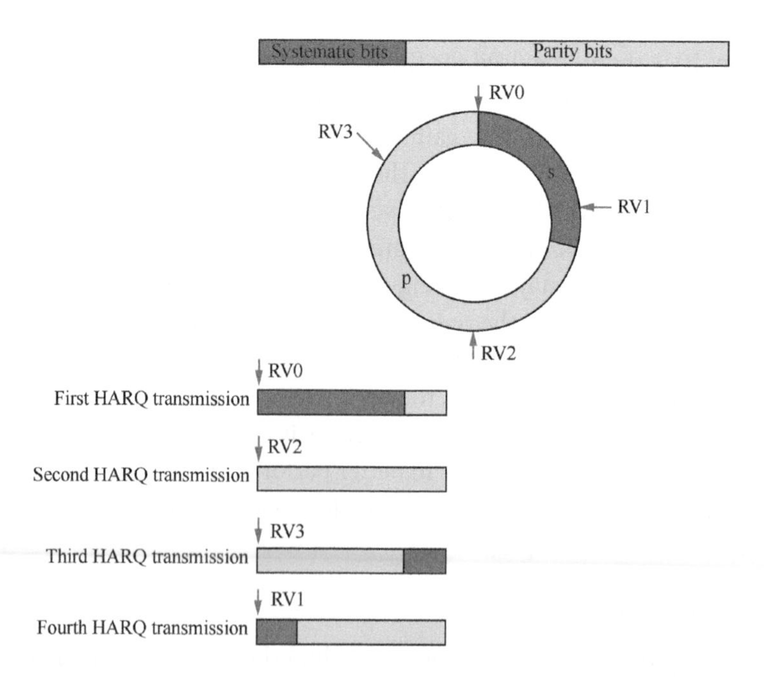

FIGURE 2.50 Illustration of circular buffer (HARQ buffer) of QC-LDPC [35].

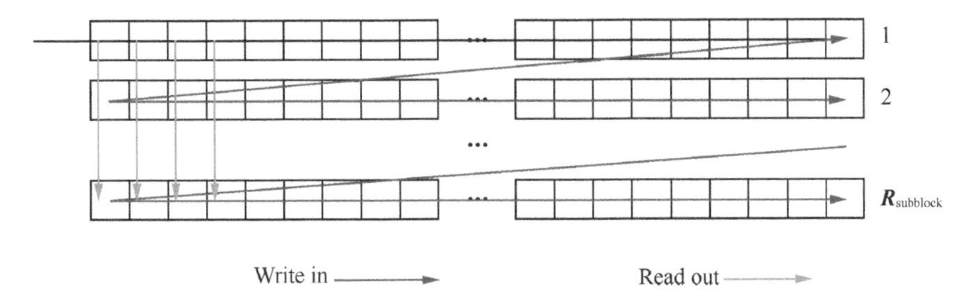

FIGURE 2.51 Row-column interleaver where the number of rows equal to the modulation order.

the interleaving, the original local concentrated (consecutive) interference is spread out into individually isolated interference, which is helpful for decoding. The benefit of interleaving is more pronounced in higher-order modulations, e.g., 16QAM, 64QAM, and 256QAM. According to [41], the interleaving is carried out in each code block individually. It means that if there are multiple code blocks, there would be no interleaving across multiple code blocks. This is to reduce the processing latency at the receiver. In 5G NR LDPC, a row-column interleaver is used to interleave each code block. In [23–25], a bit-level interleaving method for LDPC was proposed. A more detailed description of this method was discussed in [22,42,43]. Below is a brief description.

The row-column interleaver is illustrated in Figure 2.51 where the number of rows is denoted as $R_{subblock}$. Here, $R_{subblock}$ is equal to the modulation order, e.g., 4 for 16QAM, 6 for 64QAM, and 8 for 256QAM. The data is reordered based on the principle of row-by-row write-in and column-by-column read-out.

The interleaver in Figure 2.51 can be used for all the redundancy versions. For RV0, systematic bits would take the priority in interleaving to high reliable bits. As shown in Figure 2.52 [44], take 256QAM as an example, systematic bits would be mapped to the first few bits of a 256QAM symbol. It is known that in higher-order QAM modulations, the reliability of each bit is different. For instance, the first two bits of the 8 bits for a 256QAM symbol have the highest reliability, followed by the 3rd and 4th bits and then the 5th and 6th bits. The last two bits have the lowest reliability. Mapping the systematic bits to the highly reliable bits would provide better protection for the systematic bits. This has been verified by many simulations of QC-LDPC codes which show the performance benefit. Since RV0 is often used for the initial HARQ transmission, such a

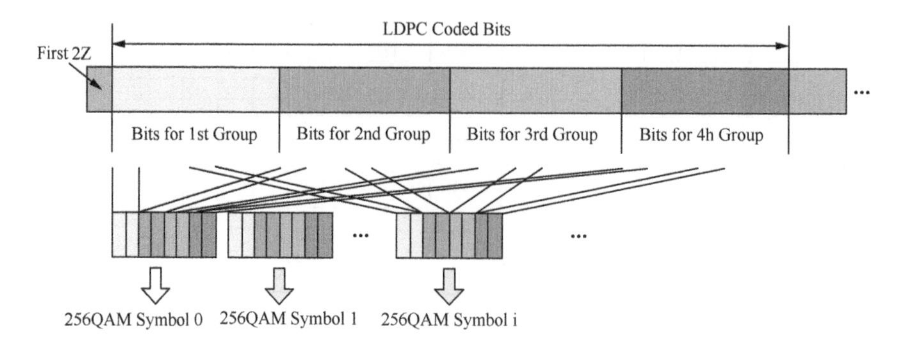

FIGURE 2.52 The method that prioritizes the systematic bits in the case of higher-order modulation [44].

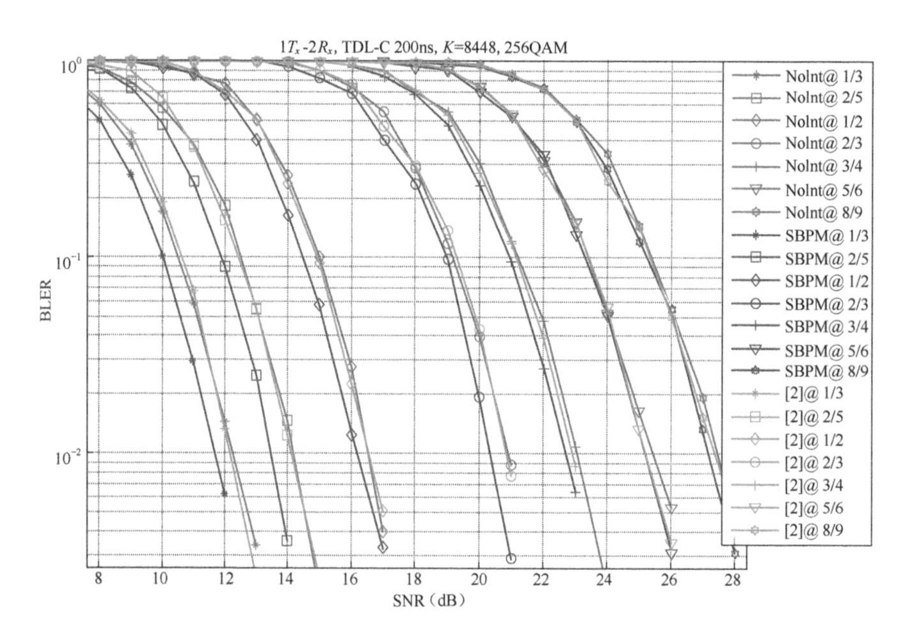

FIGURE 2.53 Performance of systematic bits prioritized mapping interleaving [44].

row-column interleaving method can improve the performance of initial transmission. Therefore, the interleaving method was also named as systematic bits prioritized mapping interleaving.

When the above systematic bits prioritized method is used, even if carried out alone, 0.5 dB performance gain can be achieved for fading channel at BLER = 10%, as seen in Figure 2.53. Due to its performance merit, the method is adopted by 3GPP.

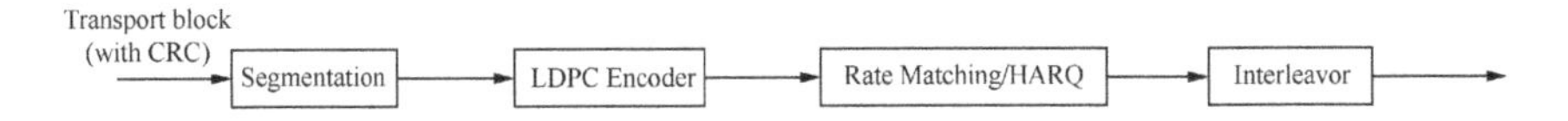

FIGURE 2.54 Channel coding processing chain at the transmitter.

2.5.6 Segmentation

Before discussing segmentation, let us first describe the location of the segmentation within the entire channel coding chain. As shown in Figure 2.54, when the physical layer receives the transport block from a medium access control (MAC) layer, a bunch of cyclic redundancy check (CRC) bits (16 or 24) would be appended to the end of each transport block. If the number of bits in the appended transport block exceeds a certain number, it has to be segmented into two or more code blocks. Then CRC bits are appended to the end of each code block. Each CRC appended code block would be individually LDPC encoded. After that, each encoded block is rate-matched, HARQ-processed, and interleaved.

Although the longer the LDPC codes and the better the performance, hardware limitations should be considered for practical designs, e.g., a long block would be segmented into smaller blocks, which is the purpose of segmentation. Segmentation may facilitate parallel decoding across multiple smaller blocks, thus reducing the decoding latency.

Performance impact should be considered during the segmentation. As Figure 2.55 [45] shows, when the block length is beyond a certain point, the performance benefit of using longer blocks diminishes. Hence, the threshold for segmentation should not be too large. On the other hand, the threshold should not be too small; otherwise, there would be too many code blocks. If the performance of the entire transport block should be kept the same, the BLER requirement for each code block would be higher, which may cause the issues of resource overhead, latency, etc.

In LTE, the maximum transport block size is 391656 [46]. In 5G NR, the maximum transport block size would be even greater, e.g., in million. According to [39], for BG1, the maximum CBS is 8448 (including 24-bit CRC). It means that segmentation would be carried out if the size of a transport block is over 8448–24=8424. For BG2, the maximum CBS is 3840 (including 16-bit CRC). It means that segmentation would be carried out if the size of a transport block is over 3840–16=3824.

It should be pointed out that different from LTE, according to [37], if there is new transmission (initial HARQ transmission) whose code rate $R_{init}>1/4$, when the code block length (excluding CRC bits) is greater

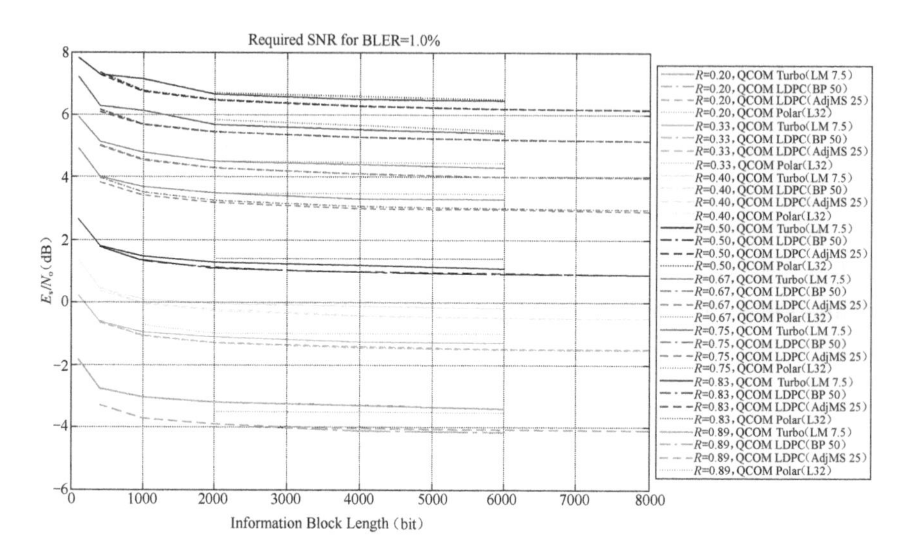

FIGURE 2.55 Performance of LDPC as a function of block length [45].

than 3824, BG1 should be used. If $R_{\text{init}} \leq 1/4$, BG2 should be used. Hence, whether to segment depends not only on the transport block size but also on the code rate of the initial transmission.

In summary, if using BG1, the maximum CBS is 8448. If using BG2, the maximum code block size is 3840.

2.5.7 Channel Quality Indicator (CQI) Table and Modulation and Coding Scheme (MCS) Table

Channel quality indicator (CQI) refers to the downlink channel quality reported by a terminal, based on the reception quality (e.g., SINR) measured over the downlink reference signal (e.g., CSI-RS). MCS refers to the modulation order and code rate (transport block size can be derived from the MCS and the number of physical resource blocks) which is used for PDSCH transmitted by a base station or PUSCH transmitted by a terminal.

Our focus in this book is the CQI tables and MCS tables for eMBB of 5G NR.

2.5.7.1 CQI Tables

The current 5G NR specification supports QPSK, 16QAM, 64QAM, and 256QAM for downlink traffic channels. After measuring the CSI-RS and estimating the channel quality, a terminal should report CQI which contains modulation order and code rate. To minimize the CQI signaling

TABLE 2.9 CQI Table with the Highest Modulation Order of 64QAM

CQI Index	Modulation	Code Rate × 1024	Efficiency
0		Out of range	
1	QPSK	78	0.1523
:	:		:
15	64QAM	948	5.5547

TABLE 2.10 CQI Table with the Highest Modulation Order of 256QAM

CQI Index	Modulation	Code Rate × 1024	Efficiency
0		Out of range	
1	QPSK	78	0.1523
:	:	:	:
15	256QAM	948	7.4063

TABLE 2.11 Sub-band Offset CQI Table

Spatial Differential CQI Value	Offset Level
0	0
1	1
2	≥2
3	≤−1

overhead while refining the granularity of CQI, two CQI tables are introduced in 5G NR, namely, one to support up to 64QAM and the other to support up to 256QAM. Both tables use 4 bits to quantize CQI. Radio resource control (RRC) signaling is employed to indicate which CQI table is to be used.

Tables 2.9 and 2.10 show the 64QAM CQI table and 256QAM CQI table, respectively. The first column in the tables is the CQI index which can take values 0, 1, …, 15, representing 16 cases, and can be indicated by 4 bits. The second column is the modulation order. The third column is the code rate. The four columns are the transmission efficiency based on the production of the modulation order and the code rate.

Tables 2.9 and 2.10 are for wide-band CQI. In 5G NR standards, a system bandwidth can be divided into multiple sub-bands and each sub-band can have sub-band CQI. To reduce the signaling overhead, sub-band CQI is defined as the offset of CQI with respect to the wide-band CQI. The offset is indicated by 2 bits. Table 2.11 is a sub-band offset CQI table where the first column represents four cases and can be signaled in 2 bits. The second column is the sub-band offset level, which is defined as

Sub-band Offset level = wideband CQI index – sub-band CQI index

2.5.7.2 MCS Tables

1. **Downlink MCS Tables.** The MCS information is indicated in downlink control information (DCI) for PDSCH transmission. MCS contains modulation order and code rate information. Similar to the case of CQI tables, two downlink MCS tables are specified in 5G NR for eMBB: one for modulation order up to 64QAM and the other for modulation order up to 256QAM. Both tables use 5 bits to indicate MCS indices. RRC signaling is employed to indicate which MCS table is to be used.

   Tables 2.12 and 2.13 show 64QAM MCS table and 256QAM MCS table for PDSCH. The first column is for the MCS index whose value can be 0, 1,…, 31, representing 32 cases, indicated by 5 bits. The second column is the modulation order, 2 for QPSK, 4 for 16QAM, 6 for 64QAM, and 8 for 256QAM. The third column is the code rate. The fourth column is the transmission efficiency derived from the modulation order and code rate. In the 64QAM MCS table, when the MCS index is 29, 30, or 31, it refers to the modulation order of HARQ retransmission of PDSCH. In the 256QAM MCS table, when MCS index is 28, 29, 30, or 31, it refers to the modulation of HARQ retransmission of PDSCH. During the retransmission, different modulation orders can be chosen to get different transmission efficiency and improve the flexibility of coding and modulation.

TABLE 2.12 Downlink MCS Table with 64QAM as the Highest Modulation Order

MCS Index I_{MCS}	Modulation Order Q_m	Target Code Rate × [1024]	Spectral Efficiency
0	2	120	0.2344
:	:	:	:
28	6	948	5.5547
29	2	Reserved	
30	4	Reserved	
31	6	Reserved	

TABLE 2.13 Downlink MCS Table with 256QAM as the Highest Modulation Order

MCS Index I_{MCS}	Modulation Order Q_m	Target code Rate × [1024] R	Spectral Efficiency
0	2	120	0.2344
:	:	:	:
27	8	948	7.4063
28	2	Reserved	
29	4	Reserved	
30	6	Reserved	
31	8	Reserved	

5 bits are used to indicate the MCS index, 1 bit more than the 4-bit CQI index. This means that the granularity of PDSCH transmission efficiency is finer than that of CQI feedback.

2. **Uplink MCS Tables.** The current 5G NR supports Pi/2-BPSK (optional), QPSK, 16QAM, 64QAM, and 256QAM for PUSCH transmission. Pi/2-BPSK is only used in the DFT-s-OFDM waveform and for those terminals that support Pi/2-BPSK. The reason for the joint use DFT-s-OFDM and Pi/2-BPSK is to reduce the PAPR of the signal and improve the uplink coverage. When Pi/2-BPSK is used, 256QAM would not be considered. Hence, Pi/2-BPSK is not included in the 256QAM MCS table.

   To reduce the MCS signaling overhead while supporting different granularities of MCS, three uplink MCS tables are specified in 5G NR for eMBB. When Pi/2-BPSK is not supported, two uplink MCS tables are used: 64QAM MCS table and 256QAM MCS table. Both use 5 bits for MCS index indication. RRC signaling is employed to indicate which uplink MCS table is to be used. These two tables are the same as the MCS tables for the downlink, e.g., Tables 2.12 and 2.13. When Pi/2-BPSK is supported, one MCS table is used, with the highest modulation order being 64QAM. 5 bits are used to indicate the MCS index, as shown in Table 2.14.

2.5.8 Determination of Transport Block Size (TBS)

At the physical layer, the uplink and downlink data are transmitted with the basic unit of transport block (TB). In LTE, the size of TB (called TBS) can be looked up from a table indexed by the number of physical resource blocks (denoted as N_{PRB}) and the index of TBS (denoted as I_{TBS}). In 5G NR standardization, both formulae and look-up tables are used to quantize the different ranges of TBS in order to provide more flexibility for scheduling.

TABLE 2.14 Uplink MCS Table Supporting Pi/2-BPSK and Up to 64QAM

MCS Index I_{MCS}	Modulation Order Q_m	Target code Rate × 1024 R	Spectral Efficiency
0	1	240	0.2344
:	:	:	:
27	6	948	5.5547
28	1	Reserved	
29	2	Reserved	
30	4	Reserved	
31	6	Reserved	

2.5.8.1 Procedure to Determine TBS for PDSCH

PDSCH transmission is scheduled by PDCCH in which cell-radio network temporary identity (C-RNTI) is used to scramble CRC bits. DCI format 1_0/1_1 is used to indicate various transmission formats of PDSCH. A look-up table type of approach was proposed in [35] to get the actual TBS. It is shown that for short TBS, such a method has an advantage over the formula-based approach. Hence, it was adopted by 5G NR standards.

If RRC signaling indicates to use the 256QAM MCS table for PDSCH and the configurable MCS indices are 0~27, or RRC signaling does not indicate to use the 256QAM MCS table for PDSCH and the configurable MCS indices are 0~28, the determination of TBS has the following steps:

1. **To Determine an Intermediate Value for Information Bits, Based on the Parameters Indicated by DCI**. Theformula is $N_{\inf o} = N_{RE} * R * Q_m * \upsilon$, where v is the number of transmission layers; Q_m is the modulation order, according to the MCS index; R is the code rate, indicated by the MCS index; N_{RE} is the total number of resource elements (REs), $N_{RE} = N_{RE} \times N_{PRB}^{XL}$, where N_{PRB}^{XL} is the number of allocated PRBs; N'_{RE} is the available number of REs in each PRB. N'_{RE} can be quantized as

   1. The number of available REs in each allocated PRB:

   $$N'_{RE} = N_{sc}^{RB} * N_{symb}^{sh} - N_{DMRS}^{PRB} - N_{oh}^{PRB},$$

   where $N_{sc}^{RB} = 12$ is the number of subcarriers in each PRB; N_{symb}^{sh} is the number of OFDM symbols in a slot that can be scheduled; N_{DMRS}^{PRB} is the number of REs for DMRS (including the overhead of DMRS CDM groups indicated in DCI format 1_0/1_1) within a scheduled time duration; N_{oh}^{PRB} is the overhead for higher layer configuration parameter Xoh-PDSCH. If Xoh-PDSCH, taking one of {0, 6, 12, 18}, is not configured, Xoh-PDSCH is set to 0°

TABLE 2.15 The Number of Available REs in a PRB

Index	N'_{RE}	Index	N'_{RE}	Index	N'_{RE}	Index	N'_{RE}	Index	N'_{RE}	Index	N'_{RE}
1	6	11	36	21	66	31	90	41	114	51	138
2	8	12	40	22	68	32	92	42	116	52	140
3	12	13	42	23	70	33	94	43	118	53	144
4	16	14	44	24	72	34	96	44	120	54	148
5	18	15	48	25	76	35	100	45	124	55	150
6	20	16	52	26	78	36	102	46	126	56	152
7	24	17	54	27	80	37	104	47	128	57	156
8	28	18	56	28	82	38	106	48	130	58	
9	30	19	60	29	84	39	108	49	132	59	
10	32	20	64	30	88	40	112	50	136	60	

TABLE 2.16 TBS Table for $N_{\text{info}} \leq 3824$

Index	TBS	Index	TBS	Index	TBS	Index	TBS
1	24	31	336	61	1288	91	3624
:	:	:	:	:	:	:	:
30	320	60	1256	90	3496		

2. After N'_{RE} is calculated, use $N_{RE} = \min(156, N'_{RE}) \times n_{PRB}$ to figure out the total number of REs.

   For example, when the number of allocated OFDM symbols $N_{symb}^{sh} = 3\sim14$, N_{DMRS}^{PRB} would be one value of {4, 6, 8, 12, 16, 18, 24, 32, 36, 48}. N_{oh}^{PRB} would be one of the values of {0, 6, 12, 18}. N'_{RE} can be calculated by following Steps ① and ②. The calculated N'_{RE} are listed in Table 2.15.

2. **To Quantize the Intermediate N_{inf} and Get the Final TBS.** The procedure is as follows:

   If $N_{\text{info}} \leq 3824$

$$N'_{\text{info}} = \max\left(24, 2^n * \left\lfloor \frac{N_{\text{info}}}{2^n} \right\rfloor\right), \text{ where } n = \max\left(3, \left\lfloor \log_2\left(N_{\text{info}}\right) \right\rfloor - 6\right);$$

To find a TBS in Table 2.16 that is no less than N'_{info}, and the closest to N'_{info} as the final TBS.

else

$$N'_{info} = \max\left(3840, 2^n \times round\left(\frac{N_{info} - 24}{2^n}\right)\right), \text{ where}$$

$n = \lfloor \log_2(N_{info} - 24) \rfloor - 5$, round(·) denotes rounding to the closest integer;

if $R \leq 1/4$

$$TBS = 8 * C * \left\lceil \frac{N'_{info} + 24}{8 * C} \right\rceil - 24, \text{ where } C = \left\lceil \frac{N'_{info} + 24}{3816} \right\rceil$$

else

if $N'_{info} > 8424$

$$TBS = 8 * C * \left\lceil \frac{N'_{info} + 24}{8 * C} \right\rceil - 24 \text{ where } C = \left\lceil \frac{N'_{info} + 24}{8424} \right\rceil$$

else

$$TBS = 8 * \left\lceil \frac{N'_{info} + 24}{8} \right\rceil - 24$$

end

end

end

2.5.8.2 *Scheduling Flexibility*

Scheduling flexibility refers to the ranges of resource parameters such as number of PRBs, number of OFDM symbols, MCS levels, etc., in order to get the same TBS for initial HARQ transmissions or retransmissions. The wider the ranges, the more flexibility would be for the scheduler. For instance, during the initial HARQ transmission, at least 18 MCS levels can be supported by a look-up table-based TBS. Scheduling flexibility is an important criterion for TBS design.

2.5.8.3 *MAC Layer Overhead Ratio*

Overhead ratio is defined as Overhead Ratio = $(TBS_j - (TBS_{j-1} - 8))/TBS_j$. Generally speaking, the overhead ratio should be kept small to avoid significant performance loss. For eMBB, the overhead ratio should be smaller than 5%. The overhead ratio is also an important criterion for TBS design.

2.5.8.4 *Design of TBS Tables*

In the procedure of TBS determination in Section 2.5.8.1, when $N'_{info} \leq 3824$, the TBS table is used to get the final TBS. In this section, the design of TBS tables is discussed, together with the performance benefits.

In communications systems, a transport block (TB) needs to be encoded before it can be transmitted. Hence, TBS should fulfill the requirement of channel coding. In 5G NR, LDPC is used for physical shared channels. The requirements for TBS are

1. TBS should be aligned at the byte level, e.g., divisible by 8.
2. No padding after the segmentation, e.g., divisible by the number of code blocks C.
3. CBS should be aligned at the byte level, e.g., TBS should be divisible by 8*C.
4. TBS should be the same for initial HARQ transmission and retransmissions.

According to the above requirements and criteria, the TBS table can be designed (or searched) using the following steps:

1. Search through all the TBS in the LTE TBS table and order them in ascending order and denote them as set A; pick those that are divisible by 8 and those that satisfy the second and the third requirements and order them in ascending order, and denote them as set B. Compare each element in set A (denoted as TBS^{A}) with the elements in set B, and pick the highest value in set B that is no greater than TBS^{A}. Put these elements into a new set: set C.
2. Screen out the elements in set C that do not satisfy overhead ratio ≤4% (denoted as TBS_j) and compare them to the elements in set B. Select an element in set B that is not greater than TBS_j and also satisfies the following constraint, replace TBS_j in set C, and form a new set: set D.

$$\left(\mathrm{TBS}_j-\left(\mathrm{TBS}_{j-1}-8\right)\right)/\,\mathrm{TBS}_j\leq 0.04$$

3. Based on the simulation results, delete or modify those TBSs of low scheduling flexibility in set D. Then get the TBS table; as seen in Table 2.16, for $\mathrm{TBS}\leq 3824$.

Simulations are conducted for the PDSCH 64QAM MCS table and the 256QAM MCS table. All the numbers of REs listed in Table 2.15 are

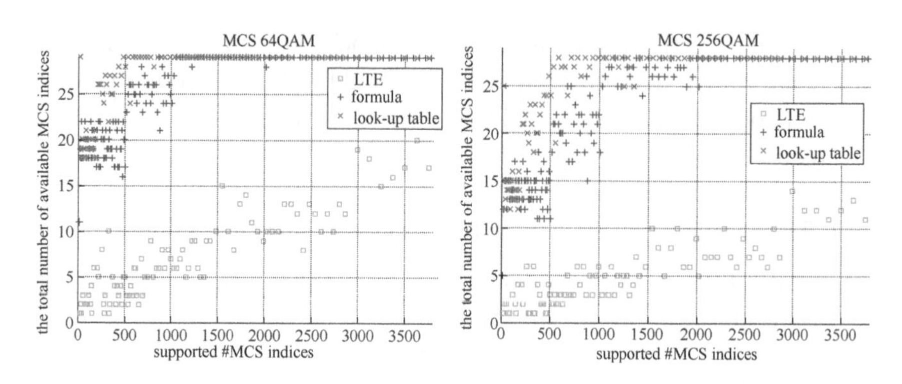

FIGURE 2.56 The number of supported MCS indices as a function of TBS, when TBS is determined by LTE TBS table, by formula, and by 5G NR TBS table, for initial HARQ transmission.

simulated. The number of layers ranges from 1 to 4. Qm corresponds to the modulation orders in the PDSCH MCS table. The number of PRBs ranges from 1 to 275. The comparison is between the TBS determined by TBS tables vs. determined by the formulae, for the initial HARQ transmission and retransmission, shown in Figures 2.56 and 2.57, respectively. It is observed that when TBS is less than 3824, if TBS of the initial transmission and retransmission is determined by the formula, the number of supported MCS indices is relatively smaller, which would limit the scheduling flexibility. By contrast, there is no such issue observed if TBS is determined by TBS tables. Hence, when TBS is less than 3824, the TBS table is used for TBS determination.

2.6 COMPLEXITY, THROUGHPUT, AND DECODING LATENCY

2.6.1 Complexity

Regarding the memory complexity, according to [47], considering the block size of information bits K=8000, the length of encoded block N=40000. For LDPC, assume 7 bits for storing channel LLR and 5 bits to store inner LLR. In the case of turbo codes, 6 bits are assumed to store the channel LLR and 9 bits for path metrics, 8 bits for extrinsic LLR, and three shifting register for two constituent convolutional codes. With all these parameter settings, the memory for LDPC is about 1.14 Mb, compared to 1.8 Mb memory for turbo. Hence, the memory requirement for LDPC is about 2/3 of turbo codes.

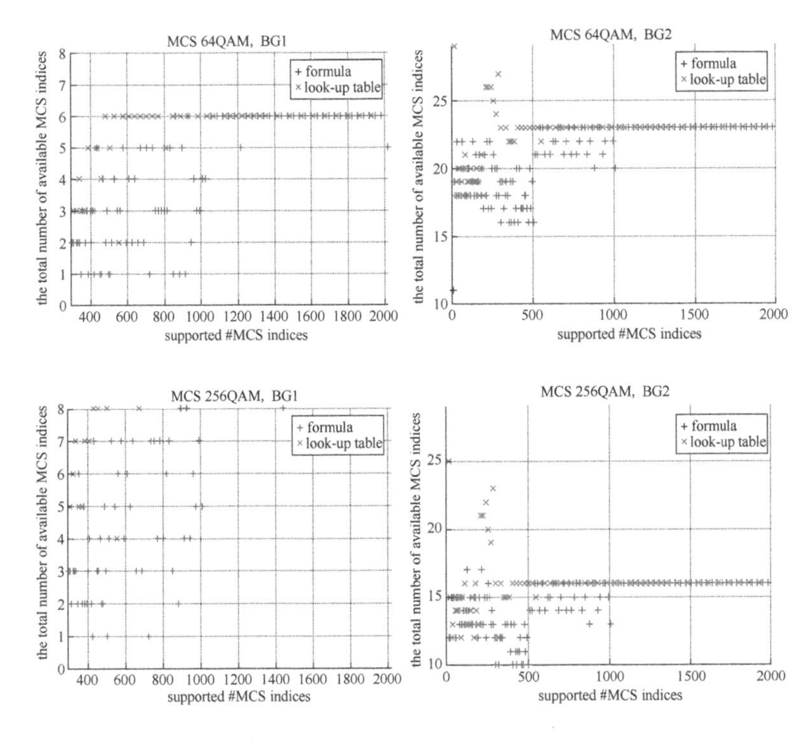

FIGURE 2.57 The number of supported MCS indices as a function of TBS, when TBS is determined by formula and by 5G NR TBS table, for HARQ retransmission.

Regarding the computation complexity, according to [47], considering the block size of information bits $K=8000$, the length of encoded block $N=40000$. For LDPC, the level of parallel processing is assumed 256. The maximum lifting factor $Z_{max}=384$. For turbo codes, eight sliding windows are assumed for the decoding. The number of additions/subtractions for LDPC is about 3072 which is not significantly lower than that for turbo codes.

2.6.2 Throughput Analysis for the QC-LDPC Decoder

2.6.2.1 Throughput of the Row-Parallel Structure

In row-parallel decoding, the update of the CNU is row by row in a serial fashion. The throughput formula is as follows:

$$T_{\text{row-parallel}} = \frac{f \cdot K}{L \cdot I} \tag{2.53}$$

where I is the number of decoding iterations. K is the block length of information bits (including CRC bits). f is the operating frequency. L is the number of layers (rows).

2.6.2.2 *Throughput of Block-Parallel Structure*

The throughput of block-parallel structure can be estimated by Eq. (2.54):

$$T_{\text{block-parallel}} = \frac{f \cdot K}{I \cdot \left\lceil \frac{N \cdot D_v}{P \cdot c} \right\rceil} \tag{2.54}$$

where I is the number of decoding iterations. K is the block length of information bits (including CRC bits). f is the clock frequency. C_y is the number of clock cycles for each decoding iteration.

According to [47], for the information block size K=8000 and block length of coded bits N=9000. Assuming I_{LDPC}=7 and $Z_{\max}$=384 for LDPC, and #sliding windows=8 and I_{Turbo}=5.5 for turbo codes, the throughput of LDPC is 12.2f bps (e.g., 12.2 Gbps @ f=1 GHz). By contrast, the throughput of turbo codes is merely 1.45f bps (e.g., 1.45 Gbps @ f=1 GHz). The throughput of LDPC is more than 8 times of turbo codes.

2.6.3 Decoding Latency

According to [47], for the information block size K=8000 and block length of coded bits N=9000. Assuming I_{LDPC}=7 and $Z_{\max}$=384 for LDPC, and #sliding windows=8 and I_{Turbo}=5.5 for turbo codes. LDPC decoding requires 658 cycles (e.g., 0.658 μs @ f=1 GHz). By contrast, turbo code requires 5500 cycles to finish the decoding (e.g., 5.5 μs @ f=1 GHz). Hence, the decoding latency of LDPC is only about 1/8 of turbo codes. The fundamental reason for this is the highly parallel processing in the LDPC decoder, e.g., Zmax=384.

Low decoding latency is very important for a self-contained subframe structure of 5G NR so that the receiver can feed back the outcome of the decoding in time, and facilitate the base station for retransmission or new transmission.

2.7 LINK-LEVEL PERFORMANCE

Link-level simulations are carried out mostly for the AWGN channel, with a focus on QPSK modulation. Log-BP with floating-point precision is assumed for LDPC decoding. The maximum number of decoding iterations is 50.

2.7.1 Short Block Length

From Figure 2.58 [48], it is observed that compared to turbo codes, the performance of 5G NR LDPC is significantly better at high code rates (e.g., 5/6) and short block lengths (e.g., 128 bits).

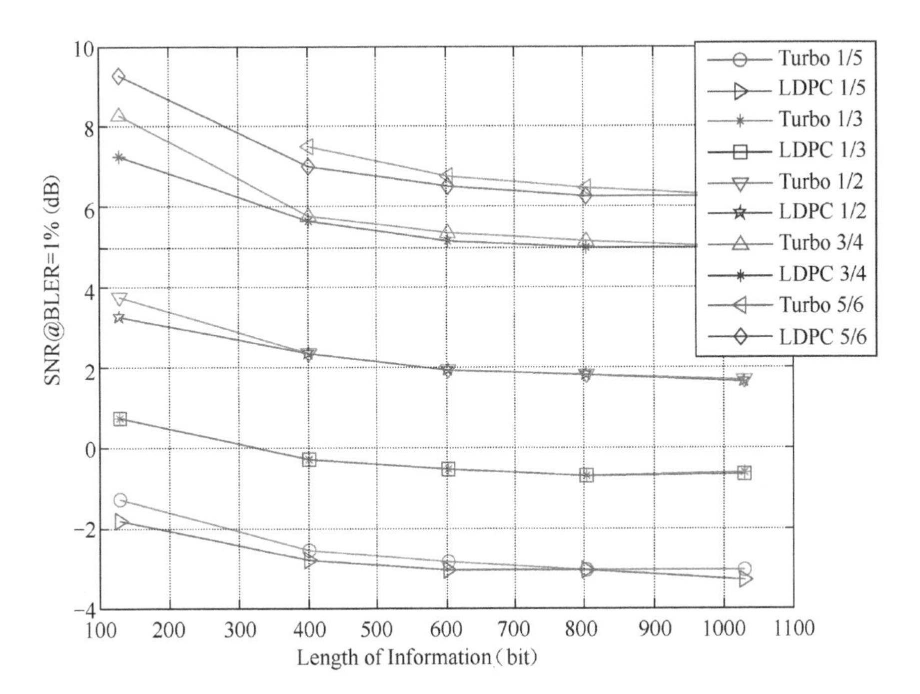

FIGURE 2.58 Performance of 5G NR LDPC for short block length [48].

2.7.2 Medium Block Length

Figure 2.59 [49] shows the required SNR to get BLER=10–2 under different block lengths (medium level) for BG1 and BG2. It is observed that for a high code rate (2/3 and 3/4) and smaller information block size (300 bits), BG2 (triangle key) noticeably outperforms BG1 (circle key). No big difference is seen in other situations.

2.7.3 Long Block Length

It can be observed from Figure 2.60 [50] that at a high code rate, the performance gain of LDPC is about 0.1~0.4 dB compared with turbo codes.

In summary, due to its lower complexity of decoding, higher throughput, lower decoding latency, and excellent link-level performance, LDPC was chosen as the channel coding scheme for traffic channels (PDSCH and PUSCH) at least for eMBB scenarios.

2.8 LDPC DESCRIBED IN 3GPP SPECIFICATIONS

LDPC has been widely used in various communications systems, for instance, WiMAX (IEEE 802.16e) [8], WiFi (IEEE 802.11n) [9], and DVB-S2 [11].

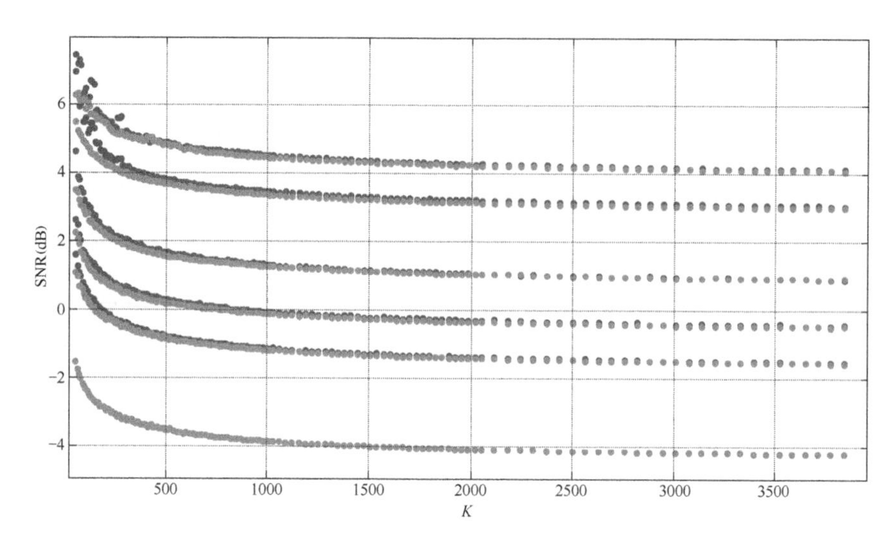

FIGURE 2.59 Required SNR for BLER$=10^{-2}$ at different block lengths and code rates [49], compared between BG1 and BG2, with a zoomed-in portion.

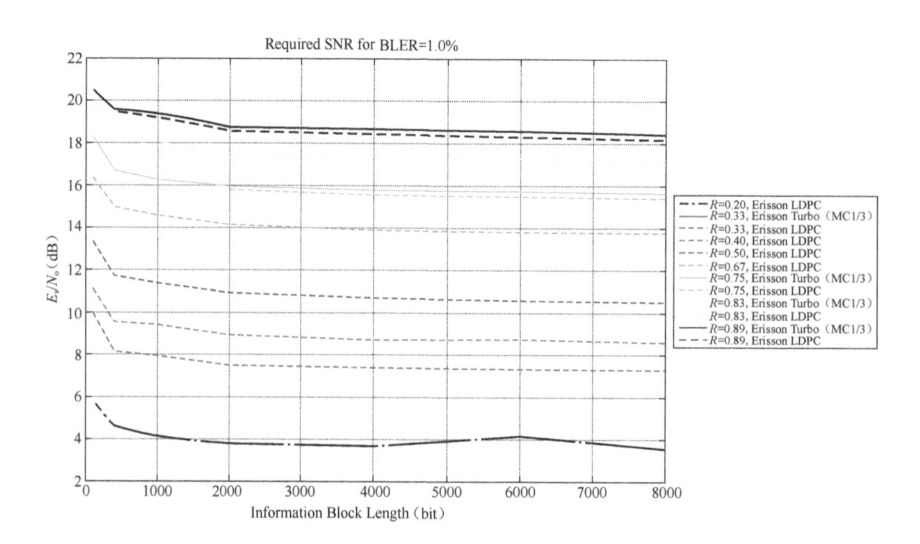

FIGURE 2.60 Performance of 5G NR LDPC for 64QAM [50].

At the beginning of LTE specification in 3GPP, LDPC was used to be one of the candidate channel coding schemes [51]. However, considering the implementation complexity, turbo codes were adopted in 4G. After a significant amount of effort in academia and the industry over the last 10 years, LDPC was finally adopted by 3GPP [52]. Next, we show how LDPC is described in 3GPP standards [39] (Table 2.17).

TABLE 2.17 LDPC Described in 3GPP Specifications

3GPP Specifications [39]	**Comments**
5.2 Code block segmentation and code block CRC attachment **5.2.2 LDPC coding**	Segmentation of LDPC into multiple segments, e.g., code blocks
The input bit sequence to the code block segmentation is denoted by $b_0, b_1, b_2, b_3, ..., b_{B-1}$, where $B > 0$. If B is larger than the maximum CBS K_{cb}, segmentation of the input bit sequence is performed and an additional CRC sequence of $L = 24$ bits is attached to each code block.	To append 24 CRC bits to each segment (code block)
For LDPC base graph 1, the maximum CBS is - $K_{cb} = 8448$.	Maximum info block size is 8448 for BG1
For LDPC base graph 2, the maximum CBS is - $K_{cb} = 3840$.	Maximum info block size is 3840 for BG2
Total number of code blocks C is determined by if $B \le K_{cb}$ $L = 0$ Number of code blocks: $C = 1$ $B' = B$	

else
$L = 24$
Number of code blocks: $C = \lceil B/(K_{cb} - L) \rceil$.
$B' = B + C \cdot L$
end if
	To determine the number of code blocks

The bits output from code block segmentation are denoted by $c_{r0}, c_{r1}, c_{r2}, c_{r3}, ..., c_{r(K_r-1)}$, where $0 \le r < C$ is the code block number, and K_r is the number of bits for the code block number r.	
Number of bits in each code block: $K' = B'/C$; For LDPC base graph 1, $K_b = 22$.	
For LDPC base graph 2, if $B > 640$ $K_b = 10$; Else if $B > 560$ $K_b = 9$; Else if $B > 192$ $K_b = 8$; else $K_b = 6$;	K_b for BG1 (the number of columns in the kernel matrix) K_b for BG2 (the number of columns in the kernel matrix)
end if	Lifting factor Z

find the minimum value of Z in all sets of lifting sizes in Table 5.3.2-1, denoted as Z_c, such that $K_b \cdot Z_c \geq K'$, and denote $K = 22Z_c$ for LDPC base graph 1 and $K = 10Z_c$ for LDPC base graph 2;

$s = 0$;

for $r = 0$ to $C - 1$

    for $k = 0$ to $K' - L - 1$

        $c_{rk} = b_s$;

        $s = s + 1$;

    end for

    if $C > 1$

        The sequence $c_{r0}, c_{r1}, c_{r2}, c_{r3}, \ldots, c_{r(K'-L-1)}$ is used to calculate the CRC parity bits $p_{r0}, p_{r1}, p_{r2}, \ldots, p_{r(L-1)}$ according to section 5.1 with the generator polynomial $g_{\mathrm{CRC24B}}(D)$.

        for $k = K' - L$ to $K' - 1$

            $c_{rk} = p_{r(k+L-K')}$;

        end for

    end if

    for $k = K'$ to $K - 1$ — Insertion of filler bits

        $c_{rk} =< NULL >$;

    end for

end for

5.3.2 LDPC coding

LDPC encoding

The bit sequence input for a given code block to channel coding is denoted by $c_0, c_1, c_2, c_3, \ldots, c_{K-1}$, where K is the number of bits to encode as defined in Section 5.2.1. After encoding the bits are denoted by $d_0, d_1, d_2, \ldots, d_{N-1}$, where $N = 66Z_c$ for LDPC base graph 1 and $N = 50Z_c$ for LDPC base graph 2, and the value of Z_c is given in Section 5.2.1.

N is the block length after encoding and it is divisible by Z_c.

For a code block encoded by LDPC, the following encoding procedure applies:

Z_c is the minimum value in the set of lifting factors°

1) Find the set index i_{LS} in Table 5.3.2-1 which contains Z_c.

2) for $k = 2Z_c$ to $K - 1$

    if $c_k \neq< NULL >$

        $d_{k-2Z_c} = c_k$;

    else

        $c_k = 0$;

        $d_{k-2Z_c} =< NULL >$;

    end if

end for

3) Generate $N+2Z_c-K$ parity bits $\mathbf{w}=\left[w_0,w_1,w_2,...,w_{N+2Z_c-K-1}\right]^T$ such that

$$\mathbf{H}\times\begin{bmatrix}\mathbf{c}\\ \mathbf{w}\end{bmatrix}=\mathbf{0}$$

, where $\mathbf{c}=\left[c_0,c_1,c_2,...,c_{K-1}\right]^T$; $\mathbf{0}$ is a column vector of all elements equal to 0. The encoding is performed in GF(2).

For LDPC base graph 1, a matrix of $\mathbf{H}_{BG}$ has 46 rows with row indices $i=0,1,2,...,45$ and 68 columns with column indices $j=0,1,2,...,67$. For LDPC base graph 2, a matrix of $\mathbf{H}_{BG}$ has 42 rows with row indices $i=0,1,2,...,41$ and 52 columns with column indices $j=0,1,2,...,51$. The elements in $\mathbf{H}_{BG}$ with row and column indices given in Table 5.3.2-2 (for LDPC base graph 1) and Table 5.3.2-3 (for LDPC base graph 2) are of value 1, and all other elements in $\mathbf{H}_{BG}$ are of value 0.

The matrix $\mathbf{H}$ is obtained by replacing each element of $\mathbf{H}_{BG}$ with a $Z_c\times Z_c$ matrix, according to the following:

- Each element of value 0 in $\mathbf{H}_{BG}$ is replaced by an all-zero matrix $\mathbf{0}$ of size $Z_c\times Z_c$;
- Each element of value 1 in $\mathbf{H}_{BG}$ is replaced by a circular permutation matrix $\mathbf{I}(P_{i,j})$ of size $Z_c\times Z_c$, where i and j are the row and column indices of the element, and $\mathbf{I}(P_{i,j})$ is obtained by circularly shifting the identity matrix $\mathbf{I}$ of size $Z_c\times Z_c$ to the right $P_{i,j}$ times. The value of $P_{i,j}$ is given by $P_{i,j}=\text{mod}(V_{i,j},Z_c)$. The value of $V_{i,j}$ is given by Tables 5.3.2-2 and 5.3.2-3 according to the set index i_{LS} and base graph.

4) for $k=K$ to $N+2Z_c-1$

$d_{k-2Z_c}=w_{k-K}$;

end for

Table 5.3.2-1: Sets of LDPC lifting size Z

Set index (i_{LS})	Set of lifting sizes (Z)
0	{2, 4, 8, 16, 32, 64, 128, 256}
:	:
7	{15, 30, 60, 120, 240}

c is the systematic bit stream

w is the parity bit stream°

$\mathbf{H}^*\mathbf{x}=0$, $\mathbf{H}$ is parity check matrix°

Design of base matrix

The meaning of"0"in the base matrix

The meaning of"1"in the base matrix

Construction of **H** for QC-LDPC

Lifting factor (lifting size)

Table 5.3.2-2: LDPC base graph 1 (H_{BG}) and its parity check matrices ($V_{i,j}$)

Base matrix of BG1 in table form

Row Index i	Column index j	Set index i_{LS} 0	1	2	3	4	5	6	7
0	0	250	307	73	223	211	294	0	135
	:	:	:	:	:	:	:	:	:
	23	0	0	0	0	0	0	0	0
:	:			:					
45	1	149	135	101	184	168	82	181	177
	6	151	149	228	121	0	67	45	114
	10	167	15	126	29	144	235	153	93
	67	0	0	0	0	0	0	0	0

Base matrix of BG2 in table form

Table 5.3.2-3: LDPC base graph 2 (H_{BG}) and its parity check matrices ($V_{i,j}$)

Row index i	Column index j	Set index i_{LS} 0	1	2	3	4	5	6	7
0	0	9	174	0	72	3	156	143	145
	:	:	:	:	:	:	:	:	:
	11	0	0	0	0	0	0	0	0
:				:					:
41	1	129	147	0	120	48	132	191	53
	:	:	:	:	:	:	:	:	:
	51	0	0	0	0	0	0	0	0

5.4.2 Rate matching for LDPC code

Rate matching. Same rate matching method for all the code blocks

The rate matching for LDPC code is defined per coded block and consists of bit selection and bit interleaving. The input bit sequence to rate matching is $d_0, d_1, d_2, ..., d_{N-1}$. The output bit sequence after rate matching is denoted as $f_0, f_1, f_2, ..., f_{E-1}$.

5.4.2.1 Bit selection

Circular buffer is used for rate matching

The bit sequence after encoding $d_0, d_1, d_2, ..., d_{N-1}$ from Section 5.3.2 is written into a circular buffer of length N for the r-th coded block, where N is defined in Section 5.3.2.

For the r-th code block, let $N_{cb} = N$ if $I_{LBRM} = 0$ and $N_{cb} = \min(N, N_{ref})$ otherwise, where $N_{ref} = \left\lfloor \frac{TBS_{\text{LBRM}}}{C \cdot R_{\text{LBRM}}} \right\rfloor$, $R_{\text{LBRM}} = 2/3$, TBS_{LBRM} is determined according to section X.X in [6, TS38.214] assuming the following:

- maximum number of layers supported by the UE for the serving cell;
- maximum modulation order configured for the serving cell;
- maximum coding rate of 948/1024;
- $\bar{N}'_{RE} = 156$;
- $n_{PRB} = n_{PRB,LBRM}$ is given by Table 5.4.2.1-1;

Maximum code rate is 0.9258

C is the number of code blocks of the transport block determined according to Section 5.2.2.

Table 5.4.2.1-1: Value of $n_{PRB,LBRM}$

Maximum number of PRBs across all configured BWPs of a carrier	$n_{PRB,LBRM}$
Less than 33	32
33 to 66	66
67 to 107	107
108 to 135	135
136 to 162	162
163 to 217	217
Larger than 217	273

Denoting by E_r the rate matching output sequence length for the r-th coded block, where the value of E_r is determined as follows:

Set $j = 0$

for $r = 0$ to $C-1$

if the r-th coded block is not for transmission as indicated by CBGTI according to Section X.X in [6, TS38.214]:

$E_r = 0;$

else

if $j \le C' - \text{mod}(G/(N_L \cdot Q_m), C') - 1$

$$E_r = N_L \cdot Q_m \cdot \left\lfloor \frac{G}{N_L \cdot Q_m \cdot C'} \right\rfloor;$$

else

$$E_r = N_L \cdot Q_m \cdot \left\lceil \frac{G}{N_L \cdot Q_m \cdot C'} \right\rceil;$$

end if

$j = j + 1;$

end if

end for

where

- N_L is the number of transmission layers that the transport block is mapped onto;
- Q_m is the modulation order;
- G is the total number of coded bits available for transmission of the transport block;
- $C' = C$ if CBGTI is not present in the DCI scheduling the transport block and C' is the number of scheduled code blocks of the transport block if CBGTI is present in the DCI scheduling the transport block.

Four RVs are defined

Denote by rv_{id} the redundancy version number for this transmission ($rv_{id} = 0$, 1, 2, or 3), the rate matching output bit sequence e_k, $k = 0,1,2,...,E-1$, is generated as follows, where k_0 is given by Table 5.4.2.1-2 according to the value of rv_{id}:

$k = 0;$

$j = 0;$

while $k < E$

if $d_{(k_0+j)\bmod N_{cb}} \neq$ <NULL>

$e_k = d_{(k_0+j)\bmod N_{cb}};$

$k = k+1;$

end if

$j = j+1;$

end while

mod( ) means that when the data fetching reaches the end, it should return to the beginning to continue the fetching

Table 5.4.2.1-2: Starting position of different redundancy versions, k_0

rv_{id}	k_0	
	Base graph 1	Base graph 2
0	0	0
1	$\left\lfloor \frac{17N_{cb}}{66Z_c} \right\rfloor Z_c$	$\left\lfloor \frac{13N_{cb}}{50Z_c} \right\rfloor Z_c$
2	$\left\lfloor \frac{33N_{cb}}{66Z_c} \right\rfloor Z_c$	$\left\lfloor \frac{25N_{cb}}{66Z_c} \right\rfloor Z_c$
3	$\left\lfloor \frac{56N_{cb}}{66Z_c} \right\rfloor Z_c$	$\left\lfloor \frac{43N_{cb}}{50Z_c} \right\rfloor Z_c$

HARQ: different starting positions for bits to be transmitted, based on different RVs

5.4.2.2 Bit interleaving

The bit sequence $e_0, e_1, e_2, ..., e_{E-1}$ is interleaved to bit sequence $f_0, f_1, f_2, ..., f_{E-1}$, according to the following, where the value of Q is given by Table 5.4.2.2-1.

for $j = 0$ to $E/Q - 1$
  for $i = 0$ to $Q - 1$
    $f_{i+j \cdot Q} = e_{i \cdot E/Q + j}$;
  end for
end for

To pick one bit for every E/Q bits, and then consecutively put these bits

Assuming there are 20 bits for 16QAM, with input order 0, 1, 2, 3, 4, …, 18 and 19, the output order is 0, 5, 10, 15, 1, 6, 11, 16, 2, 7, 12, 17, 3, 8, 13, 18, 4, 9, 14 and 19

Table 5.4.2.2-1: Modulation and number of coded bits per QAM symbol

Modulation	Q
π/2-BPSK, BPSK	1
QPSK	2
16QAM	4
64QAM	6
256QAM	8

Modulation and then mapped to the modulation symbols

6.2 Uplink shared channel

Determine BG for PUSCH

6.2.2 LDPC base graph selection

For initial transmission of a transport block with coding rate R indicated by the MCS according to Section X.X in [6, TS38.214] and subsequent retransmission of the same transport block, each code block of the transport block is encoded with either LDPC base graph 1 or 2 according to the following:

- if $A \le 292$, or if $A \le 3824$ and $R \le 0.67$, or if $R \le 0.25$, LDPC base graph 2 is used;
- otherwise, LDPC base graph 1 is used,

  where A is the payload size described in Section 6.2.1.

BG2 is for a small code block, low to medium code rate; BG1is for a long code block, high code rate. Refer to Section 2.5.1 in this book

7.2 Downlink shared channel and paging channel

Appending CRC for PDSCH

7.2.1 Transport block CRC attachment

Error detection is provided on each transport block through a CRC.

The entire transport block is used to calculate the CRC parity bits. Denote the bits in a transport block delivered to layer 1 by $a_0, a_1, a_2, a_3, ..., a_{A-1}$, and the parity bits by $p_0, p_1, p_2, p_3, ..., p_{L-1}$, where A is the payload size and L is the number of parity bits. The lowest-order information bit a_0 is mapped to the most significant bit of the transport block as defined in Clause 6.1.1 of [TS38.321].

24-bit CRC for block size larger than 3824;

The parity bits are computed and attached to the DL-SCH transport block according to Section 5.1, by setting L to 24 bits and using the generator polynomial $g_{CRC24A}(D)$ if $A > 3824$; and by setting L to 16 bits and using the generator polynomial $g_{CRC16}(D)$ otherwise.	16-bit CRC for block size larger than or equal to 3824
The bits after CRC attachment are denoted by $b_0, b_1, b_2, b_3, ..., b_{B-1}$, where $B = A + L$.	
7.2.2 LDPC base graph selection	Determine BG for PDSCH, similar to PUSCH
For initial transmission of a transport block with coding rate R indicated by the MCS according to Section X.X in [6, TS38.214] and subsequent retransmission of the same transport block, each code block of the transport block is encoded with either LDPC base graph 1 or 2 according to the following: - if $A \leq 292$, or if $A \leq 3824$ and $R \leq 0.67$, or if $R \leq 0.25$, LDPC base graph 2 is used; - otherwise, LDPC base graph 1 is used, where A is the payload size in Section 7.2.1.	BG2 is for small block length, low to medium code rate; BG1is for long block length, high code rate. Refer to Section 2.5.1 in this book

2.9 FUTURE DIRECTIONS

In the future, LDPC is expected to continuously evolve, with the following potential directions:

- To better support short block length (e.g., new optimized base graphs, BG3, BG4). For IoT and MTC scenarios, small data packets are dominant. IoT and MTC also care about energy consumption, which requires high performance.
- Multi-edge LDPC codes [21] which have a lower error floor and can be beneficial for high-reliability scenarios such as URLLC.
- Parallel concatenated Gallager code (PCGC) [53] where the inner code of the turbo code is LDPC. At the code rate of R=0.3367, PCGC can approach Shannon's limit as close as 0.4 dB.
- LDPC in the form of staircase code [54,55]. The merit of the staircase code is that its error floor is very low.

2.10 SUMMARY

In this chapter, LDPC codes are discussed which covers the basic principle, QC-LDPC, the decoding architecture of QC-LDPC, the standardization of LDPC in 5G NR, the complexity, throughput, link-level performance,

LDPC specification, and future directions. Overall speaking, LDPC has good performance, low complexity, high throughput, and low decoding latency, due to the inherent parallel processing. Certainly, LDPC is not perfect. Its codeword construction is a little complicated, which may not be suitable for short block length. Nevertheless, it is believed that as people get deeper into the research on channel coding, LDPC may find more use in practical systems.:

REFERENCES

[1] R. G. Gallager, "Low-density parity-check codes," *IRE Trans. Inform. Theory*, vol. 8, Jan. 1962, pp. 21–28.

[2] R. G. Gallager, *Low density parity-check codes*, MIT, 1963.

[3] C. E. Shannon, "A mathematical theory of communication," *Bell System Technol. J.*, vol. 27, no. 3, July 1948, pp. 379–423.

[4] R. W. Hamming, "Error detecting and Error correcting codes," *Bell System Technol. J.*, vol. 29, no. 2, April 1950, pp. 147–160.

[5] E. Prange, "Cyclic Error-Correcting Codes in Two Symbols," AFCRC-TN-57, Air Force Cambridge Research Center, 1957.

[6] R. M. Tanner, "A recursive approach to low complexity codes," *IEEE Trans. Inf. Theory*, 27, no. 5, 1981, pp. 533–547.

[7] C. Berrou, "Near Shannon Limit Error-Correcting Coding and Decoding: Turbo Codes," *Proc. IEEE Intl. Conf. Communication (ICC 93)*, May 1993, pp. 1064–1070.

[8] IEEE. 802.16e.

[9] IEEE. 802.11a.

[10] S. Y. Chung, "On the design of low-density parity-check codes within 0.0045 dB of the Shannon limit," *IEEE Commun. Lett.*, vol. 5, no. 2, 2001, pp. 58–60.

[11] ETSI, "EN 302 307 V1.3.1-Digital Video Broadcasting (DVB) Second generation framing structure, channel coding and modulation systems for Broadcasting, Interactive Services, News Gathering and other broadband satellite applications (DVB-S2)," 2013.03.

[12] F.R. Kschischang, "Factor graphs and the sum-product algorithm," *IEEE Trans. Inf. Theory*, vol. 47, no. 2, Feb 2001, pp. 498–519.

[13] A.I.V. Casado, "Informed Dynamic Scheduling for Belief-Propagation Decoding of LDPC Codes," *ICC'07. IEEE International Conference on Communications*, 2007., arXiv:cs/0702111.

[14] H. Xiao, "Graph-based message-passing schedules for decoding LDPC codes," *IEEE Trans Commun.*, vol. 52, no. 12, 2004, pp. 2098–2105.

[15] D. J. C. MacKay, "Near Shannon limit performance of low density parity check codes," *Electron. Lett.*, vol. 3, no. 6, March 1997, pp. 457–458.

[16] S. Lin, *Error control coding*, 2nd edition (translated by Jin Yan), Mechanical Industry Press, Beijing, China, June 2007.

[17] M.G. Luby, "Efficient erasure correcting codes," *IEEE Trans. Inf. Theory*, vol. 47, no. 2, 2001, pp. 569–584.

[18] D.J.C. Mackay, "Good error-correcting codes based on very sparse matrices," *IEEE Trans. Inf. Theory*, vol. 45, no. 2, 1999, pp. 399–431.
[19] T. J. Richardson, "Design of capacity-approaching irregular low-density parity-check codes," *IEEE Trans. Inf. Theory*, vol. 47, no. 2, 2001, pp. 619–637.
[20] 3GPP, R1-167532, Discussion on LDPC coding scheme of code structure, granularity and HARQ-IR, MediaTek, RAN1#86, August 2016.
[21] T.J. Richardson, "Multi-edge type LDPC codes," submitted IEEE IT, EPFL, LTHC-REPORT-2004-001, 2004.
[22] Jin Xu, Jun Xu, "Structured LDPC Applied in IMT-Advanced System," *International Conference on Wireless Communication*, 2008, pp.1–4.
[23] "LDPC technology solutions," white book of IMT-Advanced key technology study (China), ZTE, Oct. 2009
[24] IMT-A_STD_LTE+_07061 LDPC coding for PHY of LTE+ air interface.
[25] 3GPP, R1-061019, Structured LDPC coding with rate matching, ZTE, RAN1#44bis, March 2006.
[26] IEEE, Rate = 5/6 LDPC coding for OFDMA PHY, ZTE, IEEE C802.16e-05/126rl, 2005.03.09.
[27] IEEE, High girth LDPC coding for OFDMAPHY, ZTE, IEEE C802.16e-05/031rl, 2005.1.25.
[28] C. Roth, "Area, Throughput, and Energy-Efficiency Trade-offs in the VLSI Implementation of LDPC Decoders," *IEEE International Symposium on Circuits & Systems*, 2011, pp. 1772–1775.
[29] 3GPP, R1-1608971, Consideration on Flexibility of LDPC Codes for NR, ZTE, RAN1#86bis, October 2016.
[30] 3GPP, R1-1611111, Consideration on Flexibility of LDPC Codes for NR, ZTE, RAN1#87, November 2016.
[31] 3GPP, R1-1700247, Compact LDPC design for eMBB, ZTE, RAN1#AH_NR Meeting, January 2017.
[32] J. Xu, *LDPC code and its application in 4G mobile communications networks*, Master Thesis of Nanjing Post Univ., 2003.
[33] H. Wen, C. Fu and L. Zhou, *Principle of LDPC codes and application*, Electronics Technology Press, Beijing, China, Jan 2006.
[34] 3GPP, Draft Report of 3GPP TSG RAN WG1 #90 v0.1.0, August 2017.
[35] 3GPP, R1-1719525, Remaining details of LDPC coding, ZTE, RAN1#91, November, 2017.
[36] 3GPP, R1-1715732, Redundancy Version for HARQ of LDPC Codes, Ericsson, RAN1 Meeting NR#3, September 2017.
[37] 3GPP, Draft Report of 3GPP TSG RAN WG1 #AH_NR3 v0.1.0, September 2017.
[38] 3GPP, R1-1700384, LDPC HARQ design, Intel, RAN1 Ad hoc, January 2017.
[39] 3GPP, TS38.212-NR Multiplexing and channel coding (Release 15), http://www.3gpp.org/ftp/Specs/archive/38_series/38.212/.
[40] 3GPP, R1-1715664, On rate matching for LDPC code, ZTE, RAN1#AH3, September 2017.
[41] 3GPP, Draft Report of 3GPP TSG RAN WG1 #90 v0.1.0, August 2017.

[42] 3GPP, R1-1715663, On bit level interleaving for LDPC code, ZTE, RAN1#AH3, September 2017.
[43] 3GPP, R1-1713230, On interleaving for LDPC code, ZTE, RAN1#90, August 2017.
[44] 3GPP, R1-1715663, On bit level interleaving for LDPC code, ZTE, RAN1 NR Ad-Hoc#3, September 2017.
[45] 3GPP, R1-1610600, Updated Summary of Channel Coding Simulation Data Sharing for eMBB Data Channel, InterDigital, RAN1#86bis, October 2016.
[46] 3GPP, TS36.213 V14.4.0-E-UTRA Physical layer procedures (Release 14), 2017.09.
[47] 3GPP, R1-166372, Performance and implementation comparison for EMBB channel coding, Qualcomm, RAN1#86, August 2016.
[48] 3GPP, R1-1612276, Coding performance for short block eMBB data, Nokia, RAN1 #87, November, 2016.
[49] 3GPP, R1-1714555, Remaining issues for LDPC code design, ZTE, RAN1#90, August 2017.
[50] 3GPP, R1-1610423, Summary of channel coding simulation data sharing, InterDigital, RAN1#86bis, Oct.2016.
[51] 3GPP, TR25.814 -V710 -Physical layer aspects for E-UTRA, 2006.9.
[52] 3GPP, "Draft_Minutes_report_RAN1#86b_v100," October 2016.
[53] F. Wang, *Application of LDPC code in future mobile communications systems*, Ph. D thesis of Southeast Univ, Nov. 2007.
[54] B. P. Smith, "Staircase codes: FEC for 100 Gb/s OTN," *IEEE/OSA J. Lightwave Technol.*, vol. 30, no. 1, Jan. 2012, pp. 110–117.
[55] IETF, "Simple Low-Density Parity Check (LDPC) Staircase: Forward Error Correction (FEC) Scheme for FECFRAME," Internet Engineering Task Force, Request for Comments: 6816, Category: Standards Track, ISSN: 2070-1721.

CHAPTER 3

Polar Codes

Focai Peng, Mengzhu Chen and Saijin Xie

POLAR CODE WAS INVENTED by Prof. Erdal Arikan of Bilkent University in Turkey in 2008 [1]. So far, polar code is the only channel coding scheme that can reach Shannon's limit, whereas all other coding schemes can just approach Shannon's limit. Due to the relatively high complexity of its encoding and decoding, polar code did not get wide attention right after it was proposed.

In 2011, Prof. Alexander Vardy of the University of California at San Diego made a breakthrough in the decoding algorithms for polar codes [2], which made it possible for practical use of Polar codes. The structure of Chapter 3 is outlined in Figure 3.1.

3.1 ORIGIN OF POLAR CODES

This section will discuss how polar codes were first proposed [3]. According to the description by Prof. Erdal Arikan, the invention of polar codes is the result of his many years of research in channel coding.

It is known that with random codes and maximum likelihood decoding, the cut-off rate R_0 determines the corresponding bit error rate $P_e = 2^{-NR_0}$, where N is the block length. When the channel rate R is less than R_0, the average bit error rate $\overline{P_e} < 2^{-N(R_0-R)}$, if using random codes and maximum likelihood (ML) decoding. The original motivation of polar codes is to improve the cut-off rate.

In the case of sequential decoding, it is impossible to calculate the cut-off rate for convolutional codes, because the cut-off rate R_0 can only be achieved for random codes and ML decoding. In his early works, Arikan studied sequential decoding for multiple access channels in his Ph.D.

DOI: 10.1201/9781003336174-3

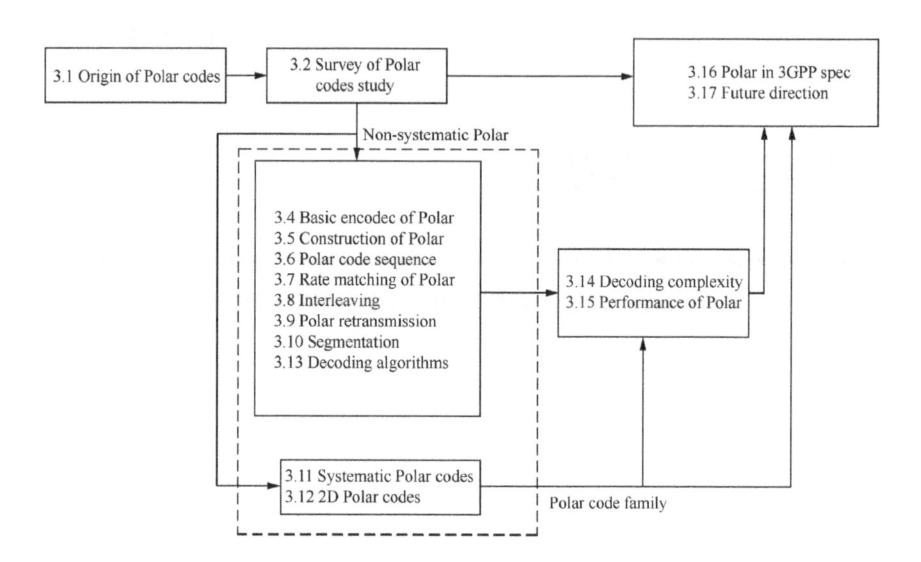

FIGURE 3.1 Structure of Chapter 3.

thesis [4]. Between 1985 and 1988, he studied the upper bound of the channel cut-off rate for sequential decoding [5]. Sequential decoding is a path searching algorithm along a tree structure. A major issue with sequential decoding is that its computation complexity is a random variable. When the SNR is low, it is generally believed that sequential decoding is not suitable. By denoting the channel cut-off rate computed based on sequential decoding as R_{Comp}, Arikan proved that the practical viable channel cut-off rate will be $R_0 \leq R_{\text{Comp}}$. In practice, a sequential decoding is a guess that a decoder guesses with a probability in which the codeword is transmitted. From 1994 to 1996, Arikan studied the number of guesses and channel cut-off rates of sequential decoding [5].

Between 2004 and 2006, Arikan tried to use channel combining and channel splitting to improve the channel cut-off rate [6]. To split a discrete memoryless channel (DMC) W into two correlated sub-channels: W_1 and W_2, the relationship of the channel capacities before and after the splitting is $C(W_1)+C(W_2)\leq C(W)$ where $C(*)$ denotes channel capacity. The relationship of the channel cut-off rates before and after the splitting is $R_0(W_1)+R_0(W_2)>R_0(W)$, where $R_0(*)$ denotes the channel cut-off rate. This means that the sum capacity of the two sub-channels would not exceed the capacity of the original channel. However, the sum cut-off rate is increased by channel splitting [7]. Channel splitting and channel combining are illustrated in Figures 3.2 and 3.3, respectively. At first, Arikan used the idea of channel combining/splitting for inner code (as W in Figure 3.2).

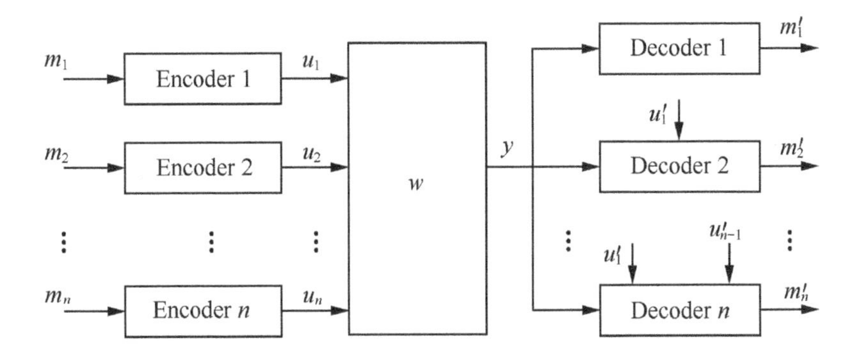

FIGURE 3.2 An illustration of channel splitting and SC [6].

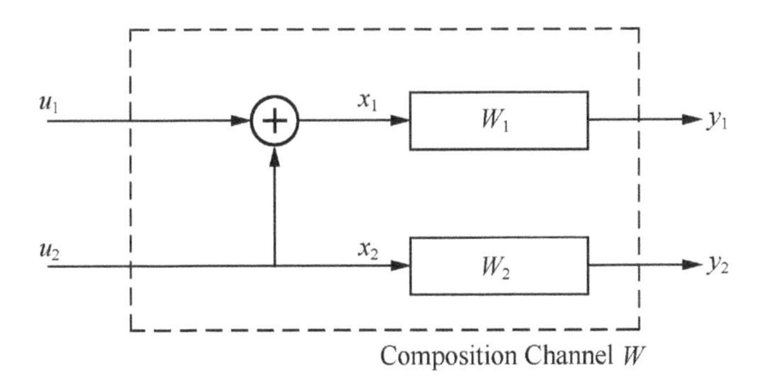

FIGURE 3.3 An illustration of channel combining.

Later on, he wondered whether by purely relying on channel combining/ splitting, the inner code can work as LDPC or turbo code, without outer code (e.g., the encoder in Figure 3.2).

In 2009, Arikan confirmed his conjecture: such inner code can result in better performance than turbo codes. This inner code is polar code. It has been proven that the polar code can reach Shannon's limit for binary erasure channel (BEC) and binary discrete memoryless channel (B-DMC).

3.2 SURVEY OF THE POLAR CODE STUDY

Due to the relatively high complexity of decoding, the polar code did not get immediate attention right after it was proposed, even though there was a continuing study on it. In 2009, N. Hussami studied the performance of Polar codes [8] and provided several ways to improve the performance, for instance, to select different frozen bits. Some of his simulation

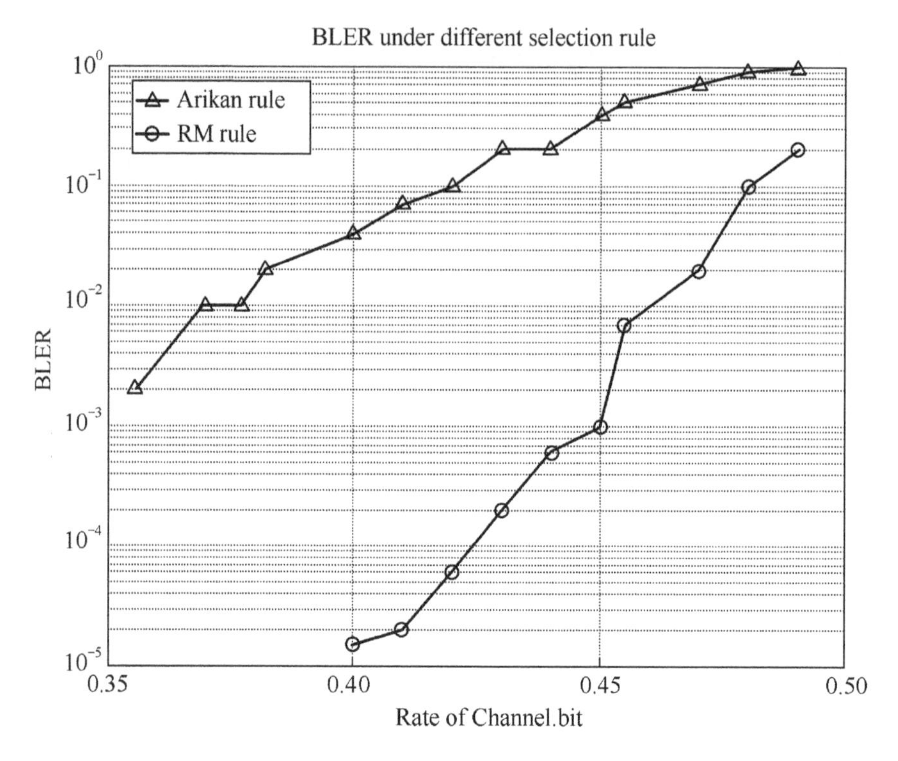

FIGURE 3.4 Performance of polar code using different methods for channel selection [8] ($N=1024$).

results are shown in Figure 3.4. It is seen that the block error rate (BLER) can be reduced if the selection of sub-channels is based on maximizing Hamming's distance, compared to the original method by Arikan.

In 2009, R. Mori studied the performance of the polar code using the density evolution method [9]. In 2010, A. Eslami of the University of Massachusetts studied the BER performance of the polar code [10]. In the case of encoded block length = 8192 bits and code rate = 1/2, the BER performance is shown in Figure 3.5 [10]. It is found that by using the guess method proposed, 0.2 dB gain can be obtained at BER = 10^{-5}.

In 2010, E. Şaşoğlu studied a 2-user polar code for binary input multiple access channels [11]. The result shows that the computation complexity of the 2-user polar code is the same as that of the single user, which is $O(n \bullet \log(n))$. The BLER of two users is also the same as that of the single user which is $O(e^{-n^{1/2-\varepsilon}})$.

Between 2010 and 2011, Prof. A. Vardy of the University of California at San Diego studied the capacity of polar code as well as the list decoding and hardware architecture of the decoding, which results in a big

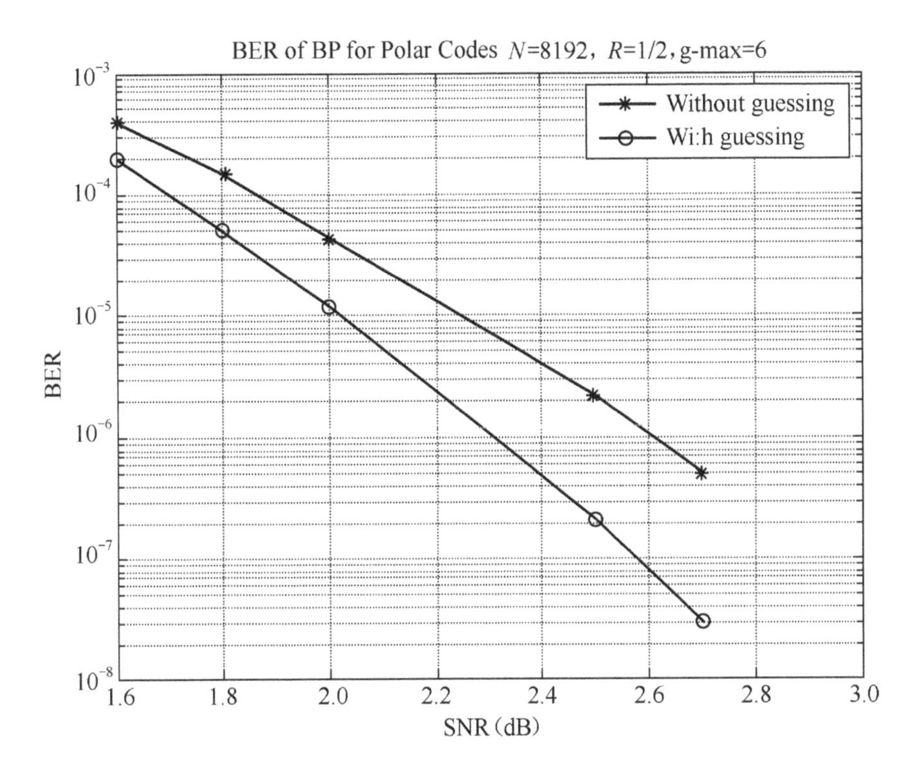

FIGURE 3.5 BER performance of polar code, with and without guessing [10].

breakthrough in the decoding algorithm of Polar codes [2,12–14]. Since then, more universities, institutes, and companies started to look into Polar codes.

In 2012, V. Miloslavskaya at Univ., St. Petersburg, Russia designed a new transformation kernel that is different from the Arikan kernel [15]. As Figure 3.6 shows, the new proposed kernel of size 64×64 can offer a 0.5 dB performance gain over the Arikan kernel.

In 2013, H. Si of the University of Texas studied the layered polar code for fading channels and the performance of his proposed coding scheme reaches the capacity of faded DMC [16]. In 2014, P. Giard tried implementing a soft decoding algorithm of polar code on a common-purpose CPU [17] and achieved 200 Mpbs throughput which is higher than the peak throughput of 150 Mbps required by the 2-antenna turbo decoder in Release 8 LTE. This paves the way for commercial use of Polar codes.

In his study of a shortened polar code [18] in 2015, V. Miloslavskaya demonstrated a 0.4 dB performance gain of polar code compared to LDPC at BLER = 1%, as illustrated in Figure 3.7 when the information block size is 432 and the encoded block size is 864 (reduced from the original $N = 1024$).

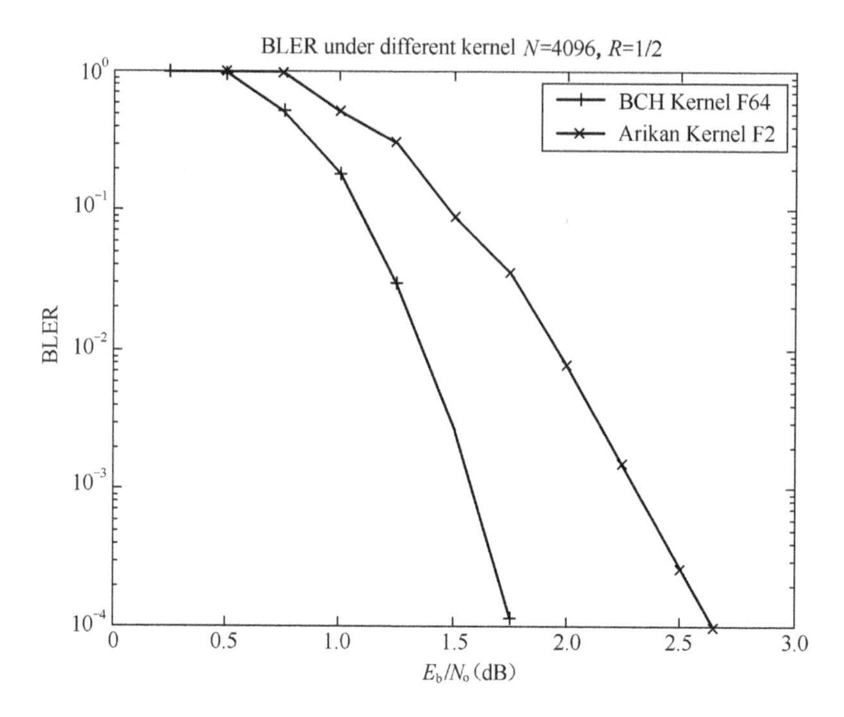

FIGURE 3.6 Performance of F64 kernel compared to Arikan kernel [15].

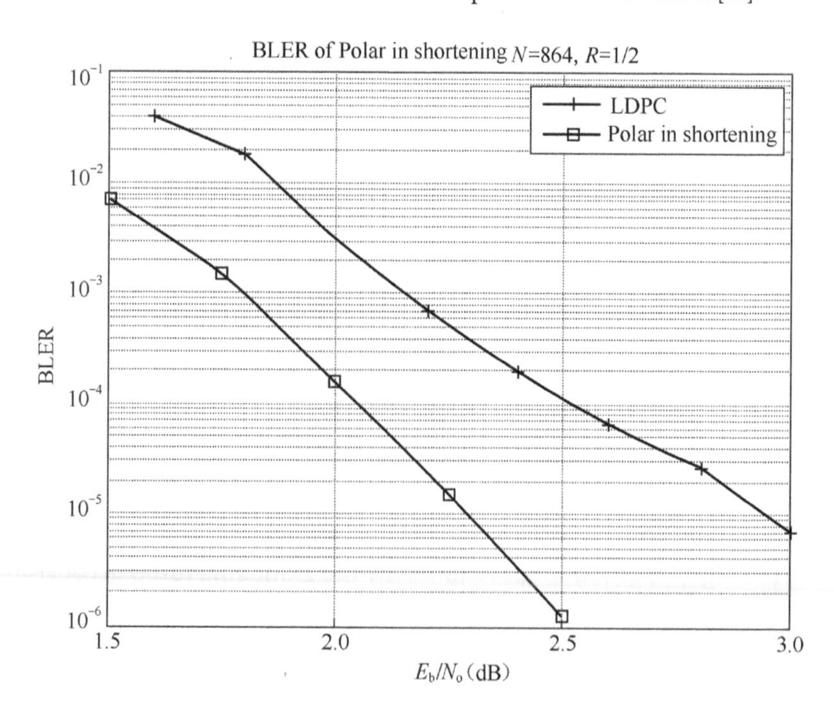

FIGURE 3.7 Performance of a shortened polar code compared to LDPC [18].

In March 2016, the study of 5G NR was kicked off in 3GPP [19]. Between October and November of 2016, a number of technical contributions to 3GPP [20–29] demonstrated the advantages of polar codes in performance and implementation complexity for short block length and low code rate. Because of these, polar code was adopted as the channel coding scheme for physical control channels of 5G NR [3]. In addition, the sequence order for uplink control information was proposed [31] and adopted by the standards. In January 2017, it was agreed in 3GPP to remove the bit-reversal module in the original polar code [32]. In May 2017, polar code was extended to the physical broadcast channel (PBCH) [33].

In China, Beijing University of Posts and Telecommunications is one of the earliest universities to start the research of polar codes where Prof. K Niu and his team studied the encoding and decoding algorithms for polar codes [34,35]. At Xidian University, Prof. Y. Li and her team studied the construction of a polar code based on Gaussian approximation and the related decoding methods [36,37]. At the Harbin Institute of Technology, Prof. X. Wang's team studied the encoding and decoding of Polar codes [38].

3.3 BASIC PRINCIPLE OF POLAR CODES

Briefly, the principle of Polar codes can be summarized as three steps of the process [44]: channel combining, channel splitting, and channel polarization. Among them, channel combining and channel polarization are carried out during the encoding, and channel splitting is conducted during the decoding.

3.3.1 Basic Channels

Strictly speaking, a channel coding scheme is usually designed for specific channel(s) so that the performance can be optimized, e.g., approaching the capacity of the channel(s). Such a coding scheme may perform well in other types of channels. When Arikan designed polar code, the target channel is a binary discrete memoryless channel (B-DMC), for instance, BEC and binary symmetric channel (BSC). Considering the B-DMC W:X ——> Y illustrated in Figure 3.8. Its input random variable X can only take two values {0, 1}. The output random variable Y can take arbitrary real numbers. The channel transfer probability is $P = W(y|x),\ x \in X, y \in Y$.

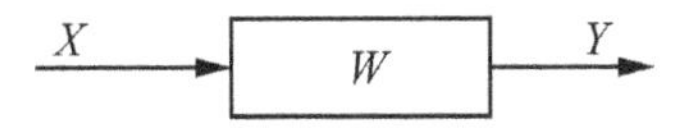

FIGURE 3.8 An illustration of B-DMC W:X ——> Y.

In the case of a binary channel such as BEC and BSC, when the probabilities of taking {0, 1} are equal, the capacity can be defined as the mutual information as:

$$C(W)=I(X;Y)=\sum_{y\in Y}\sum_{x\in X}\frac{1}{2}W(y|x)\log\frac{W(y|x)}{\frac{1}{2}W(y|0)+\frac{1}{2}W(y|1)} \tag{3.1}$$

If the log is 2-based (e.g., the capacity is represented as the number of bits), we can see $0\le C(W)\le 1$. When $C(W)=1$, W at this time is called noiseless channel (very good channel); When $C(W)=0$, W at this time is called pure noise channel (useless channel). The purpose of such "polarization" is to convert the commonly observed channels into two sets of channels with extreme reliabilities. The capacity calculation may be complicated. However, the Bhattacharyya parameter can be used to examine the capacity range. When channel W is used to transmit a single bit, Bhattacharyya parameter $Z(W)$ provides a way to estimate the upper bound of error probability when using maximum a posterior probability (MAP) detection.

$$Z(W)\overset{\Delta}{=}\sum_{y\in Y}\sqrt{W(y|0)W(y|1)} \tag{3.2}$$

For the rest of the chapter, we will primarily use the Bhattacharyya parameter for capacity analysis to explain the principle of polar codes, especially for B-DMC.

$$C(W)\ge\log\frac{2}{1+Z(W)} \tag{3.3}$$

$$C(W)\le\sqrt{1-Z(W)\bullet Z(W)} \tag{3.4}$$

3.3.2 Channel Combining

Figure 3.9 shows an example of two BEC channel W_1 and W_2 combined into channel W. where the notation "+" is modulo-2 addition. Since u_1 and u_2 are chosen from {0, 1} and independently and identically distributed, the capacity of the combined channel is

$$C(W)=I(U;Y) \tag{3.5}$$

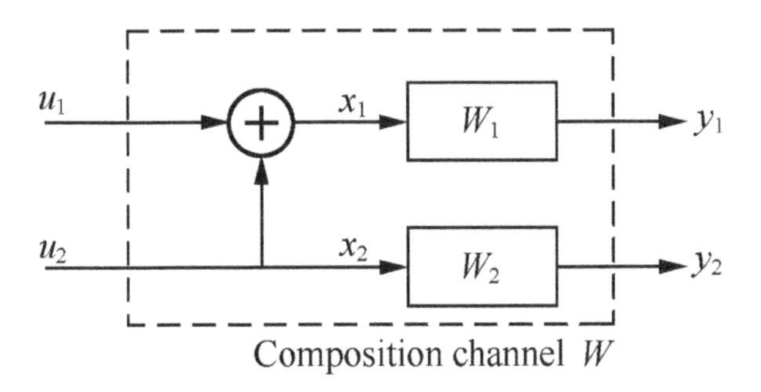

FIGURE 3.9 Illustration of channel combining [6].

where $U=[u_1, u_2]$, $Y=[y_1, y_2]$.

$$C(W)=I(X;Y) \tag{3.6}$$

where $X=[x_1, x_2]$.

$$C(W)=I(x_1;y_1)+I(x_2;y_2)=C(W_1)+C(W_2)=2\bullet C(W_1)=2\bullet C(W_2) \tag{3.7}$$

It is seen that for BEC, the capacity of the combined channel W is equal to the sum capacity of W_1 and W_2. Similarly, if there are N channels combined, the following will be held

$$C(W)=N\bullet C(W_1) \tag{3.8}$$

It means that the capacity abides by the conservation principle for the BEC channel: the capacity of a combined channel is equal to the sum capacity of sub-channels.

Assuming in Figure 3.9 the erasure probability $p=1/2$, the capacity W_1 and W_2 channels of $p=1/2$. After u_1 passes the combined channel, the capacity of the corresponding sub-channel is $1-(2*p-p*p)=1/4$. After u_2 passes the combined channel, the capacity of the corresponding sub-channel is $1-p*p=3/4$. The total capacity of the two sub-channels after the combining is $1/4+3/4=1$ which is equal to the sum capacity of the sub-channels before the combining: $2*(1/2)=1$.

From the above calculation, it is found that the capacity of the sub-channel carrying u_1, denoted as W^-, is lower than the original independent channel W_1, i.e., worse than the original channel. The capacity of the sub-channel carrying u_2, denoted as W^+, is higher than the original independent channel W_2., i.e., better than the original channel.

For the AWGN channel, the total capacity will not change before and after the channel combining. However, there is no analytical formula to quantify the capacity. The following empirical formula [46] may be used as an approximation, e.g., compared to BEC, the bad channel becomes a little better and the good channel becomes a little worse.

$$C_{AWGN}^{-} = C_{BEC}^{2} + \delta \tag{3.9}$$

where $\delta = \frac{|C_{BEC} - 0.5|}{32} + \frac{1}{64}$.

$$C_{AWGN}^{+} = 2C_{BEC} - C_{BEC}^{2} - \delta \tag{3.10}$$

Compared to the case of combining two independent sub-channels (W_1 and W_2) into the synthesized channel W, as shown in Figure 3.9, we can form a larger combined channel by combining on top of the combined channel W in Figure 3.9, as shown in Figure 3.10. Here the bit-reversal module is removed (if bit reversal were implemented, the positions of u_2 and u_3 would be swapped, while the positions of u_1 and u_4 would be unchanged).

In Figure 3.10, there are four BEC channels W_1, W_2, W_3, and W_4, where the erasure probability of each BEC is assumed p=1/2. Hence, the capacity of each BEC is 1−p=1/2, and the total capacity is 4*1/2=2. After the channel

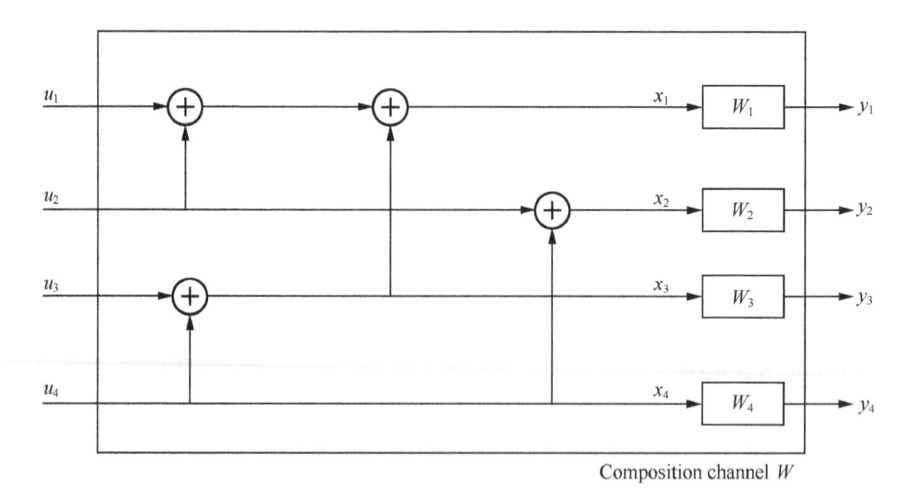

FIGURE 3.10 An illustration of combining four sub-channels [44].

combining, the capacity of the sub-channel carrying u_1 is 1/16. The capacity of the sub-channel carrying u_2 is 7/16. The capacity of the sub-channel carrying u_3 is 9/16. The capacity of the sub-channel carrying u_4 is 15/16. The total capacity is 2, the same as the sum capacity of the original four BEC.

The above channel combining method can be generalized as combining two sets of length-*N*/2 synthesized channels recursively into one set of length-N synthesized channels, as illustrated in Figure 3.11.

3.3.3 Channel Splitting

Channel splitting is to split the mixture of multiple input information $y_1^N = (y_1, y_2, y_3, \ldots\ldots, y_N)$ into multiple independent input information $u_1^N = (u_1, u_2, u_3, \ldots\ldots, u_N)$. Considering the decoding of a F2 polar code generated bits as illustrated in Figure 3.12. First, the decoder is to separate ul from $Y^N = Y^2 = (y_1, y_2)$. Then, the decoder would separate u_2 from (u_1, y_1, y_2) to finish the channel splitting of $Y^N \longrightarrow> U^N, N = 2$. Hence, in general, the channel before the splitting is Y^N, and the channel after the splitting is U^N. The splitting is to get extract U^i from (Y^N, U^{i-1}), that is, $(Y^N, U^{i-1}) \longrightarrow> U^i$. The entire channel splitting is essentially a successive cancelation (SC) decoding.

In Figure 3.12, the channel capacity of the channel to be de-composited is $C(W_{Befor}) = I(Y^N, U^N) = I(Y^2, U^2)$, where $C(W_{Befor}) = I(Y^2, u_1) + I(u_1; Y^2, u_2) = 2 \bullet C(W)$. That is, the total channel capacity $2C(W)$ is kept unchanged. Similarly, the channel capacity of the channel to be

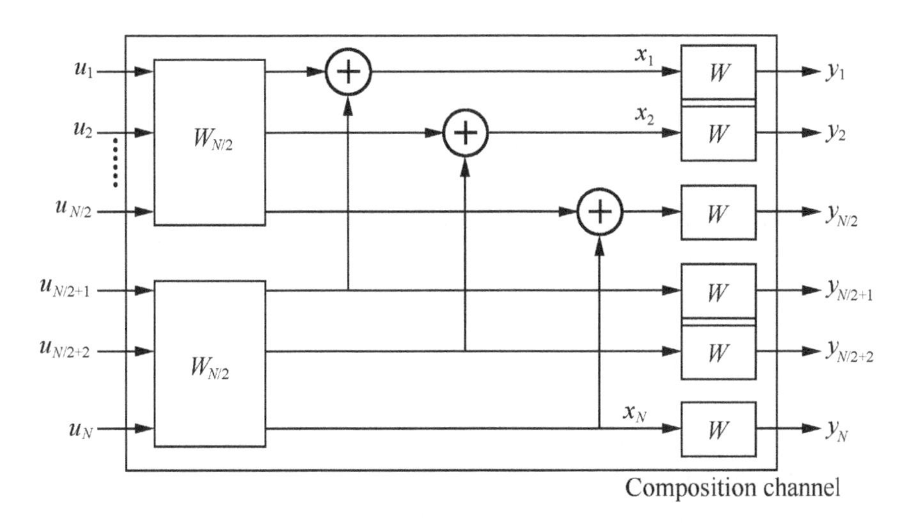

FIGURE 3.11 An illustration of recursive combining of two sets of length-*N*/2 synthesized sub-channels into a set of length-*N* synthesized channels [44].

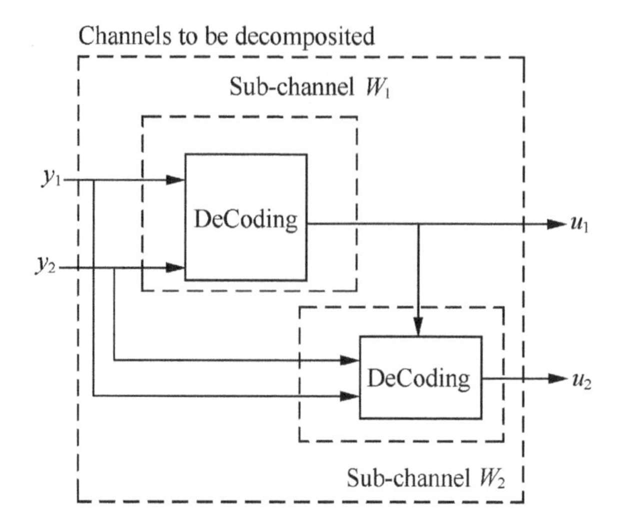

FIGURE 3.12 Illustration of channel splitting.

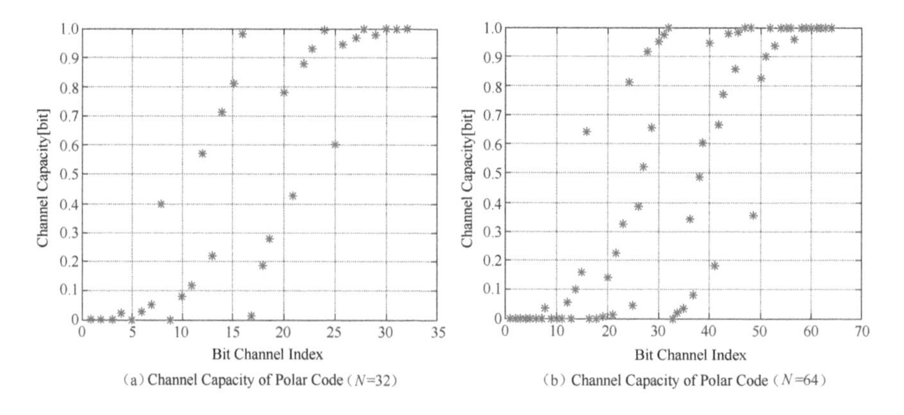

FIGURE 3.13 Channel polarization when $N=32$ and 64.

de-composited with a larger code length is also kept unchanged. That is, the channel capacity is always conservative for any code length.

3.3.4 Channel Polarization

Channel polarization refers to the trend that as the mother code length $N=2^n$ increases, the channel capacities of sub-channels (serial numbers of input bits of polar code) become polarized into two extremes: noiseless channels (e.g., channel capacity=1) and pure noisy channel (e.g., channel capacity=0) [47]. As observed from Figures 3.13–3.15 and Table 3.1, assuming a BEC channel with erasure probability $p=1/2$, when the mother

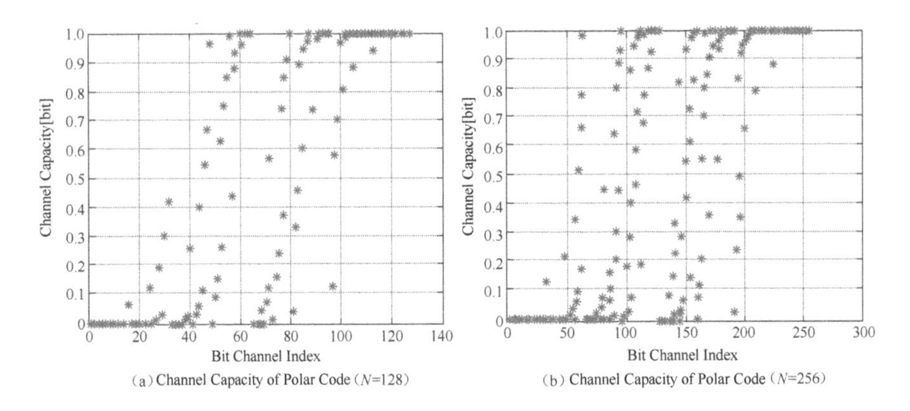

FIGURE 3.14 Channel polarization when $N=128$ and 256.

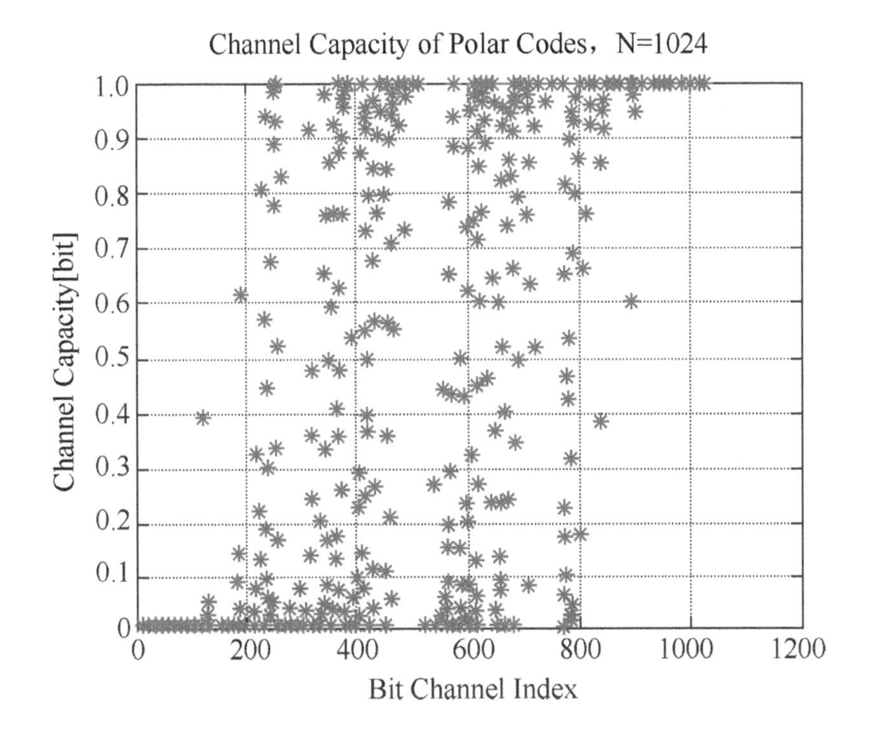

FIGURE 3.15 Channel polarization when $N=1024$.

code length $N \geq 512$, the polarization effect is already quite significant, e.g., more than 80% of sub-channels are already polarized. By taking advantage of channel polarization, we can use high-capacity sub-channels to transmit the data, while set other sub-channels as frozen bits (known, e.g., 0).

TABLE 3.1 Percentages of Polarized Sub-Channels Under Different Mother Code Lengths (δ=0.05; Noiseless Sub-Channels Whose Capacities >1−δ, Pure Noisy Sub-Channels Whose Capacities <δ)

Mother code length *N*	**8**	**16**	**32**	**64**	**128**	**256**	**512**	**1024**	**2048**	**4096**
Percentage of capacity <δ(%)	12.5	25	25	31	33	36	40	41	42.6	44
Percentage of capacity >1−δ (%)	12.5	25	25	31	33	36	40	41	42.6	44
Percentage of polarization (%; sum above two)	25	50	50	62	67	72	80	82	85.2	88

3.4 BASICS OF ENCODING AND DECODING OF POLAR CODES

3.4.1 Basics of Encoding

Polar code encoder is to select the suitable sub-channels (e.g., the locations of input bits in the entire mother code block) for data, put frozen bits to bad sub-channels, and then perform logic operations. Typically, sub-channels of high capacity (or reliability) would be used to carry the data. As shown in Figure 3.16, for N=8 and R=1/2, polar codes u_4, u_6, u_7, and u_8 should be selected for carrying the data. Since the number of coded bits to be transmitted may not be the same as the mother code length of polar code (for instance, here we like to transmit 12 bits instead of 8 bits), a certain bit selection mechanism needs to be carried out after the encoding (to be discussed in Section 3.7).

Assuming that there are four information bits $[u_4, u_6, u_7, u_8]=[0,1,0,1]$ to be encoded, the other four sub-channels are set to frozen bits. After the first level of encoding, the bits are [0, 0, 0, 0, 1, 1, 1, 1] (from the top to the bottom). After the second level of encoder, the bits are [0, 0, 0, 0, 0, 0, 1, 1]. After the third level of encoding (the final round), the obtained encoded bits are $[x_1, x_2, x_3, \ldots\ldots, x_8]=[0, 0, 1, 1, 0, 0, 1, 1]$.

3.4.2 Basics of Polar Code Decoding

The encoding procedure of polar codes can be represented as $x = u \bullet G$. When right-multiplying G^{-1} to both sides of the equation, we can get $u = x \bullet G^{-1}$, where G^{-1} is the inverse matrix of G in the binary domain (GF(2)). Since $G^{-1} = (F^{\otimes n})^{-1} = (F^{-1})^{\otimes n} = F^{\otimes n} = G$, where F is the Arikan kernel $F = \begin{bmatrix} 1 & 0 \\ 1 & 1 \end{bmatrix}$,

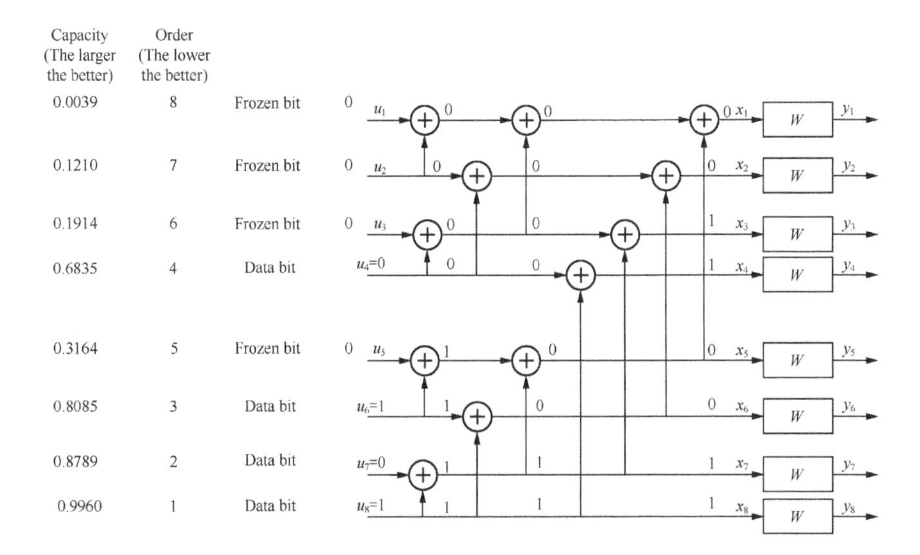

FIGURE 3.16 An illustration of a polar code encoder with $N=8$ and $R=1/2$.

we have $u=x\bullet G^{-1}=x\bullet G$, meaning that the polar code decoding is similar to polar code encoding, e.g., estimate the received signal y as x. Then, perform polar encoding on x to get the original information bit u. More specific procedures are as follows.

The basic decoding algorithm of polar code is SC [44,45] which includes two steps: in the first step, treat the $N/2$ received data with larger indices as the random noise, and try to decode the $N/2$ received data with small indices; In the second step, treat the $N/2$ decoded data with small indices as the known bits, and try to decode the $N/2$ received data with larger indices.

Take a G4 polar code as an example, shown in Figures 3.17–3.19. Assuming the modulation is BPSK, e.g., the mapping from the bits to the modulation symbols is y = 2x–1 (bit "0" mapped to "1", bit "1" mapped to "+1"). The received data are y_1, y_2, y_3, y_4 to be converted to the log-likelihood ratio (LLR) via Eq. (3.11).

$$x=LLR(y)=\ln\frac{W(y|0)}{W(y|1)} \tag{3.11}$$

where $W(y \mid 0)$ denotes the probability of receiving y when the transmitted bit is "0". $W(y \mid 1)$ denotes the probability of receiving y when the transmitted bit is "1". ln (*) denotes the natural logarithm, e.g., with e as the base.

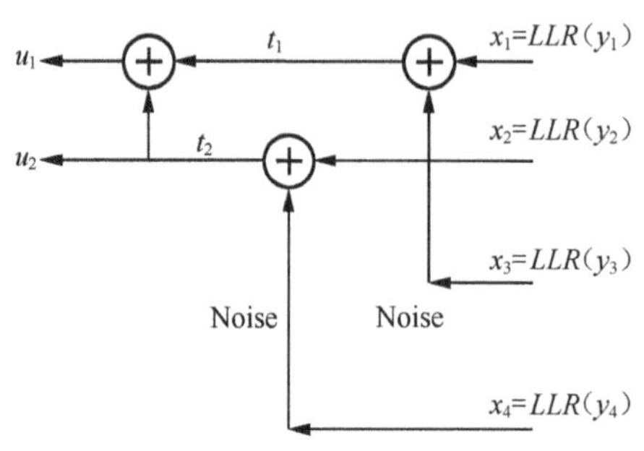

Step 1. To treat the lower part (x_3 and x_4) as noise, then to compute t_1 and t_2

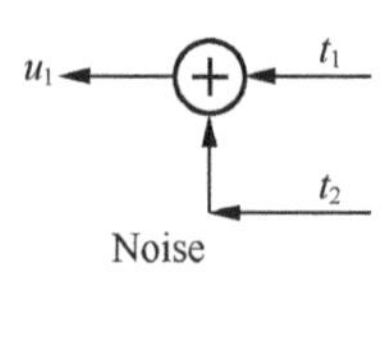

Step 2. To treat the lower part (t_2) as noise, then to compute u_1

FIGURE 3.17 Illustration of decoding of $N=4$ polar code (Step 1 and Step 2).

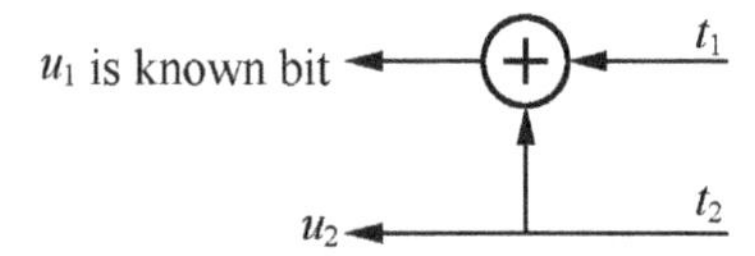

Step 3. To treat the upper part (u_1) as known bit, then to decode u_2

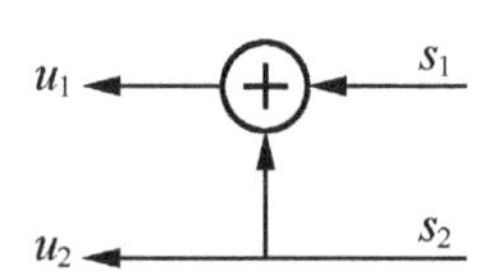

Step 4. To compute $s_1=u_1+u_2$ and $s_2=u_2$

FIGURE 3.18 Illustration of decoding of $N=4$ polar code (Step 3 and Step 4).

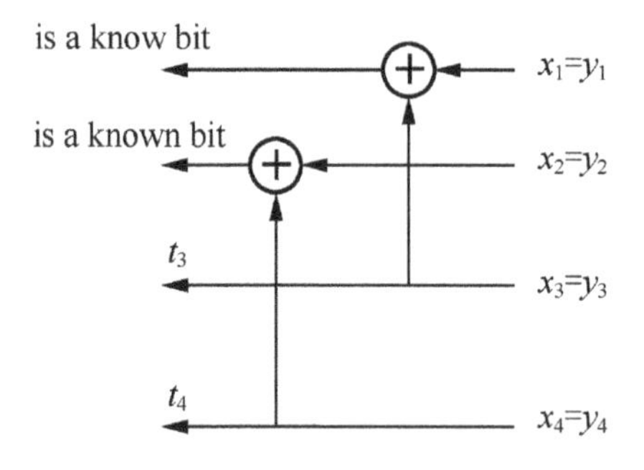

Step 5. To treat the upper part (i.e.,s1 and S2)as know bit,then to compute t3 and t4.

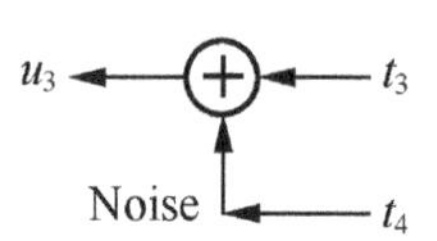

Step 6. To treat the lower part (i.e.,t4)as noise,then to compute U3.

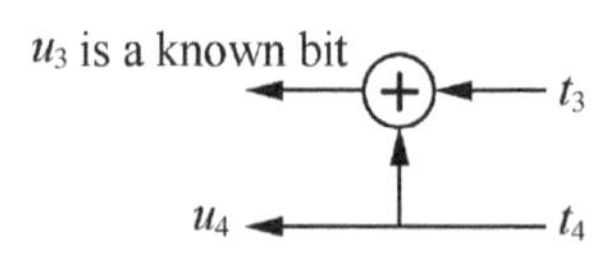

Step 7. To treat the upper part (i.e., u3)as known bit,then to compute U4.

FIGURE 3.19 Illustration of decoding of $N=4$ polar code (Step 5 to Step 7).

$W(y \mid 1)$ can be written as

$$W(y|1) = \frac{1}{1+e^{-2y/\delta^2}} \tag{3.12}$$

where δ^2 is the variance of noise. $W(y \mid 0)$ can be calculated by $W(y \mid 0) = 1 - W(y \mid 1)$.

Assuming that u_1, u_2, u_3, and u_4 carry information bits, e.g., $R=1$, the decoding can be carried out as follows:

Step 1. To treat the data of the lower half part (e.g., x_3 and x_4) as noise and calculate temporary variables t_1 and t_2, based on the data of the upper half part (e.g., x_1 and x_2). Here, "to treat it as noise" is from the aspect of encoding (or data transmission), that is,

$$\begin{aligned} x_1 &= t_1 + n_1 \\ x_2 &= t_2 + n_2 \end{aligned} \tag{3.13}$$

where t_1 and t_2 are useful data, while n_1 and n_2 are noise. During decoding, in order to remove the impact of noise, we need to carry out the following:

$$\begin{aligned} t_1 &= x_1 - n_1 \\ t_2 &= x_2 - n_2 \end{aligned} \tag{3.14}$$

Here "–" is not the regular subtraction. In fact, it is the box-plus operation, represented as

$$t_1 = x_1 - n_1 = f(x_1, n_1) = x_1 \boxplus n_1 = \ln\frac{1+e^{x_1+n_1}}{e^{x_1}+e^{n_1}} \tag{3.15}$$

Step 2. to treat the lower half (e.g., the temporary variable t_2) as noise, and use box-plus operation $\boxplus$ to determine u_1.

Step 3. to treat the upper half (e.g., u_1) as the known bit and determine u_2. That is, based on whether u_1 takes 0 or 1, as well as the values of temporary variables t_1 and t_2, to decode u_2.

Step 4. to calculate another set of temporary variables $s_1 = u_1 + u_2$ (here refers to mod 2 addition e.g., $\oplus$) and $s_2 = u_2$.

Step 5. to treat the upper part (e.g., the temporary variables s_1 and s_2) as the known bits and calculate another set of temporary variables t_3 and t_4.

Step 6. similar to Step 2, to treat the lower part (e.g., the temporary variable t_4) as noise and decode u_3.

Step 7. similar to Step 3, to treat the upper part (e.g., u_3) as the known bit and decode u_4.

Assuming that the received data are $y_1=-1.61$, $y_2=-0.06$, $y_3=-1.89$, and $y_4=1.69$. By using $x=\text{LLR}(y)$, we get $x_1=3.22$, $x_{2=}0.12$, $x_3=3.78$, and $x_4=-3.38$.

Step 1. treat the lower-half part of the information (e.g., x_3 and x_4) as noise, and calculate the temporary variables t_1 and t_2 based on the upper half part (e.g., x_1 and x_2). That is, $t_1=f(x_1, x_3)=x_1 \boxplus x_3=2.77$, $t_2=f(x_2, x_4)=x_2 \boxplus x_4=-0.11$.

Step 2. treat the lower half part (e.g., temporary t_2) as noise and decode u_1. That is, $z_1=f(t_1, t_2)=t_1 \boxplus t_2=-0.09$. If using Eq. (3.16) (if it is a frozen bit, no need to make a decision, just to set to the value of the frozen bit) to decide z_1, we get $u_1=1$. Note that this is an error which is reasonable because SNR is assumed 0 dB, code rate $R=1$ (no redundancy bits) and the mother code length $N=4$ is not long enough, leading to the insignificant polarization effect.

$$u_i = \begin{cases} 0, & \text{if LLR}_i \geq 0 \\ 1, & \text{else} \end{cases} \tag{3.16}$$

Step 3. treat the upper half part (e.g., u_1) as the known bit and decode u_2. That is, according to the value of u_1 (0 or 1) process the temporary variables t_1 and t_2, and then decode u_2. Based on the rule of the polar code encoding, if $u_1=0$, then after the encoding, we have $t_1=t_2$, Then, $z_2=t_1+t_2$. By performing hard decision decoding on z_2, we will get u_2. If $u_1=1$, then after the encoding, we have $t_1=1-t_2$ (after BPSK mapping, there is $t_1+t_2=0$). Then, $z_2=t_2-t_1$. By using hard decision decoding on z_2, we will get u_2. Put them together, $z_2=t_2+(1-2^*u_1)^*t_1$ (also called g operation). Substituting the values of u_1, t_1 and t_2, we obtain $z_2=-0.11+(1-2^*1)^*\ 2.77=-2.88$. By doing a hard decision on z_2, we get $u_2=1$, which is correct in this example.

Step 4. compute another set of temporary variables $s_1=u_1+u_2$ and $s_2=u_2$. Based on the results from Step 2 and Step 3, we get $s_1=0$ and $s_2=1$.

Step 5. treat the upper half part (e.g., the temporary variables s_1 and s_2) as the known bits and calculate another set of temporary variables t_3 and t_4. Similar to Step 3, we get $t_3=x_3+(1-2*s_1)*x_1=3.78+(1-2*0)*3.22=7.00$, $t_4=x_4+(1-2*s_2)*x_2=-3.38+(1-2*1)*0.12=-3.50$.

Step 6. similar to Step 2, treat the lower half part (e.g., the temporary variable t_4) as noise and decode u_3. That is, $z_3=f(t_3,t_4)=t_3 \boxplus t_4=-3.47$. By using hard decision decoding, we get $u_3=1$which is an error in this example.

Step 7. similar to Step 3, to treat the upper half part (e.g., u_3) as the known bit and decode u_4. That is $z_4=t_4+(1-2*u_3)*t_3=-3.50+(1-2*1)*7.00=-10.50$. By using hard decision decoding on z_4, we get $u_4=1$ which is correct in this example.

The above decoding process is successive, e.g., first to decode u_1, followed by u_2, then u_3, and finally u_4, When decoding the later bit (e.g., u_2), the impact of the previous bit (e.g., u_1) is fully taken into account. Therefore, the impact of previous bits is gradually reduced. The entire process is called SC. For long mother length blocks, the above process needs to be repeated. More details about the polar code decoding are to be discussed in Section 3.13.

3.5 POLAR CODE CONSTRUCTION

Polar code construction refers to designing the entire signal processing chain of polar code at the transmitter to facilitate decent operation. Generally speaking, it can include error detection, generation matrix generation, sequence order, rate matching, and interleaving (pre-encoder interleaving and post-encoder interleaving), as shown in Figure 3.20. Pre-encoder interleaving is only used for 5G NR downlink (not for uplink), whereas post-encoder interleaving is only used for 5G NR uplink (not for downlink). In addition, there is a sub-block interleaving during rate matching. For PBCH, there is a de-interleaving procedure. In this section, we focus on error detection, pre-encoding interleaving, and encoding matrix generation.

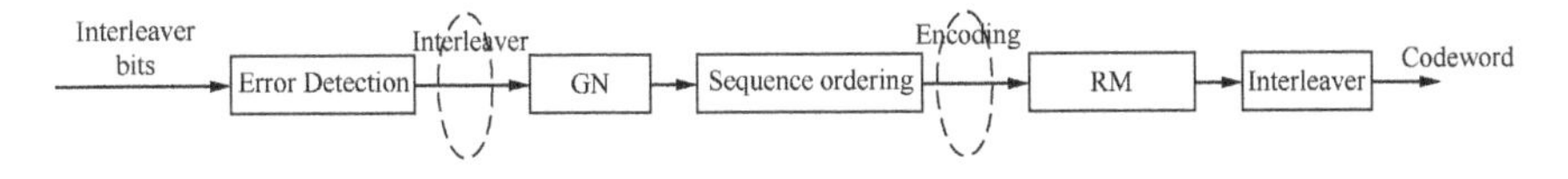

FIGURE 3.20 Construction of the polar code (block diagram of the entire signaling processing chain at the transmitter).

3.5.1 Error Detection

3.5.1.1 CRC-Aided Polar Codes (CA-Polar)

Cyclic redundancy check (CRC) aided encoding is a widely used technology. Its purpose is to detect whether a code block has been successfully decoded. Compared to the application of CRC to turbo codes and LDPC codes, CRC as well as the parity check (to be discussed in Section 3.5.1.2) plays a very important role in polar code development, in the sense that during the successive cancelation with list decoding (SCL), CRC can effectively prune those potential paths that violate CRC check. As seen in Figure 3.25 (in Section 3.5.1.3), assuming the list depth is $L=2$ and the left path [0, 0, 1, 1] cannot pass CRC check, whereas the right path [1, 0, 0, 0] can pass CRC check. Hence, the left path will be pruned and the right path will be kept.

CRC would introduce overhead, especially for polar codes used in control channel or channel information encoding because of the relatively short block length of control channels. Hence, special care would be taken when selecting the length or the number of bits for CRC and parity check.

In addition, CRC can also be considered as an outer code of polar code, even though CRC bits are not like Reed–Muller code that can be actually decoded.

Quite often, at the encoder, the result of CRC calculation would be appended to the end of originally encoded block. To the original encoded block, this CRC is an overhead and should be minimized. However, to keep the false alarm rate low, CRC length should not be too short. Also, when using the SCL decoding algorithm, the list depth or list size (L) effectively increases the false alarm (equal to $\log_2(L)$ bits).

In the physical downlink control channel (PDCCH) of LTE, there are 16 CRC bits. In NR-PDCCH of 5G NR, there are CRC 24 bits [48] with the generator polynomial as CRC24(D) = [D24 + D23 + D21 + D20 + D17 + D15 + D13 + D12 + D8 + D4 + D2 + D + 1]. To avoid the ambiguity issue of downlink control information (DCI) (carried in NR-PDCCH) bit length, CRC shift registers are initialized to be all "1" [49]. Considering that the generation of CRC can be implemented by either a left-shift register or a right-shift register, the CRC results would be different even when the CRC shift registers are initialized to be all "1". Hence, further restrictions on the implementation ought to be specified.

For the channel coding of 5G NR uplink control indication (UCI), bit repetition is used for 1-bit UCI. For 2-bit UCI, the simple encoder is used, which is similar to parity check code. For instance, 2 bits, i.e., C0 and C1, are encoded to get C2, where C2 = C0 + C1. For the UCI length between 3 and 11 bits, Reed–Muller (RM) code is used without CRC [49]. For the UCI length equal to or greater than 12 bits, polar code would be used and the maximum length of information bits is 2048*5/6=1706 [49]. According to [33] and [49], for 12- to 19-bit UCI, 6-bit CRC should be used with the generator polynomial g(D)=D6+D5+1, accompanied by CA-PC-polar code (to be discussed in Section 3.5.1.2). For UCI length equal to or greater than 20 bits, 11-bit CRC should be used with the generator polynomial as g(D)=D11 + D10 + D9 + D5 + 1, accompanied by the CA-polar code.

As shown by the simulation illustrated in Figure 3.21 [34], after the CRC-aided feature is introduced, the performance of Polar codes can be improved by 0.7 dB at BLER=1%. Simulation results in [38] show that when 24-CRC bits are used for mother code length N=1024, the code rate of 1/2, and list size L=8, polar codes can provide 0.3 dB performance gain at BLER=1%.

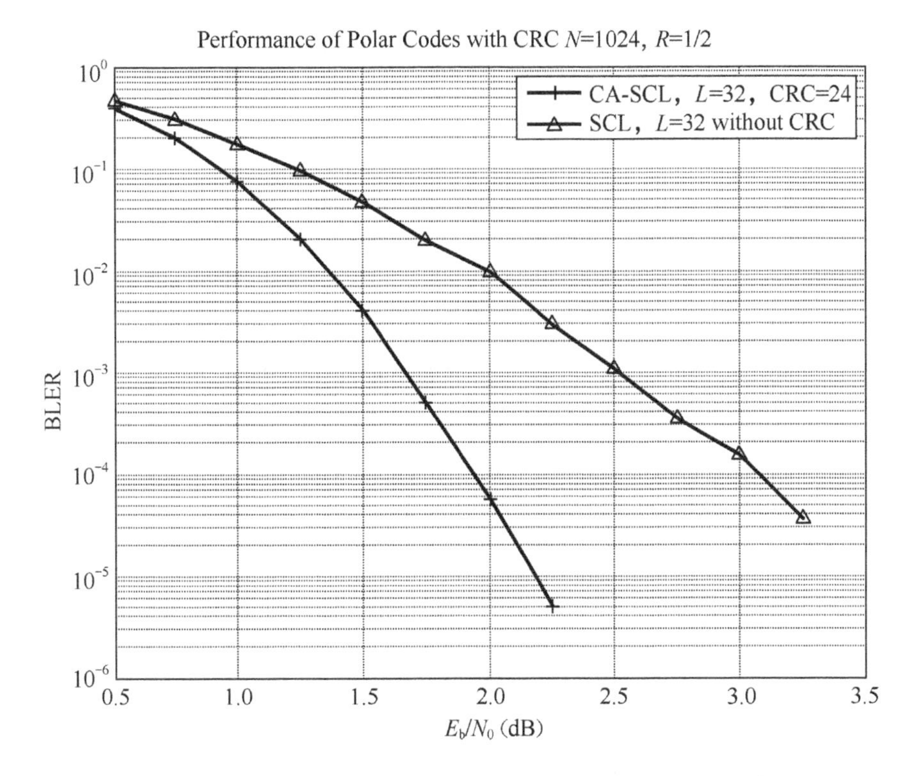

FIGURE 3.21 Performance of polar codes with and without CRC aid [34].

3.5.1.2 Parity Check Polar Code (PC-Polar) vs. CA-PC-Polar

Parity check (PC) bits can also be used to check whether polar code decoding is successful. PC bits can be considered as a degraded form of CRC. Parity check allows the decoder to prune the invalid paths and terminate early. At the different stages of polar code encoding (for a mother code length of $N=2^n$, there are n levels or stages of encoding), PC bits can be calculated, respectively. The following is an example showing how to construct PC-polar code for mother code length $N=23=8$ (three original information bits), as illustrated in Figure 3.22.

Step 1. select two positions to carry the PC bits (e.g., u_5, u_8) with the criterion to facilitate the operation (e.g., make it easier to insert the second-stage PC bit);

Step 2. select three positions to carry the information bits (u_4, u_6, u_7) with the criterion of prioritizing high-reliability bits;

Step 3. calculate the first-stage PC bit (as seen in the upper left of Figure 3.22). In 3GPP, a shift register is used to generate PC bits [49,52];

Step 4. set $u_1=u_2=u_3=u_8=0$, and set u_5 as the first-stage PC bit, and then calculate the temporary results t_1, t_2, t_3, t_4, t_5, t_6, t_7;

Step 5. calculate the second-stage PC bit (the lower left of Figure 3.22);

Step 6. set u_8 as the second-stage PC bit and calculate x_1–x_8.

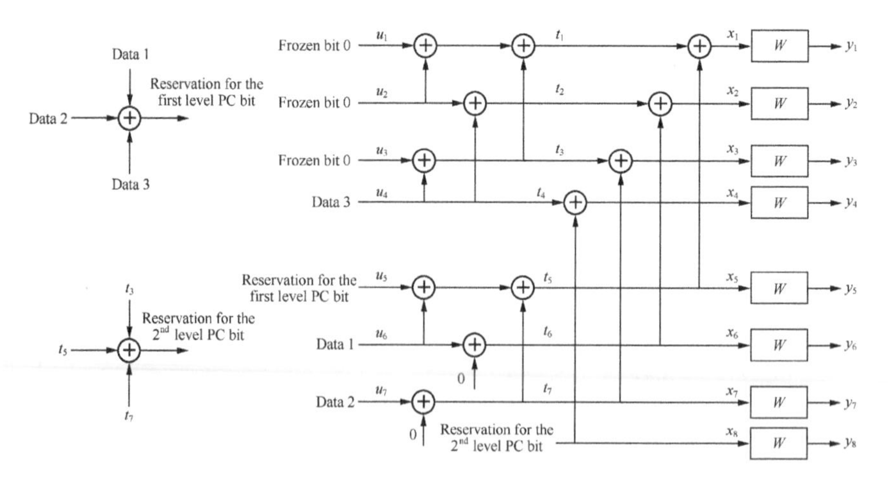

FIGURE 3.22 An illustration of constructing the PC-polar code.

So far, the construction of this example of the PC-polar code is completed. More discussion about PC-polar codes can be found in [50–52].

The decoding of PC-polar is illustrated in Figure 3.23 which corresponds to the encoding procedure shown in Figure 3.22. During the decoding, if t_3, t_5, t_7, and the second-stage PC bit are decoded, then perform modulo-2 summation over t_3, t_5, and t_7 and compare with the second-stage PC bit. If the parity check fails, then try a different decoding path (e.g., prune) or simply terminate this decoding; If the parity check passes, e.g., $t_3+t_5+t_7=\text{PC2}$, then to presume the successful decoding of t_3, t_5, t_7 and prune the right path and keep the left path, then continue the subsequent decoding of u_4, u_5, u_6, u_7, and to perform modulo-2 summation over u_4, u_6, u_7, and compare with u_5 (here u_5 is the first-stage PC, e.g., PC1). If the parity check fails, then try a different path or declare decoding failure. If the parity check passes, e.g., $u_4+u_6+u_7=u_5$, then presume the successive decoding.

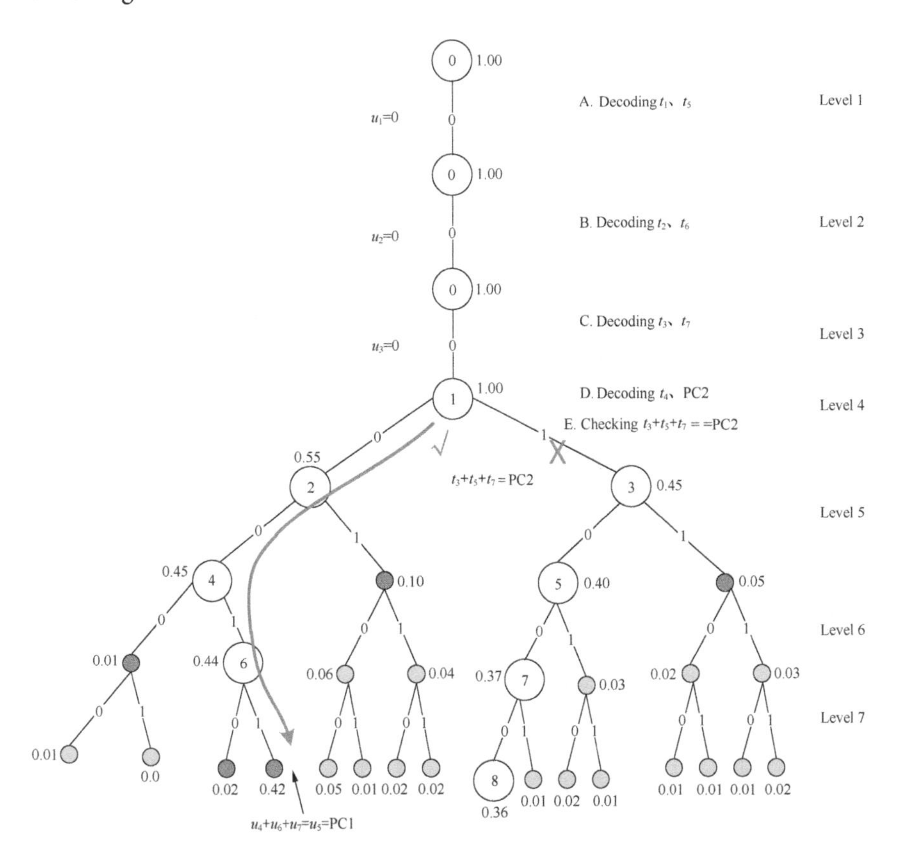

FIGURE 3.23 An illustration of the PC-polar code.

Compared to the above description and CA-polar code in Section 3.5.1.1, it is seen that for CA-polar, only after all the data have been decoded would the receiver know whether there is an error in the decoded bits. However, for PC-polar code, different stages of PC bits can be used to detect whether there is an error in the decoded bits.

By combining CRC bits and PC bits with polar code, we will get CA-PC-polar. That is, first append CRC bits to the original information bits, and then perform polar encoding. During each stage, PC bits are added. The decoding process is similar to the above as of PC-polar code, with the only difference being that a CRC check would be carried out in the end. The entire decoding would be declared only if the CRC check passes.

3.5.1.3 Distributed CRC-Aided Polar Code (Dist-CA-Polar) and Pre-Encoder Interleaving

As discussed before, in CA-polar, CRC bits are consecutively appended after the end of the original information block. Alternatively, CRC bits can also be inserted into the middle of the information block, which is called Dist-CA-polar [53,54]. It is noticed that the PC bits of CA-PC-polar are also distributed. However, their difference is that: PC bits require multiple times of computing and should be placed according to different stages, whereas the positions of CRC bits have nothing to do with polar code construction. In fact, distributing CRC bits over the original information bits is essentially interleaving. Hence, such interleaving is often called pre-encoder interleaving (as opposed to the interleaving after polar code encoding). The interleaving pattern of pre-encoder interleaving can be found in [54]. Because the interleaving would scramble the information embedded in the consecutively placed NR-PBCH, a de-interleaver (e.g., pre-interleaver) is defined so that after the pre-encoder interleaving, the information bits of NR-PBCH are still consecutive. The pattern of de-interleaver can be found in Table 7.1.1-1 of [112]. The construction of Dist-CA-polar is illustrated in Figure 3.24.

Distributed CRC bits can help to prune the path during the decoding, as shown in Figure 3.25. Note that if it is not distributed CRC (e.g., localized CRC bits), then the CRC bits cannot be used for pruning the paths. Instead, the localized bits can select the lists at the end of polar code decoding.

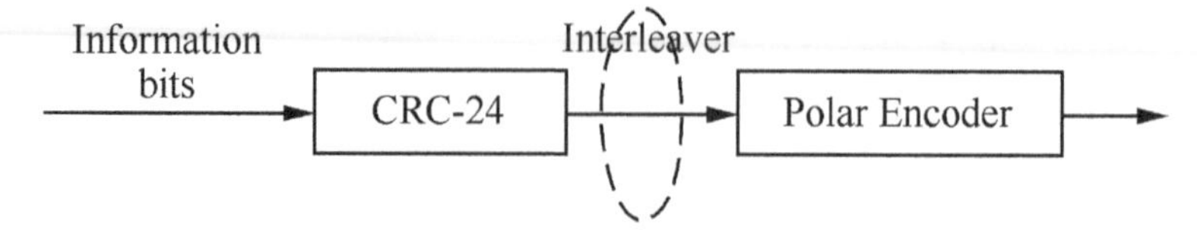

FIGURE 3.24 An illustration of Dist-CA-polar (pre-encoder interleaving).

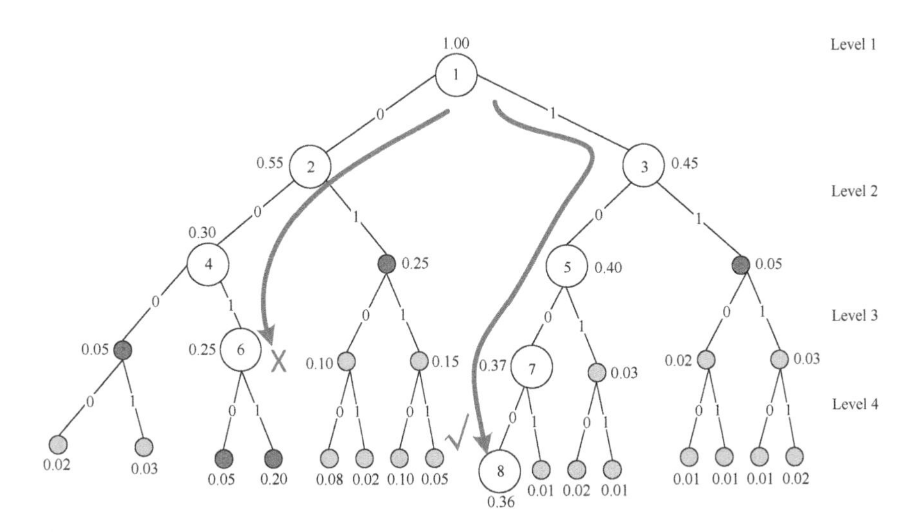

FIGURE 3.25 An illustration of CRC bits used for pruning the path that fails the CRC check [34].

During the discussion of distributed CRC in 3GPP [48], various design principles and methods were proposed:

- By carrying out the row swap and column swap for the generation matrix of CRC G, we can get an upper triangle matrix G', as illustrated in Eq. (3.17):

$$\begin{pmatrix} 1 \dots 0 & g_{0,0} \cdots & g_{0,K-2} \\ \vdots \;\; \vdots & & \\ 0 \dots 1 & g_{n-2,0} \cdots & g_{n-2,K-2} \end{pmatrix} \rightarrow G' = \begin{pmatrix} g_{0,0} & g_{0,2} & \cdots & g_{0,K-1} \\ g_{1,0} & g_{1,2} & \cdots & g_{1,K-1} \\ \vdots & \vdots & \cdots & \vdots \\ g_{d(0),0} & \vdots & & \vdots \\ 0 & g_{d(1),1} & \cdots & \vdots \\ 0 & 0 & \cdots & \vdots \\ \vdots & \vdots & \cdots & \vdots \\ 0 & 0 & \cdots & g_{d(K-1),K-1} \end{pmatrix} \tag{3.17}$$

- When some bits are decoded, early termination can be conducted by checking CRC, which is very important for downlink control channel decoding (since there are many DCI formats and aggregation levels, requiring multiple times of blind decoding).

- Reduce the false alarm rate (FAR) as low as possible. This requires carefully selecting the positions of CRC bits, based on certain analyses and numerous simulations.
- Use one interleaver for distributed CRC. In theory, the optimization of interleavers needs to fit different block lengths and code rates. However, from practical engineering and complexity of standardization points of view, it is preferable to reduce the number of interleavers, and the performance can be ensured over certain ranges of block lengths and code rates. For downlink physical control channel (PDCCH), the payload size is usually below 140 bits. After appending 24 CRC bits, the total would rarely exceed 164 bits. Hence an interleaving pattern of 224 bits long would be sufficient. The interleaving function can be achieved by row swap and column swap. The actual interleaving matrix specified in 5G NR can be found in Table 5.3.1.1-1 of [112].

Both NR-PDCCH and NR-PBCH use CA-polar for encoding and the CRC bits are distributed.

3.5.1.4 Hash-Polar Code

The construction of hash-polar code [55] is illustrated in Figure 3.26. The first step is to append CRC bits. Then segment the CRC appended information block into a number of sub-blocks of equal size (e.g., 32 bits). Then perform a hash function to each sub-block in the ascending/descending order of integer numbers. In the end, put the original information block, CRC, and hash numbers (in bit form) into the polar encoder. Hash-polar is equivalent to CA-polar, plus CRC, and can be considered as a type of Dist-CA-polar.

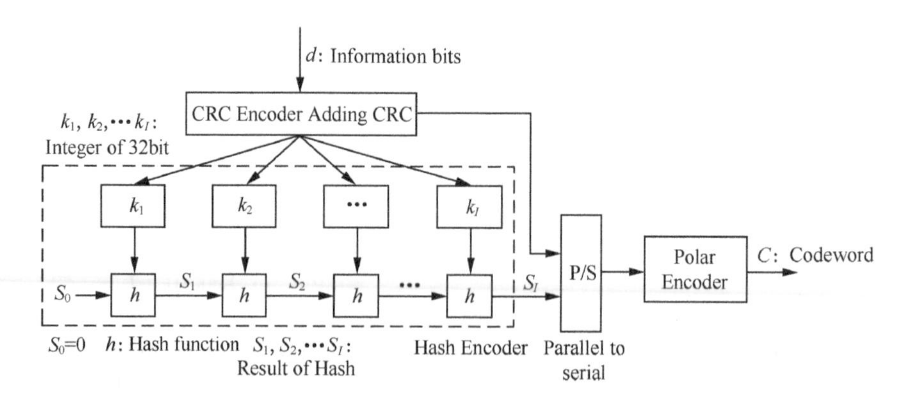

FIGURE 3.26 An illustration of hash-polar code construction [55].

3.5.2 Generation of Encoder Matrix

The encoder matrix of polar code refers to the matrix that is used for the calculation of codeword during the encoding (e.g., G in $x=u\bullet G$). The encoder matrix is often stored beforehand without the need to change. For a polar code of mother code length N = 2n, its encoder matrix can be calculated from the Arikan kernel $F=\begin{bmatrix}1 & 0\\ 1 & 1\end{bmatrix}$, which is $G_N=F^{\otimes n}$, where $\otimes$ denotes Kronecker power (tensor product). For instance, $G_2=F=\begin{bmatrix}1 & 0\\ 1 & 1\end{bmatrix}$,

$$G_4=F^{\otimes 2}=F\otimes F=\begin{bmatrix}1000\\ 1100\\ 1010\\ 1111\end{bmatrix}.$$

In the current 5G NR standards, in order to reduce the decoding complexity and latency, the maximum value of N is limited to 512 for downlink and to 1024 for uplink. Considering the minimum code rate $R_{\min}=1/8$ and DCI is at least 24 bits (to include CRC bits), typical values of N for downlink would be 32, 64, 128, 256, and 512. The size of UCI is at least 12 bits (adding 6-bit CRC to get 18 bits so that polar code can be used). The minimum code rate is also $R_{\min}=1/8$. Hence, the typical values of N for uplink would be 32, 64, 128, 256, 512, and 1024. All the above calculations mean that the number of encoder matrices for 5G NR is quite small.

Considering picking four rows: fourth, sixth, seventh, and eighth from $G_8=\begin{bmatrix}1,0,0,0,0,0,0,0\\ 1,0,0,0,1,0,0,0\\ 1,0,1,0,0,0,0,0\\ 1,0,1,0,1,0,1,0\\ 1,1,0,0,0,0,0,0\\ 1,1,0,0,1,1,0,0\\ 1,1,1,1,0,0,0,0\\ 1,1,1,1,1,1,1,1\end{bmatrix}$ that have the largest Hamming distance. These four rows (with the highest reliability) form a sub-matrix $G_P(8,4)=\begin{bmatrix}1,0,1,0,1,0,1,0\\1,1,0,0,1,1,0,0\\1,1,1,1,0,0,0,0\\1,1,1,1,1,1,1,1\end{bmatrix}$. It is found that $G_P(8,4)$ is in fact the generation matrix of RM code. This means that RM is a generalized polar code, which is called by Arikan as a close cousin in [56]. Certainly, their decoding algorithms are quite different.

3.5.2.1 Small Nested Polar Code

A small nested polar code can be used to transmit different types of information [57,58] with the main purpose of using a small polar codeword to send a small amount of data and using the entire big polar codeword to send a large amount of data. The principle of the small nested polar code is illustrated in Figure 3.27. Assuming that the length of the big polar code has the mother code length $N=8$. The mother code length of a small polar code is $N'=2$. Then, pick a bit for every $N/N'=4$ bits of the big polar codeword to form small Polar codewords. A small polar code can transmit a type of information, e.g., SS block index (which contains 3~6 bits). A big polar code is used to transmit another type of information. If the channel condition is good enough, the receiver can independently decode small polar without waiting for the decoding of the entire big polar code to be completed.

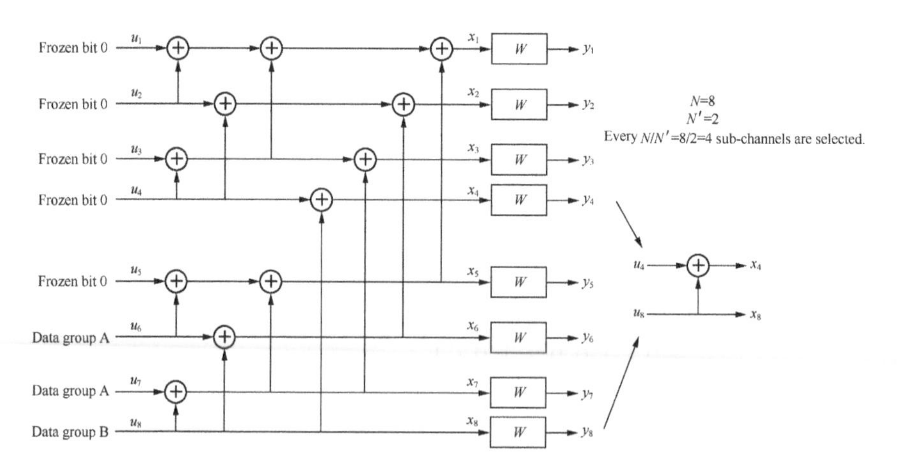

FIGURE 3.27 An illustration of the small nested polar code [57,58].

Because a small polar code requires setting some good sub-channels (e.g., u_4) to be frozen bits (otherwise, independent decoding is not possible due to the code rate $R=1$), the small polar code would degrade the performance of the big polar code. It is shown in [57] that the performance degradation is about 0.25 dB. To further improve the performance of the small polar code, $N'=64$ is used in [58] which is twice of $N'=32$ in [57], together by adding CRC.

3.5.2.2 Partial Code

The concept of partial code (sub-code or incomplete code) [59] has some similarities with the small nested polar code. For G8 polar as shown in Figure 3.27, u_1–u_6 would carry one type of information, whereas u_7 and u_8 would carry another type of information. As long as the SNR is high enough, the receiver is able to independently decode the sub-code without waiting for the entire polar code to be decoded. Since the high reliable bit u_7 is set to the frozen bit (otherwise the independent decoding would not be possible), the sub-code would degrade the performance of the entire polar code.

3.5.2.3 Polar Code of Arbitrary Length

The mother code length proposed by Arikan is a power of 2, e.g., $N=2^n$. More specifically, $G_N = F^{\otimes n}$ where $F=\begin{bmatrix} 1 & 0 \\ 1 & 1 \end{bmatrix}$. As argued in [60], the length of information block is arbitrary for a practical communications system, which requires that the code length can be adjusted based on the information block size, e.g., to construct a polar code with a more flexible mother code length and code rate. In the current method, the polar code length is adjusted by puncturing part of the codeword whose mother code length is a power of 2. At the receiver, the LLR of these punctured bits should be set to 0 and 1 with equal probability when carrying out the decoding. While this method can achieve a variable length of polar code, the decoding error probability is increased. Hence, a better way to construct variable length polar codeword was proposed in [60] where the mother code length is not a power of 2. The method can be illustrated below where $N=7$.

Step 1. represent the mother code length N in binary scale and get $N=(B_1B_2B_3......B_M)$ where $M=\text{ceil}(\log_2(N))$. For $N=7$, we have $M=\text{ceil}(\log_2(7))=3$. Hence $N=(B_1B_2B_3)=(111)$. In this example, $B_1=1$, $B_2=1$, $B_3=1$.

Step 2. calculate the number of "1" in the binary scale of N, e.g., $P = B_1 + B_2 + B_3 + \ldots\ldots + B_M$ and the corresponding serial numbers S1, S2, S3, ……, SP. For N=7, we have $P=B_1 + B_2 + B_3 = 1+1+1=3$, S1=1, S2=2, S3=3.

Step 3. generate P matrices whose sizes are $2^{(S1\text{-}1)}$, $2^{(S2\text{-}1)}$, $2^{(S3\text{-}1)}$, ……, $2^{(SP\text{-}1)}$, respectively. That is to apply (S1-1), (S2-1), (S3-1), ……, (SP-1) to "n" of $G_N = F^{\otimes n}$. For N=7, to set n=(S1-1)=0, n=(S2-1)=1, n=(S3-1)=2, respectively. That is $G_1 = F^{\otimes 0} = [1]$, $G_2 = F^{\otimes 1} = \begin{bmatrix} 1 & 0 \\ 1 & 1 \end{bmatrix}$,

$$G_4 = F^{\otimes 2} = \begin{bmatrix} 1 & 0 & 0 & 0 \\ 1 & 1 & 0 & 0 \\ 1 & 0 & 1 & 0 \\ 1 & 1 & 1 & 1 \end{bmatrix}.$$

Step 4. perform the combining of these P matrices, e.g., starting from the smallest matrices and then to the larger matrices. For N=7, first combine G1 and G2 to get G3, and then combine with G4, and finally get G7. The combination of matrices is carried out as the following. Assuming two square matrices A and B whose sizes are K*K and L*L, respectively, where K≥L, the combined matrix C is $C_{(K+L)\times(K+L)} = \begin{bmatrix} A & 0_{K\times L} \\ B \;\; 0_{L\times(K\text{-}L)} & B \end{bmatrix}$

For N=7, $C_{(2+1)\times(2+1)} = G_3 = \begin{bmatrix} G_2 & 0_{2\times 1} \\ G_1 \;\; 0_{1\times(2\text{-}1)} & G_1 \end{bmatrix} = \begin{bmatrix} 1 & 0 & 0 \\ 1 & 1 & 0 \\ 1 & 0 & 1 \end{bmatrix}$,

$$C_{(4+3)\times(4+3)} = G_7 = \begin{bmatrix} G_4 & 0_{4\times 3} \\ G_3 \;\; 0_{3\times(4\text{-}3)} & G_3 \end{bmatrix} = \begin{bmatrix} 1 & 0 & 0 & 0 & 0 & 0 & 0 \\ 1 & 1 & 0 & 0 & 0 & 0 & 0 \\ 1 & 0 & 1 & 0 & 0 & 0 & 0 \\ 1 & 1 & 1 & 1 & 0 & 0 & 0 \\ 1 & 0 & 0 & 0 & 1 & 0 & 0 \\ 1 & 1 & 0 & 0 & 1 & 1 & 0 \\ 1 & 0 & 1 & 0 & 1 & 0 & 1 \end{bmatrix},$$

where $0_{K\times L}$ is an all-zero matrix with K rows and L columns, $0_{L\times(K\text{-}L)}$ is an all-zero matrix with L rows and K-L columns.

Since the computation order would be different, the above-generated matrix may not be unique [60]. For instance, first combine G_1 and G_4 to get G_5, and then combine G_2 and G_5 to get G_7.

From the above equations, it is found that the matrix combination is essentially equivalent to the Kronecker product based on the kernel F: $G_{2N} = F^{\otimes(1+\log_2 N)} = G_N \otimes F = \begin{bmatrix} G_N & 0 \\ G_N & G_N \end{bmatrix}$, or more generally Kronecker product of two matrices: $G_{M \bullet N} = G_M \otimes G_N = \begin{bmatrix} g_{1,1} \bullet G_N, g_{1,2} \bullet G_N, & \ldots\ldots, & g_{1,M} \bullet G_N \\ g_{2,1} \bullet G_N, g_{2,2} \bullet G_N, & \ldots\ldots, & g_{2,M} \bullet G_N \\ \vdots \quad\quad \vdots & , \ldots\ldots, & \vdots \\ g_{M,1} \bullet G_N, g_{M,2} \bullet G_N, & \ldots\ldots, & g_{M,M} \bullet G_N \end{bmatrix}$

where $g_{i,j}$ are the elements of the square matrix G_M, $1 \le i \le M$, $1 \le j \le M$ and M is the number of rows in G_M. A more detailed discussion can be found in [60].

3.6 POLAR CODE SEQUENCE

3.6.1 Basic Concept

A polar code sequence is also called the determiner of good channels, e.g., to determine the bit indices for encoded bits before polar encoding. A polar code sequence can follow the order from the least reliable to the most reliable (the default method in 3GPP, also used in this book), or from the most reliable to the least reliable (which is convenient for counting). As seen in Figure 3.28, the polar code sequence is {0, 2, 1, 3} for the length-4 mother code.

From the above polar code sequence, it is seen that if the number of input information bits is only one (the output codeword has 4 bits),

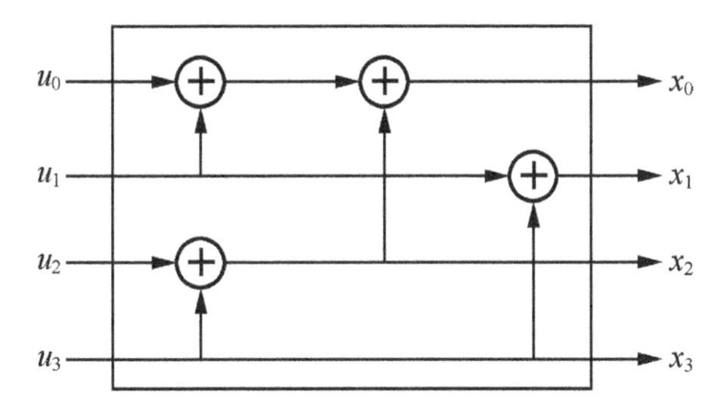

FIGURE 3.28 An illustration of the G4 polar code in 3GPP.

we would select the best sub-channel {3} (called information set [1]), e.g., set u_3 to be the input information bit and set sub-channel {0, 2, 1} to be frozen bits (e.g., "0"). Similarly, if the number of input information bits is 2, we would select sub-channel {1, 3}, e.g., u_1 and u_3 to be the input information bits, and set sub-channel {0, 2} as the frozen bits. In the same way, if the number of input information bits is 3, we would select sub-channel {2, 1, 3}, e.g., u_2, u_1, and u_3 to be the information bits, and set {0} as the frozen bit.

Note that since there is no bit inverse (BIV) [61] in the polar code in 3GPP, the polar code sequence in 3GPP is different from the channel selection method by Arikan [1].

From the polar code sequence {0, 2, 1, 3} and Figure 3.38, it is seen that:

- The first input bit (u_0) corresponds to the worse (the least reliable) sub-channel. Reliability here refers to the decoding probability of this bit, e.g., 1-BER, or the channel capacity, or the SNR.
- The last input bit (u_3) corresponds to the best (the most reliable) sub-channel.
- The sub-channels corresponding to the middle input bits (u_1 and u_2) have moderate quality between the best and the worse.

In fact, the above conclusion can be generalized to other Polar codes generated by the polar code kernel $F = \begin{bmatrix} 1 & 0 \\ 1 & 1 \end{bmatrix}$. For the BEC channel, the reliability of the third input bit (u_2) is slightly higher than that of the second input bit (u_1). Hence, the G4 polar code sequence can also be {0, 1, 2, 3}.

The polar code was originally designed for BEC where the capacities of sub-channels after polarization can be exactly computed. However, for the general binary input discrete memoryless channel (B-DMC), the capacity calculation of polarized channels is quite complicated.

Based on the SC principle, the probability density evolution method can be used recursively to calculate the capacity of the original bits in the case of polarization [62–64]. Using $a_N^{(i)}$ to represent the probability density function of the sub-channel $L_N^{(i)}(Y_1^N, 0_1^{i-1})$ sending "0" information, the computation of PDF is

$$a_{2N}^{(2i)} = a_N^{(i)} * a_N^{(i)} \tag{3.18}$$

$$a_{2N}^{(2i-1)} = a_N^{(i)} \otimes a_N^{(i)} \tag{3.19}$$

where "*" refers to the convolutional product of variable nodes and "⊗" refers to the convolutional product of check nodes. $a_1^{(1)} = a_W, a_W$ is the probability density function of the LLR of the original sub-channel sending "0".

After $a_N^{(i)}$ is computed, Eq. (3.20) can be used to calculate the integral over (-ti 0] to get the error probability of the information set. $P(A_i)$. Thus, the reliability of the sub-channel is $1-P(A_i)$.

$$P(A_i) = \int_{-\infty}^{0} 2^{-\prod(x=0)} a_N^i(x)dx \tag{3.20}$$

Note that the above convolution operations are all defined in the real-number domain. Practical calculations have certain quantization precisions. It is known that convolution operations are in general very sensitive to quantization errors. Hence two approximation methods can be considered: upgrading and downgrading quantization, to get the upper bound and lower bound of bit error probabilities [62–64]. Upgrading and downgrading quantization can convert the corresponding sub-channels into a set with reduced symbol domains, therefore reducing the computation complexity.

Even with the probability density evolution method of reduced complexity, the computation burden is still heavy, in the case of the most widely used AWGN channel, e.g., $L_1^{(i)}(y_i) \sim N(\frac{2}{\sigma^2}, \frac{4}{\sigma^2})$ (assuming the information bits are all 0). If Gaussian distribution can be used to approximate the information passing during the decoding, the computation can be significantly reduced.

Normally, the sequence design is SNR-dependent. It means that at different levels of SNR, the optimized polar code sequences may not be exactly the same. However, from a practical engineering point of view, it is desirable to have a single sequence, albeit with certain performance degradation compared to using multiple sequences. Careful adjustments can be carried out to tweak the orders of certain elements of the sequence so that one sequence can work reasonably well over a large range of SNR.

Also, the polar code sequences obtained by different methods (or criteria) may be different. Several well-known polar code sequences are discussed below.

3.6.2 Description of Several Polar Code Sequences

3.6.2.1 Row-Weight (RW) Sequences

For each row of the polar code generation matrix G, calculate the RW (the sum of all the elements in a row), and order the row weights from the lowest to the highest (a bigger number means more reliable). Then the row indices corresponding to the RWs form the RW sequence. Take the G_8 polar code as an example, its generation matrix is

$$G_8 = \begin{bmatrix} 1,0,0,0,0,0,0,0 \\ 1,1,0,0,0,0,0,0 \\ 1,0,1,0,0,0,0,0 \\ 1,1,1,1,0,0,0,0 \\ 1,0,0,0,1,0,0,0 \\ 1,1,0,0,1,1,0,0 \\ 1,0,1,0,1,0,1,0 \\ 1,1,1,1,1,1,1,1 \end{bmatrix},$$

the weights of each row are 1, 2, 2, 4, 2, 4, 4, 8. Then the RW sequence of G_8 polar code is {0, 1, 2, 4, 3, 5, 6, 7}. It is found that when $N=8$, the sequence obtained by RW calculation is exactly the same as the sequence calculated from the BEC channel, which means that when designing the sequence, simple calculation sometimes can get a good sequence.

From the principle of polar encoding $x=u \bullet G$, it is seen that when a row has high weights, the input bit would span over more output bits. Equivalently, this bit has more number of transmissions, leading to higher reliability of this bit.

3.6.2.2 Column-Weight (CW) Sequence

Similar to the RW sequence, the column-weight sequence can be obtained by calculating the CW. Taking the above G_8 polar code generation matrix as an example, the CW sequence can be determined as follows.

For each column of the generation matrix, sum over all the elements to get the CW (the smaller the value, the more important the column): 8, 4, 4, 2, 4, 2, 2, 1. Hence, the CW sequence is {0, 4, 2, 1, 6, 5, 3, 7}. It is seen that the sequence obtained by CW calculation is exactly the same as the sequence calculation from the BEC channel.

3.6.2.3 Polarization Weight (PW) Sequence

PW sequence was provided in [65]. The PW can be calculated as follows:

$$W_i = \sum_{j=0}^{n-1} B_j * 2^{j/4} \tag{3.21}$$

where W_i is the PW of the i-th input bit (e.g., i-th sub-channel). i is the index of input bit (for a polar code of length $N=2^n$ where n is the number of bits to represent the integer N, $i=0, 1, 2, \ldots, N-2, N-1$). B_j represents the j-th bit of the integer i in binary scale, e.g., $i = B_{n-1}B_{n-2}\ldots\ldots B_2B_1B_0$, $j=0, 1, 2, \ldots, n-2, n-1$.

After computing the polarization weights, arrange the weights from the highest to the lowest order, and then locate the serial number of the original input bits. For the polar code with $N=2^3=8$ bits, their polarization weights are {0, 1.4142, 1.1892, 2.6034, 1.0000, 2.4142, 2.1892, 3.6034}. Hence the PW sequence is {0, 4, 2, 1, 6, 5, 3, 7}.

Considering G_8 encoder structure in Section 3.4.1 and G_8 generation matrix in Section 3.6.2.1, if we convert the indices of eight sub-channels (e.g., indices of input bits) 0, 1, 2, 3, 4, 5, 6, 7 into binary scale: 000, 001, 010, 011, 100, 101, 110, 111. If a binary-scaled sub-channel index contains "1", this sub-channel would be applied to the sub-channels whose decimal scale indices are smaller (e.g., the information would be propagated to less reliable sub-channels, leading to stronger protection). If a binary-scaled sub-channel index contains "0", this sub-channel would be polluted by the sub-channels whose decimal scale indices are bigger (leading to weaker protection). The more number of "1" in the binary scale index of a sub-channel, the higher the capacity of this sub-channel is. By contrast, the more number of "0" in the binary scale index of a sub-channel, the lower the capacity of this sub-channel. Furthermore, the "1" on more significant bits are more important than "1" on less significant bits.

By adding "1" in the binary scale indices of sub-channels, the numbers of "1" in each sub-channel becomes 0, 1, 1, 2, 1, 2, 2, 3. Then covert those numbers of "1" per sub-channels into a power of 2 to get 1, 2, 2, 4, 2, 4, 4, 8. It is found that the above power of 2 numbers are row weights. The above conversion can be represented as

$$RW(B_1B_2B_3) = \sum_{j=1}^{3} (B_j \bullet 2^{B_j}) \tag{3.22}$$

where $B_1B_2B_3$ is the binary scale indices of sub-channel.

In Eqs. (3.21) and (3.22), power of 2 is used. In fact, other values for the base can be considered, for instance, the base for the natural log, e.g., e. Since the value of Bj is either "0" or "1", Eq. (3.22) can be simplified as

$$RW(B_1B_2B_3) = \sum_{j=1}^{3}(B_j \bullet 2) \tag{3.23}$$

It is noticed that when using Eq. (3.23) (e.g., $\sum_{j=1}^{n}(B_j \bullet 2) = 2 \bullet \sum_{j=1}^{n} B_j$), the row weights calculated for the first sub-channel ($B_1B_2B_3 = 000$) and the last sub-channel ($B_1B_2B_3 = 111$) are slightly smaller than the actual row weights $2^{\sum_{j=1}^{n} B_j}$. In another word, we may consider $\sum_{j=1}^{n} B_j$as the log-2 domain of the actual RW. However, such discrepancy would not affect the ordering, e.g., as seen in Section 3.6.1, the first sub-channel is still the worse, and the last sub-channel is still the best.

Considering the "1" in higher positions of binary scale indices are more important than the "1" in lower positions of binary scale indices, different scaling factors (via computer simulations to find suitable values) may be added to the bits in different positions. For instance, for j, the following formula can be used:

$$RW(B_1B_2B_3) = \sum_{j=1}^{3}(B_j \bullet j \bullet 2) \tag{3.24}$$

The above formula can be generalized to polar code of the mother code length of $N = 2^n$:

$$RW(B_1B_2B_3 \ldots\ldots B_{n-1}B_n) = \sum_{j=1}^{n}(B_j \bullet j \bullet 2) \tag{3.25}$$

3.6.2.4 Mutual Information Based Density Evolution (MI-DE) Sequence

Simply speaking, mutual information here refers to the correlation between the input and the output of a channel, $C = I(U; X)$, e.g., channel capacity. As discussed in Section 3.6.1, the higher the channel capacity, the more reliable the channel. Density evolution here refers to the evolution of the probability density function of information bits or sub-channels [62].

For a channel with X and Y as the input and the output, respectively, the mutual information (e.g., channel capacity C) of input X and output Y[66] is

$$C = I(X;Y) = H(Y) - H(Y|X) \tag{3.26}$$

where $H(Y)$ is the entropy of the output information, $H(Y) = -\sum_y \Pr(y)\log_2(\Pr(y))$. $H(Y|X)$ is the entropy of input being X and output being Y: $H(Y|X) = -\sum_x \sum_y \Pr(x)\Pr(y|x)\log_2(\Pr(y|x))$

For the BEC channel, when X is evenly distributed over two binary values {−1, +1}, its capacity $C = 1-p$, where p is the probability of erasure. For the AWGN channel, the situation is more complicated.

Considering BPSK modulated symbols pass through an AWGN channel and get $Y = X + n$, where X is evenly distributed over two binary values {−1, +1}, n is the zero-mean Gaussian noise with a variance of σ^2. The LLR is $2*Y/\sigma^2$ [66]. Note that LLR is a Gaussian random variable with the mean value $\pm 2/\sigma^2$ and the variance $4/\sigma^2$. The mutual information of input X and output Y (e.g., channel capacity) [66] is

$$J\left(\sqrt{4/\sigma^2}\right) = J(2/\sigma) = I(X;LLR) = I(X;Y) \tag{3.27}$$

where the function $J(.)$ can be approximated using the empirical formula [66]:

$$J(\sigma) = \begin{cases} a\bullet\sigma^3 + b\bullet\sigma^2 + c\bullet\sigma, \ 0 \le \sigma \le 1.6363 \\ 1 - e^{d\bullet\sigma^3 + e\bullet\sigma^2 + f\bullet\sigma + g}, \ 1.6363 < \sigma < 10 \\ 1, \ \sigma \ge 10 \end{cases} \tag{3.28}$$

where a = 0.0421061, b = 0.209252, c = 0.00640081, d = 0.00181491, e = 0.142675, f = 0.0822054, and g = 0.0549608. This means that for the AWGN channel, it is difficult to calculate the mutual information using the exact formula.

The basic idea of the MI-DE sequence [67] can be explained by using Figure 3.29 where a G2 polar code is shown. In the BEC channel,

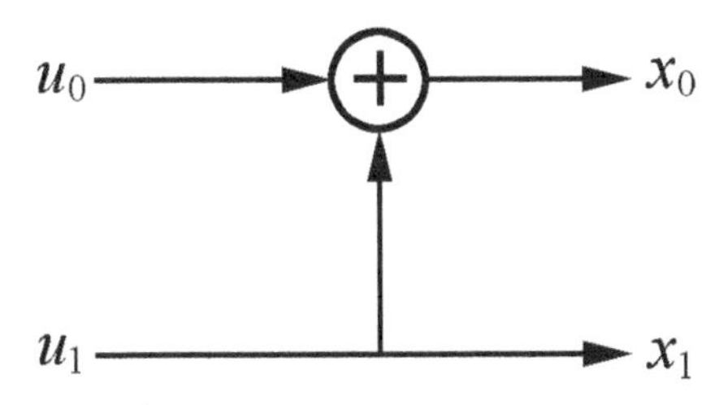

FIGURE 3.29 An illustration of the G2 polar code.

assuming the erasure probability is P. Then the capacity of u_0 is $C = 1-(2P(1-P)+P{\bullet}P) = (1-P) \bullet (1-P)$, The capacity of u_1 is $C = 1 - P{\bullet}P$. Since $1 - P{\bullet}P > (1-P) \bullet (1-P)$, the sequence of G2 polar is {0, 1}.

The capacity formula in BEC is rather simple. But for AWGN channels, there is no exact equation [67]. Instead, an approximation can be used. The empirical formula is provided in [67]: the capacity of u_0 is $C = C_{bec} + \delta$ and the capacity of u_1 is $C = C_{bec} - \delta$, where C_{bec} is the capacity of BEC channel and $\delta = 1/64 + (abs(1 - P - 0.5)/32)$. Since δ is quite small, the G2 polar code sequence is still {0, 1} for the AWGN channel.

The above method can be generalized to a longer polar code sequence. For instance, the obtained G64 polar code sequence is {0, 1, 2, 4, 8, 16, 32, 3, 5, 6, 9, 10, 17, 12, 18, 33, 20, 34, 24, 7, 36, 11, 40, 13, 48, 19, 14, 21, 22, 35, 25, 37, 26, 38, 28, 41, 15, 42, 49, 44, 50, 23, 52, 27, 56, 39, 29, 30, 43, 45, 51, 46, 53, 54, 57, 58, 31, 60, 47, 55, 59, 61, 62, 63} [67] (note that the order is reversed in the reference)

3.6.2.5 Combined-and-Nested (CN) and Optimized Combined-and-Nested (O-CN)

The design of CN sequences and O-CN sequences [68–70] for polar code starts from the mother code length of $N=64$. First, construct (N, K) = (64, 1), (64, 2), (64, 3),……, (64, 63) polar code. Then refine the sequence based on the density evolution method to get the polar sequence of mother code length $N=64$. Similarly, construct the polar code sequence of mother code length $N=128$ while maintaining the bit order from 0 to 63 to be the same as the sequence for mother code length $N=64$, in order to ensure nested property. By doing so, polar code sequences of longer mother code lengths can be constructed, as illustrated in Figure 3.30.

Compared to CN sequences, the optimization in O-CN sequences [68–70] is reflected in two design criteria and universal partial order (UPO) [69,70]:

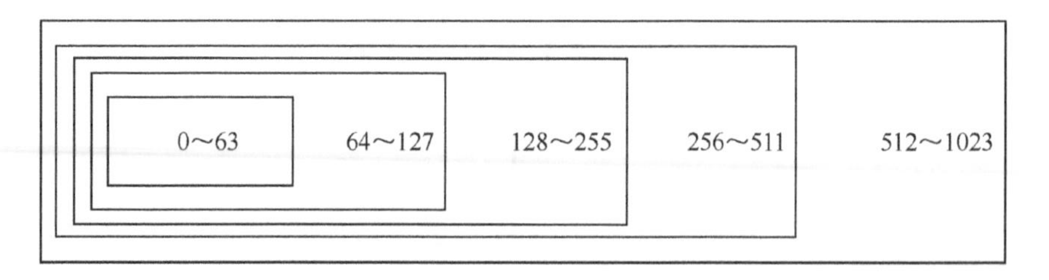

FIGURE 3.30 Illustration of CN sequences [68].

- If the binary scale indices of two input bits differ only by 1 bit (e.g., their Hamming distance is 1), the natural order is used. For instance, 12 (1100 in binary scale) and 14 (1110 in binary scale) differ by only one bit. Hence, 12 would be a precedent to 14, e.g., 12 has lower reliability and 14 has higher reliability.
- If the binary scale indices of two input bits differ by 2 bits (e.g., their Hamming distance is 2), and the binary form of the first bit is {0, 1} and the binary form of the second bit is {1,0}, the natural order is used. For instance, 11 (1011 in binary scale) and 13 (1101 in binary scale) differ by 2 bits. Hence, 11 would be a precedent to 13, e.g., 11 has lower reliability and 13 has higher reliability.

For UPO, three criteria are provided in [77] as illustrated in the Hasse graph in Figure 3.31.

- The reliability of the sub-channel corresponding to the input bit index in (a, b, c, 0) binary scale is lower than the reliability of the sub-channel corresponding to the input bit index in (a, b, c, 1) binary

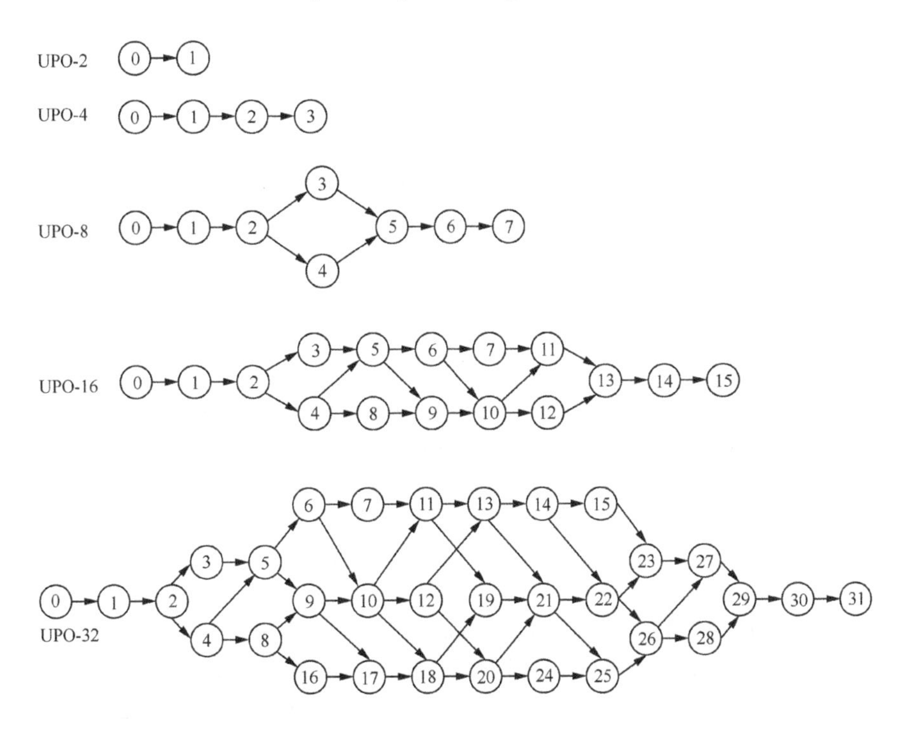

FIGURE 3.31 Hasse graph of UPO [77].

scale. For instance, for UPO-8 in Figure 3.31, the reliability of #4 sub-channel ("0100") is lower than that of #5 sub-channel ("0101").

- The reliability of the sub-channel corresponding to the input bit index in (a, 0, 1, b, c) binary scale is lower than the reliability of the sub-channel corresponding to the input bit index in (a, 1, 0, b, c) binary scale. This rule can be represented in simpler form #"01"<#"10": For instance, for UPO-32 in Figure 3.31, the reliability of #13 sub-channel ("01101") is lower than that of #22 sub-channel ("10110").
- Combine the above two rules, for UPO-32 in Figure 3.31, the reliability of #10 sub-channel ("01101") is lower than that of #22 sub-channel ("10110").

For the Hasse graph of UPO in Figure 3.31, the reliabilities of the lower sub-channels are lower than the upper sub-channels. For instance, for UPO-8, the reliability of #4 sub-channel is lower than that of #3 sub-channel. Hence, the polar code sequence for $N=8$ is {0, 1, 2, 4, 3, 5, 6, 7} and the polar code sequence for $N=16$ is {0, 1, 2, 4, 8, 3, 5, 9, 6, 10, 12, 7, 11, 13, 14, 15}.

3.6.2.6 SCL-Like Sequence

SCL-like sequence is based on a listed successive cancelation algorithm [71,72]. First, a reference sequence of a certain length (e.g., 32 bits) is generated based on a certain criterion, e.g., PW. Then, by using certain means (e.g., computer simulations) to obtain the SNR of this reference sequence under certain conditions, e.g., BLER=1%. Next, to obtain the channel reliability based on e.g., Gaussian approximation. Then increase the sequence length (e.g., from 32 bits to 64 bits) and get a new sequence. Based on the channel reliability to get L new good sequences (where L is a positive integer). Then, the "best" sequence is selected according to the performance (e.g., required SNR @ BLER=1%) of the new sequences via, for instance, computer simulation. Repeat the above procedure to get longer sequences (e.g., 256 bits).

By using this method, polar sequence of $N=128$ bits can be obtained: {0, 1, 2, 4, 8, 16, 3, 32, 5, 6, 64, 9, 10, 17, 12, 18, 33, 20, 34, 65, 7, 24, 36, 11, 66, 40, 13, 19, 14, 68, 48, 72, 21, 35, 22, 80, 25, 96, 37, 26, 38, 41, 67, 28, 42, 70, 15, 49, 69, 73, 44, 50, 82, 23, 52, 74, 27, 81, 76, 39, 56, 97, 29, 43, 84, 98, 30, 45, 88, 71, 51, 100, 46, 53, 104, 77, 75, 54, 112, 57, 85, 78, 83, 58, 90, 99, 31, 60, 86, 101, 89, 106, 92, 47, 102, 55, 105, 79, 59, 87, 113, 108, 61, 91, 114, 62,

116, 120, 103, 93, 107, 94, 109, 115, 110, 117, 63, 118, 121, 122, 95, 124, 111, 119, 123, 125, 126, 127}.

3.6.2.7 Merged Design

When the base graphs (BG) of LDPC were designed in 3GPP, the BGs finally adopted by the standards were the result of joint designs by multiple companies. Similarly, a merged design [73] was proposed by five companies as a consolidated solution. Note that there are two sequences in the merged solution, e.g., one of which was independently pushed by a member of this five-company group.

In the merge sequence of length N=1024, if the elements with indices less than 512 are punctured, the order of the rest of the 512 elements would be the same.

3.6.3 Properties of Sequences

3.6.3.1 Online Computation-Based (OCB)

OCB refers to that the sequence can be computed, without the need to be stored. The merit of OCB is to save memory or reduce memory. However, OCB would increase the amount of computation. When the computation is extensive, it would bring certain latency in encoding and decoding.

For some sequences in Section 3.6.2 (RW sequences), online computation is feasible. However, for long mother code length (e.g., N=1024), online computation is not suitable. It will be discussed in Section 3.6.5 that the polar code sequence adopted by 3GPP [48] can only be stored beforehand and does not require computation online. Hence, the OCB property is not mandatory.

3.6.3.2 Nestedness

Nested property refers to that for a sequence of mother code length N, when the elements whose indices are equal or greater than $N/2$ are punctured, the remaining elements (if not re-ordered) are the sequence of mother code length $N/2$. In another word, the sequence of mother code length $N/2$ is nested in the sequence of mother code length N. The merit of nest sequences is that a single look-up table is enough, e.g., no need for new computations, nor for re-ordering.

If most of the elements of a sequence with mother code length $N/2$ are in the same order as the sequence with mother code length N, then it is called quasi-nested. The merit of quasi-nested is that better decoding performance can be obtained by adjusting the ordering of some elements.

3.6.3.3 Symmetry

Symmetry means that in the sequence of mother code length N, the first $N/2$ elements are symmetric to the last $N/2$ elements. That is, $S(i)=N-1-S(N-1-i)$, where $i = 0, 1, 2, \ldots, N-1$, $S(i)$ is an element in the sequence. Symmetry property can be used to reduce the memory size.

Quasi-symmetry means that in the sequence of mother code length N, most of the first $N/2$ elements are symmetric to the last $N/2$ elements. Quasi-symmetry property allows certain adjustment of ordering to improve the decoder's performance.

3.6.3.4 UPO and Gaussian Universal Partial Order (GUPO)

UPO property is already discussed in the previous section for O-CN sequences. This property is not required by 3GPP [74]. To boost the chance of a merged solution, the UPO property was emphasized [76]. In fact, for $N=512$, the merged solution does not have UPO property. For $N=1024$, the part of the sequence, 512~1023, has UPO property.

GUPO is described in [77] where it is believed that GUPO and UPO are consistent, but the performance of GUPO is more consistent.

Among the above properties, the most important is the nested-ness, which is also required by 3GPP [75].

3.6.4 Criteria for Polar Sequence Down-Selection

In 3GPP, the criteria [73,74] to down-select the polar sequences are error block count, SNR spacings, and "win" counts in SINR-BLER simulation.

3.6.4.1 Error Block Count

Error block count refers to that when conducting BLER vs. SNR simulations, in order to get consistent performance, the number of error blocks should satisfy a certain number. For instance, to test the required SNR for BLER=1%, the error block count should be 1000. Hence, the total number of blocks to be simulated should be 1000/(1%)=105. If the required error block count is high, e.g., 104, it would cost a lot of simulation time.

3.6.4.2 SNR Spacing in BLER vs. SNR Simulation

Quite often for channel coding specification, the performance difference between different solutions in 3GPP is small, e.g., ~0.1 dB. In order to clearly differentiate the performance, small spacing of SNR, e.g., 0.1 or 0.2 dB, is used.

To evaluate BLER=10% to BLER=0.1%, a wide range of SNR needs to be simulated, which would cost a lot of simulation time when the SNR spacing is small.

3.6.4.3 WinCount

WinCount refers to that under the same BLER, the same information block length, if the required SNR for Sequence A is lower than that of Sequence B by a certain threshold, then Sequence A gets one win. Note here the threshold depends on the information block size and list size (L) for polar decoding. Once all the simulation cases are completed, each sequence would get a total WinCount.

In 3GPP, the final sequence was selected based on the total win count [48]. Since the AWGN channel is assumed in all simulations, the polar code sequence adopted in the standards is for the AWGN channel. Let us examine how different this sequence is from the sequence for BEC as shown in Section 3.3.4.

By comparing Figures 3.32 and 3.33, it is observed that the polar code sequence for the BEC channel has a similar trend as the polar code

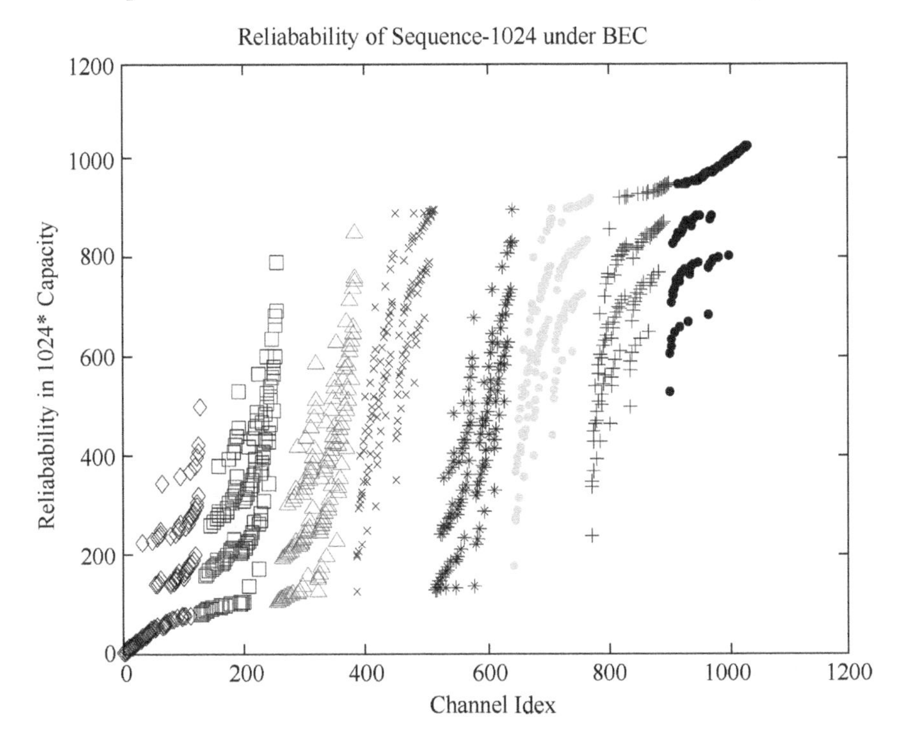

FIGURE 3.32 Reliability of polar code sequence (N=1024) for BEC channel.

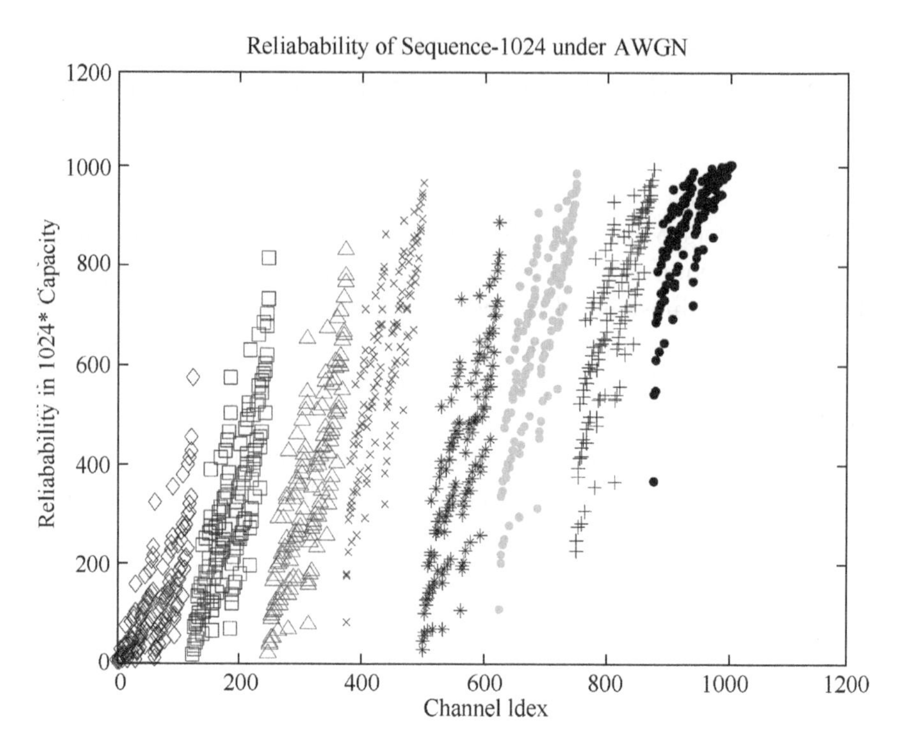

FIGURE 3.33 Reliability of polar code sequence ($N=1024$) for AWGN channel.

sequence for AWGN. However, the polar code sequence for BEC exhibits a more significant layering effect (especially, for sub-channels with large indices, several "waterfall" can be observed). By contrast, a less signification layering effect is observed in the polar code sequence for the AWGN channel. In another word, the polarization effect of the BEC channel is more obvious compared to the AWGN channel.

3.6.5 Merged Solution and Final Selection

From the perspective of standardization, the most important part of polar code encoding is the channel selection, e.g., polar code sequence. Fierce competitions were observed between companies in 3GPP. On August 4, 2017, there were a total of 7 polar sequences submitted to 3GPP [73,74]. On August 22, 2017, the winner was announced based on the total WinCount.

The current polar code in the specification has the mother code length $N=1024$. In the future, if the size of UCI increases, it is possible to extend the mother code length to $N=2048$ or even bigger.

3.6.6 Pre-Frozen Bits for Rate Matching

Pre-frozen [78] refers to assigning certain input bits (e.g., sub-channels) as frozen bits. This may have a certain effect on the polar code sequence. The main impact includes [69]:

- The input bits corresponding to the punctured bits (e.g., bits that are not to be transmitted) would be pre-frozen. For instance, assuming that the number of information bits $K=96$, the mother code length $N=256$. The number of bits allowed for transmission $M=250$. Hence, the first to sixth bits after encoding should be punctured. Then, the first to sixth input bits would be pre-frozen.
- In the case of puncturing, the following input bits should also be pre-frozen.
    - If $M \geq (3*N/4)$, the first bit to the (ceil($3*N/4-M/2$)+1)-th input bits should be pre-frozen, where ceil(.) is to round up to the closest integer. Using the above parameters, since $M=250 \geq (3*N/4)=192$, the 1st to 68th input bits should be pre-frozen. However, as Figure 3.34 shown, the 64th input bit has a very good sub-channel and ranks 63rd for $N=256$ (the best is 256)

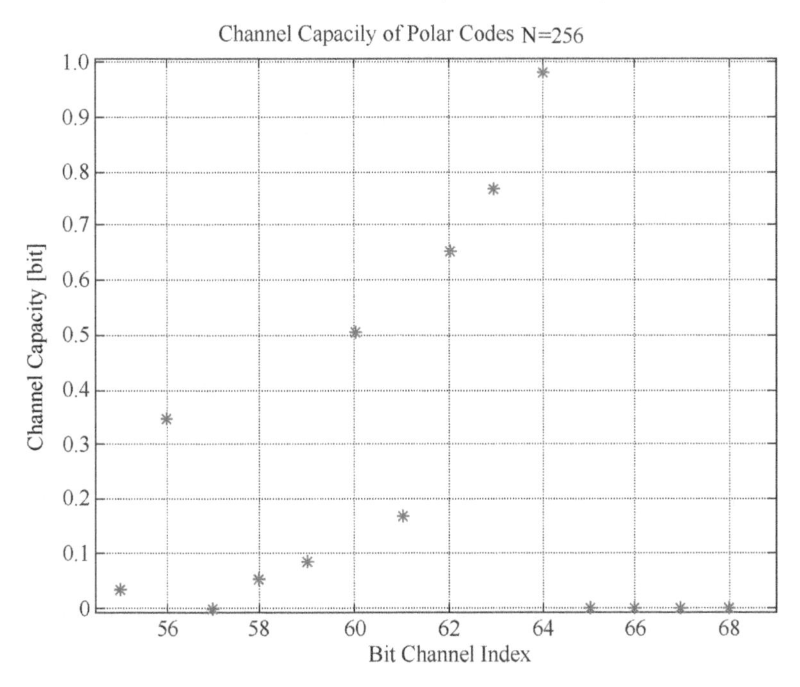

FIGURE 3.34 Sub-channels in which 64th bit has a good channel.

- If $M<(3*N/4)$, then the first to the (ceil($9*N/16-M/4$)+1)-th bits should be pre-frozen.

3.7 RATE MATCHING FOR POLAR CODES

Rate matching [57,58] refers to carrying out the selection of the encoded bits (e.g., increasing or reducing the number of coded bits) after encoding to fit the corresponding physical resources. For example, for NR-PBCH, it spans three OFDM symbols and 40, 96, and 240 subcarriers. The total physical resource of NR-PBCH contains $40+96+240=576$ resource elements (REs). After deducting 1/4 of overhead for the reference signal, there are 432 REs to carry the PBCH payload. PBCH uses QPSK. This physical resource can carry $432*2=864$ bits. The maximum mother code length in the downlink is $N=512$. Hence, 512 encoded bits should fit 864-bit resources. This is the rate matching.

Rate matching contains two operations. The first operation is interleaving of sub-blocks, which is to be discussed in Section 3.8.3. The second operation is to pick bits in certain ways, which is to be discussed in this section.

The following parameters are to be used for rate matching:

K: length of information block (including CRC bits or PC bits) after segmentation

M: number of bits corresponding to the physical resources

$R_{\min}=1/8$: minimum code rate, to be used to select the mother code length $N_r=2$^(ceil(log2($K/R_{\min}$)))

$N_{\max}$: maximum mother code length, 512 for downlink, 1024 for uplink

$N_{\min}$: minimum mother code length, fixed to be 32 (e.g., 2^5)

N_{dm}: no less than M, but is power of 2, e.g., $N_{dm}=2$^(ceil($\log_2(M)$))

N_m: another candidate of mother code length. $N_m=N_{dm}/2$ if $M\leq(9/8)*N_{dm}/2$ and $K/M<(9/16)$; $N_m=N_{dm}$ if otherwise.

The final length of the mother code is $N=\max(\min(N_r, N_m, N_{\max}), N_{\min})$. The rate matching procedure is shown in Figure 3.35. It should be pointed out that the following parameters "9/8", "9/16" and "7/16" are the result of a large number of computer simulations in order to find suitable values.

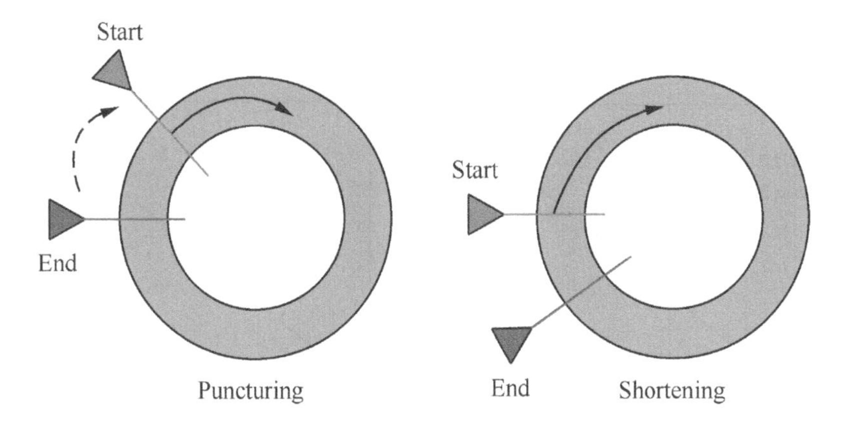

FIGURE 3.35 Illustration of rate matching for polar code [75].

Step 1. write the encoder bits into the circular buffer

Step 2. no further action is needed for rate matching if $M=N$

Step 3. puncturing should be carried out if $M<N$ and the code rate $K/M \leq (7/16)$, e.g., to pick the bits with indices from $(N-M+1)$ to the last (e.g., N) for transmission

Step 4. shortening should be carried out if $M<N$ and the code rate $K/M > 7/16$, e.g., to pick the bits from the beginning to M-th bit.

Step 5. repetition should be carried out if $M>N$, e.g., in addition to picking the N bits, the 1st to mod$((M-N), N)$ th bits should also be picked.

3.8 INTERLEAVING

The purpose of interleaving is to randomly reshuffle the orders of the bits in an information block in order to reduce the correlations between adjacent bits and be more robust to bursty errors. Compared to the pre-encoder interleaving (distributed CRC) described in Section 3.5.1.3, the pre-interleaving specifically for NR-PBCH as well as the small-scale interleaving during the rate matching (because only part of the encoded bits participate), the interleaving (also called post-encoder interleaving) discussed in this section has a large scale and can have a bigger impact on latency.

3.8.1 Interleaver with Triangle Shape

In the interleaver with isosceles right triangle shape [79], assuming the length of the edges of the right angle is P (the maximum size of interleaver

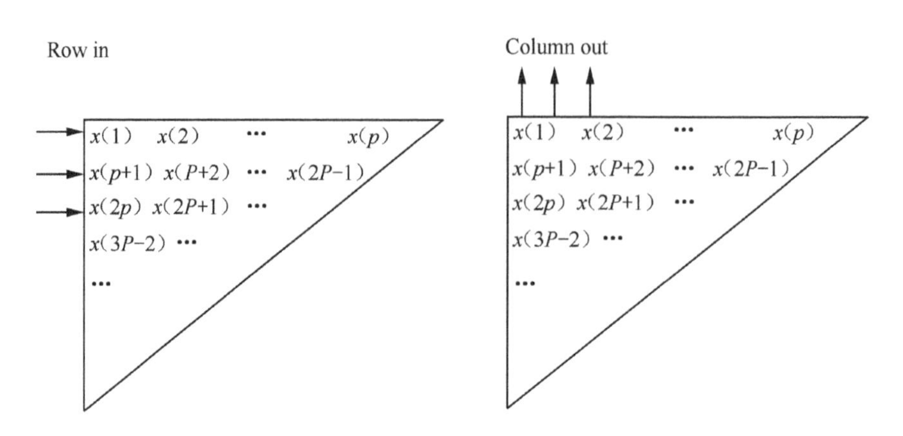

FIGURE 3.36 Illustration of triangle interleaver [79].

is set to 8192 [49] in 3GPP 5G NR, e.g., $(P*P/2) \leq 8192$, that is, $P \leq 128$). M is the number of bits after rate matching. Hence the edge length P should satisfy $P*(P+1)/2 \geq M$. As illustrated in Figure 3.36, the data are written into the interleaver row by row. If $P*(P+1)/2 > M$, NULL would be inserted. The data are read out from the interleaver column by column. If NULL is read out, NULL will be discarded.

Triangle interleaver has two important properties:

- For consecutive data before the interleaving, their distances after the interleaver become P, $P-1$, $P-2$, $P-3$,...... which are no longer the same.
- Each column has a different length so the transposed mode of each column is different.

Triangle interleave is used for UCI. As seen from Figure 3.37, its performance is quite close to that of random interleaver and noticeably better than without interleaver.

3.8.2 Double Rectangular Interleaver

Double rectangular interleaver (parallel rectangular interleaver) [80] is illustrated in Figure 3.38. The data block of M bits is divided into two sub-blocks having $M_1 = 1 + \text{ceil}(M/2)$ bits and $M_2 = M - M_1$, respectively. Then, M_1 bits are written into a rectangular interleaving of depth (e.g., #columns) = 5 in a row-by-row fashion. M_2 bits are written into another rectangular interleaver of depth = 11. NULL would be filled in if needed. After that, data are read out from these two rectangles in a column-by-column fashion.

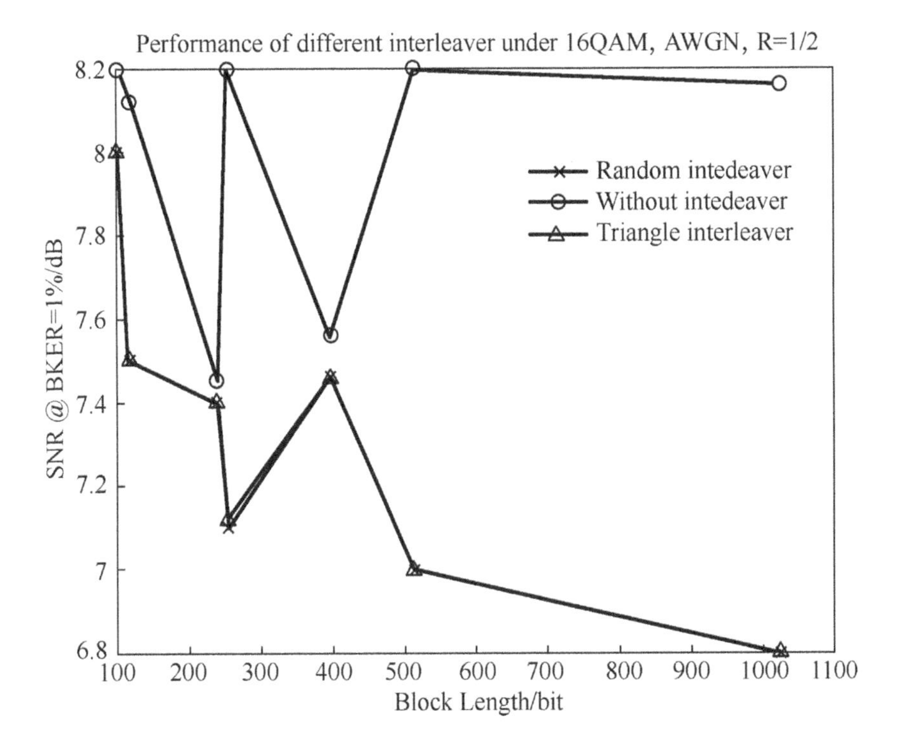

FIGURE 3.37 Performance of triangle interleaver [79].

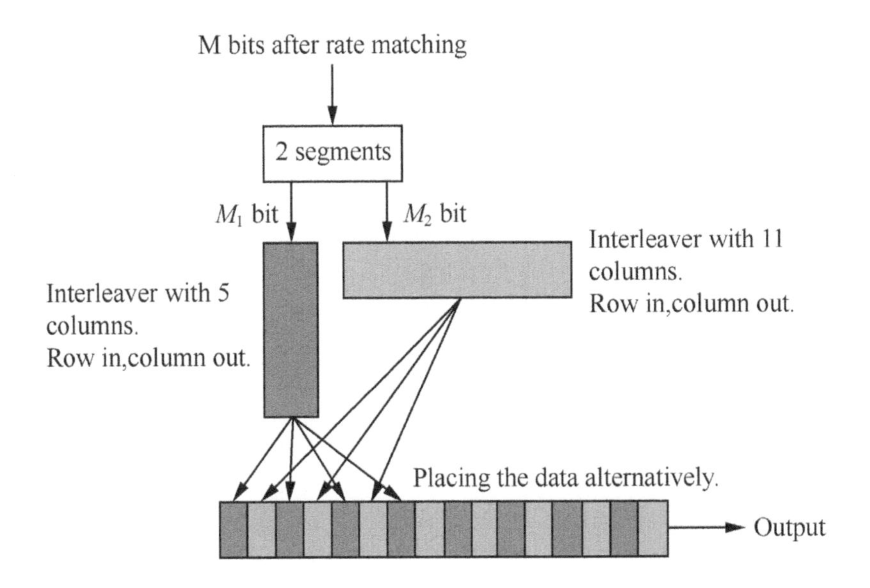

FIGURE 3.38 Illustration of a double rectangular interleaver [80].

NULL would be discarded if there is. In the end, these data are written to an interleaver of M bits in an alternate fashion.

The performance of double rectangular interleaver is quite close to that of random interleaver [80]. It was decided in 3GPP that no post-encoder interleaver is to be introduced for downlink. Hence, double rectangular interleaver has not been adopted by 5G NR so far.

3.8.3 Interleaving in Rate Matching

As per the G8 polar code illustrated in Figure 3.39, each equation consists of a different number of variables (e.g., CW). It is desirable to keep the number of variables low and have less complicated equations when solving the equations (e.g., decoding). This helps to quickly solve the equations. In another word, encoded bits with less pollution, e.g., x_8 ("clean" without "pollution"), x_7 ("polluted" once), x_6 ("polluted" once), and x_4 ("polluted" once).

In Figure 3.39, from the top to the bottom, the encoded bits are grouped into four parts (each having two encode bits). The first part is heavily polluted, followed by the upper half of the second part, then the lower half of the third part. Then the lower half of the second part. The fourth part is the least polluted.

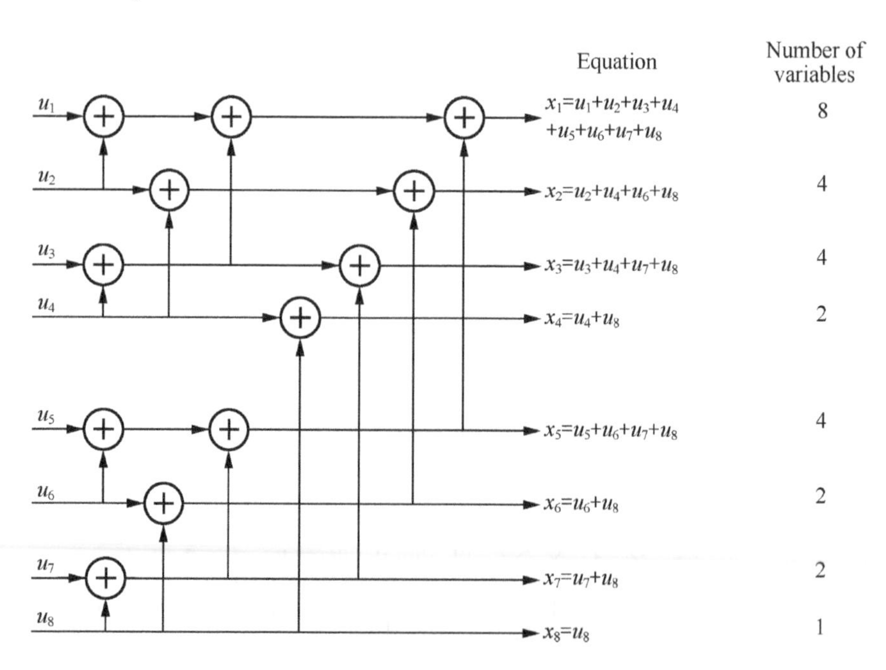

FIGURE 3.39 Equations for polar code generation.

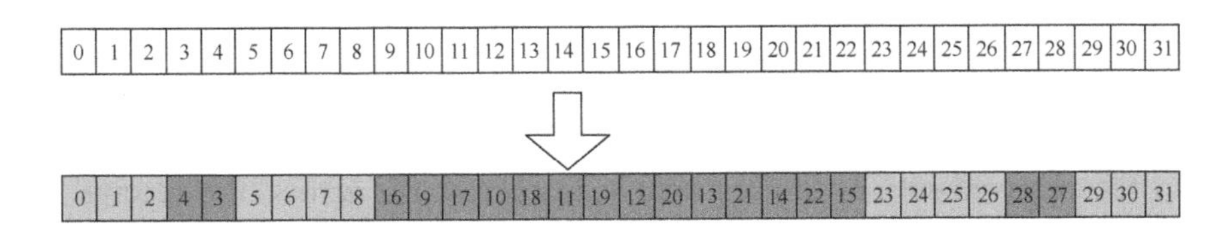

FIGURE 3.40 An illustration of sub-block interleaver [59,78].

Based on the above, an interleaver for rate matching was proposed in [81] where the encoded bits are partitioned into four parts. The first and fourth parts would be kept unchanged. The second and third parts are interlaced. In [82], bit reverse order (BRO) interleaver was proposed. In [83,84], the polar code sequence was used for the interleaver for rate matching. Different interleavers for sub-block were proposed in [85–88]. For instance, at the output of N encoded bits, every $N/32$ consecutive bits form a sub-block, resulting in 32 sub-blocks. The interleaving is applied to these 32 sub-blocks.

Take a polar code of $N=32$ as an example. The column weights are {32, 16, 16, 8, 16, 8, 8, 4, 16, 8, 8, 4, 8, 4, 4, 2, 16, 8, 8, 4, 8, 4, 4, 2, 8, 4, 4, 2, 4, 2, 2, 1}. Here it is desirable to swap the fourth and fifth elements, the 28th and 29th elements, and the middle element. After these conversions, it is better to put the weights into two categories: relatively heavier weights and relatively lighter weights, with the goal of sub-block interleaving: equations can be simpler.

During the rate matching, the first step of the operation is sub-block interleaving. The interleaver pattern is {0, 1, 2, 4, 3, 5, 6, 7, 8, 16, 9, 17, 10, 18, 11, 19, 12, 20, 13, 21, 14, 22, 15, 23, 24, 25, 26, 28, 27, 29, 30, 31} as shown in Figure 3.40.

It is seen from Figure 3.40 that the three sub-blocks on each side do not change their positions. The 14 sub-blocks in the middle change their positions. Assuming the mother code length $N=32$, meaning that each sub-block contains only one bit. Then, it becomes a bit interleaver. If the mother code length $N=64$, each sub-block contains two bits. Then it can be considered as an interleaver for a group of bits. Within a group, the 2 bits still follow the natural order.

3.9 POLAR CODE RETRANSMISSION

Retransmission refers to retransmitting information block. During retransmission, different redundant bits can be transmitted. The receiver can perform the decoding based on the retransmitted redundant information as well as the information previously received. HARQ is a combination of forward correction coding (FEC) and automatic retransmission quest (ARQ). The receiver would first rely on FEC to correct the potential errors in the received bits. If FEC is unable to correct the errors, the receiver would send "detection failure" to the transmitter. Upon knowing "detection failure", the transmitter would retransmit the information block, but with different coding parameters.

When polar code is used for physical control information such as DCI/UCI, no HARQ is needed. However, when polar code is used for NR-PBCH, the situation is a little different [89]. The receiver of NR-PBCH (presumably the terminal devices) would not indicate to the base station transmitter whether the detection is successful or failed. Nevertheless, NR-PBCH can be transmitted multiple times (up to 16). Hence, it is possible that the decoding of NR-PBCH can be finished earlier than 16.

As illustrated in Figure 3.41, the encoded bits are partitioned into four segments, each with a starting position denoted as P1, P2, P3, and P4. During the first transmission, it would start from P1 and last till the end. In the second transmission, it would start from P2 and then wind around

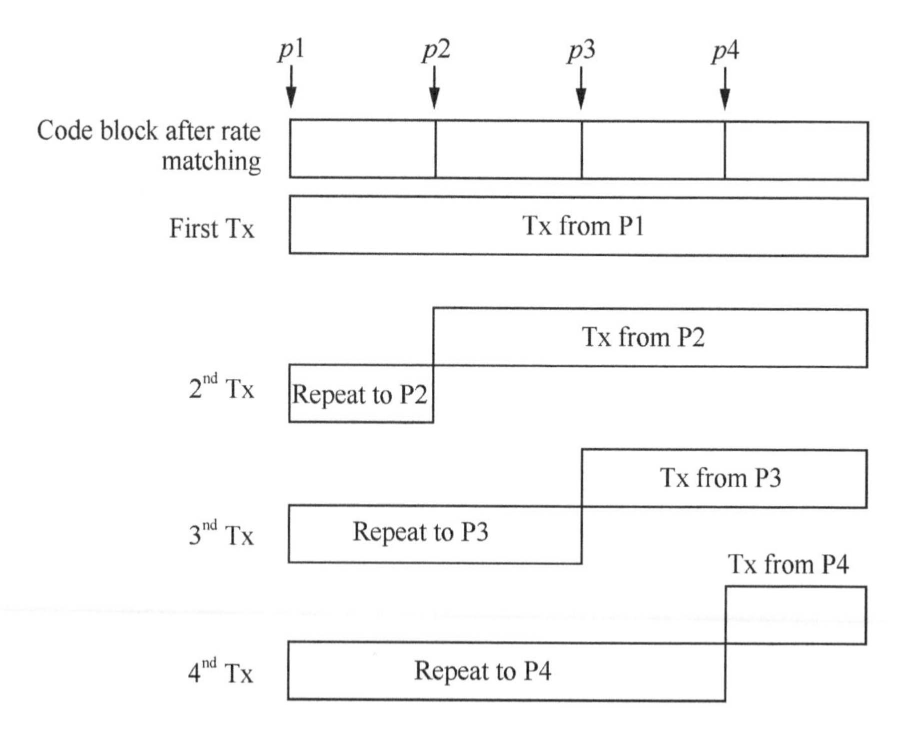

FIGURE 3.41 Illustration of HARQ [89].

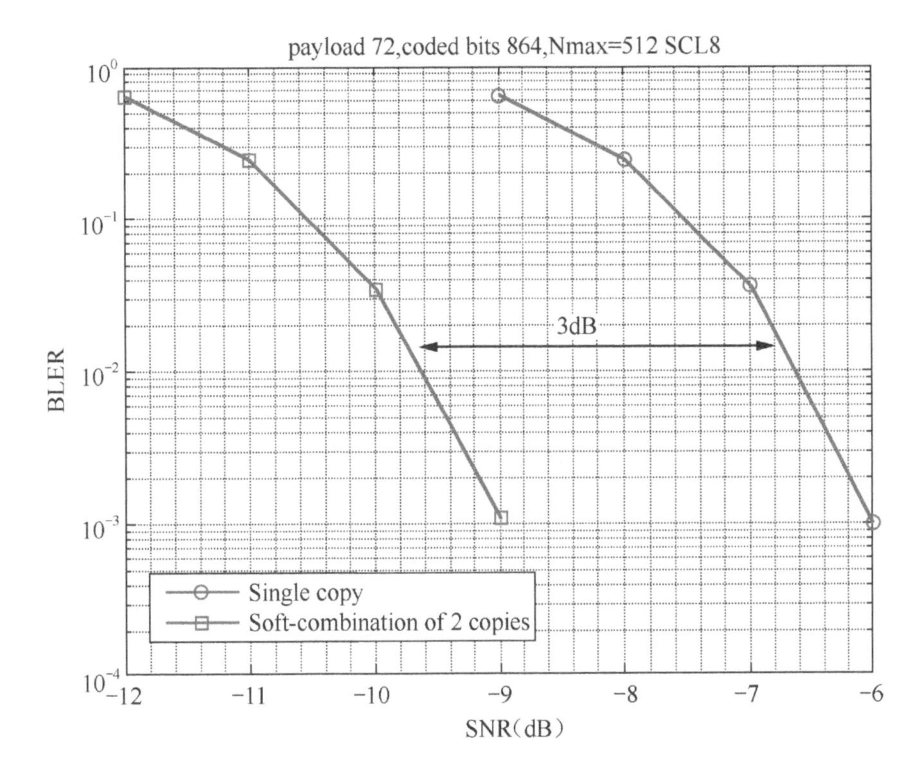

FIGURE 3.42 Combining gain due to HARQ [90].

back to P2. In the third transmission, it would start from P3 and then wind around back to P3. In the fourth transmission, it would start from P4 and wind around back to P4.

In Figure 3.41, the number of bits for each (re)transmission is the same. When the information block size is the same, different modulation orders (e.g., 16QAM to QPSK), code rates (e.g., 1/2 to 1/3), and different numbers of encoded bits (e.g., 500–200 bits) are possible during (re)transmissions. The receiver can combine the data across multiple transmissions to improve the decoding performance. As Figure 3.42 shows, 3 dB gain is observed by one retransmission (total of two transmissions).

3.10 SEGMENTATION

Segmentation [92,93] refers to segmenting the original information block into multiple sub-blocks, before carrying out polar code encoding for each sub-block. Segmentation is not always required, e.g., it is only needed when the information block size exceeds a certain length and there is enough physical resource allocated.

Polar code is used for control channel coding for 5G. In the downlink, the maximum information block sizes of NR-PDCCH and NR-PBCH are 140 and 56 bits, respectively [49], not large enough for segmentation. In the uplink, with carrier-aggregation, the size of UCI (including ACK/NACK and CSI) may exceed 500 bits. Considering that the relatively low code rate is required to maintain good coverage of UCI and the maximum mother code length $N_{max}=1024$, as well as the "good" channel ratio=41%, UCI longer than N_{max}*ratio=1024*0.41=420 bits would require segmentation.

To improve the performance of a long UCI codeword, segmentation for the single-CRC-based polar code was proposed [92] where the information bits and CRC bits are equally divided into two parts and Polar coded separately. At the receiver, based on path-metric (PM), select several paths from the decoding results of two parts, respectively, and then concatenate the paths from the two parts. Next, carry out a CRC check in order to get the final decoded bits.

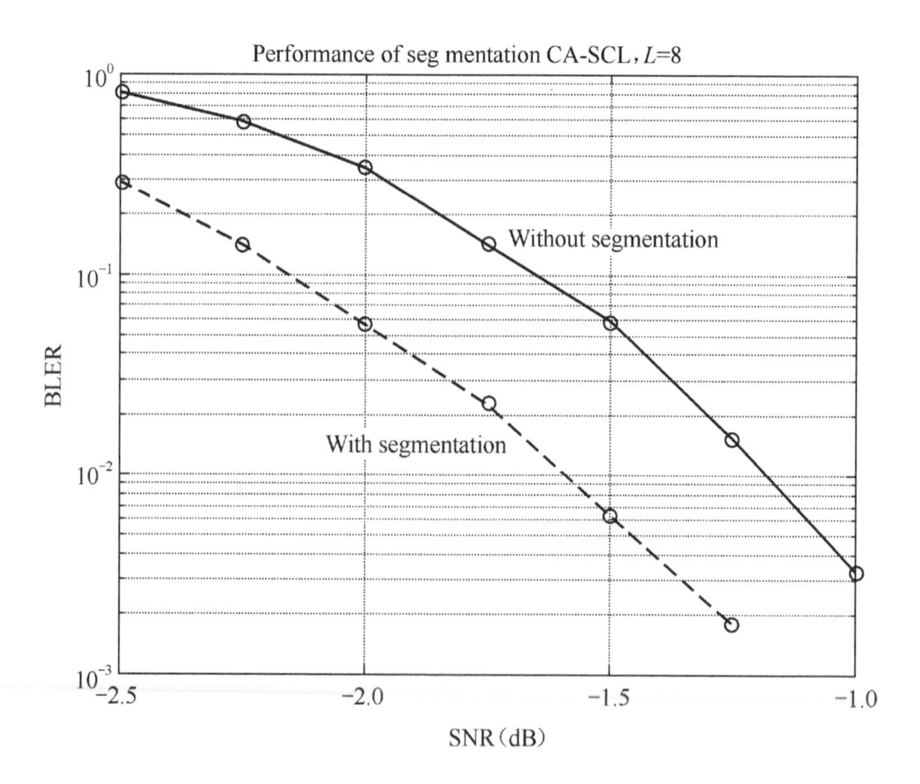

FIGURE 3.43 Performance comparison between segmentation and no segmentation [92].

In the segmentation solution of [92], UCI has 500 bits and the number of encoded bits is 2000 (code rate=1/4). When using segmentation (partitioned into two segments), use two polar encoders, each being (1024, K). Compared to no segmentation (use 1 polar encoder (1024, K) and repeat to 2000 bits), the gain is about 0.5 dB, as shown in Figure 3.43.

Although the above segmentation solution has significant performance benefits, concatenation of decoding results of two polar decoders is needed. To simplify the operation, polar code segmentation based on dual CRC was proposed in [93] where each segment individually appends CRC bits to their information bits, to achieve two independent decodings.

Polar code segmentation based on dual CRC was adopted by the standards. Furthermore, only when the length of UCI (excluding CRC) is no less than 360 bits and the number of encoded bits (to be supported by physical resources) M is no less than 1088, would the information block be segmented into two sub-blocks, each to be appended with its own CRC. If K is an odd number, "0" would be filled at the beginning of the original UCI [91].

3.11 SYSTEMATIC POLAR CODES

The original polar code proposed by Arikan [1] is non-systematic. It is known that any linear code can be converted into systematic form, which applies to polar code as well. Later on, Arikan proposed systematic polar code [94] and 2D polar code [96], which enriches the polar code family. In this section, we will discuss about systematic polar code. In the next section, 2D polar code will be discussed.

A linear code can be generally represented as $x=u \bullet G$, where both $\mathbf{x}$ and $\mathbf{u}$ are defined in GF(2). For polar code, the bits to be encoded $\mathbf{u}$ include two parts: the variable part (bits to be transmitted) u_A and the fixed part (frozen bits) u_{A^C}, that is, $u=(u_A \;, u_{A^C})$. The codeword $\mathbf{x}$ can be represented as $x=u_A \bullet \mathrm{G_A} + u_{A^C} \bullet G_{A^C}$, where $\mathrm{G_A}$ is the sub-matrix formed by the rows in G that corresponds to u_A. G_{A^C} is the sub-matrix formed by the rows in G that correspond to u_{A^C}. Here "+" is a modulo-2 addition. $A \subset \{1, 2, 3, \ldots\ldots, N\}$, $A^C \subset \{1, 2, 3, \ldots\ldots, N\}$. For instance, consider $\mathrm{G_4} = \begin{bmatrix} 1 & 0 & 0 & 0 \\ 1 & 1 & 0 & 0 \\ 1 & 0 & 1 & 0 \\ 1 & 1 & 1 & 1 \end{bmatrix}$, assuming $A=\{2, 3\}$, then $A^C=\{1, 4\}$, $\mathrm{G_A} = \begin{bmatrix} 1 & 1 & 0 & 0 \\ 1 & 0 & 1 & 0 \end{bmatrix}$ and $\mathrm{G}_{A^C} = \begin{bmatrix} 1 & 0 & 0 & 0 \\ 1 & 1 & 1 & 1 \end{bmatrix}$.

Next, we separate the codeword **x** into two parts x_B and x_{B^C}, e.g., $x=(x_B\ ,x_{B^C}$, where $B\subset\{1, 2, 3, ..., N\}$, $B^C\subset\{1, 2, 3, ..., N\}$. Then we have $x_B=u_A\bullet G_{AB}\ +\ u_{A^C}\bullet G_{A^C B}$ 和 $x_{B^C}=u_A\bullet G_{AB^C}\ +\ u_{A^C}\bullet G_{A^C B^C}$, where

G_{AB} is the sub-matrix made up of elements G_{ij} in generation matrix G that satisfy $i\in A$ and $j\in B$. The definitions of three other sub-matrices are similar.

Still use the above G_4, G_A, and G_{A^C} as examples. Assuming $B=\{2, 3\}$, then $B^C=\{1, 4\}, G_{AB}=\begin{bmatrix}1&0\\0&1\end{bmatrix}, G_{A^C B}=\begin{bmatrix}0&0\\1&1\end{bmatrix}, G_{AB^C}=\begin{bmatrix}1&0\\1&0\end{bmatrix}$, and $G_{A^C B^C}=\begin{bmatrix}1&0\\1&1\end{bmatrix}$.

According to [94], if and only if A and B have the same number of elements and are invertible, can a systematic polar encoder be converted from a non-systematic polar encoder by using a pair of parameters $(B,\ u_{A^C})$. First, compute u_A as $u_A=(x_B-u_{A^C}\bullet G_{A^C B})\bullet(G_{AB})^{-1}$. Then, substitute u_A into $x_{B^C}=u_A\bullet G_{AB^C}\ +\ u_{A^C}\bullet G_{A^C B^C}$ to get x_{B^C}. Next, combine x_B and x_{B^C} together to get the final codeword $x=(x_B\ ,\ x_{B^C})$.

Still using the above G_4, A, B as examples, assuming $u=\{u_1,\ u_2,\ u_3,\ u_4\}$, then $u_A=\{u_2,u_3\}$ and $u_{A^C}=\{u_1,u_4\}$, $x_B=u_A\bullet G_{AB}\ +\ u_{A^C}\bullet G_{A^C B}=\{x_2,x_3\}=$

$$\{u_2,u_3\}\bullet\begin{bmatrix}1&0\\0&1\end{bmatrix}+\{u_1,u_4\}\bullet\begin{bmatrix}0&0\\1&1\end{bmatrix}=\{u_1+u_2,\ u_3+u_4\},\quad x_{B^C}=u_A\bullet G_{AB^C}\ +$$

$$u_{A^C}\bullet G_{A^C B^C}=\{x_1,x_4\}=\{u_2,u_3\}\bullet\begin{bmatrix}1&0\\1&0\end{bmatrix}+\{u_1,u_4\}\bullet\begin{bmatrix}1&0\\1&1\end{bmatrix}=\{u_1+u_2+u_3+u_4,u_4\}.$$

Setting frozen bits to be 0, e.g., $u_{A^C}=\{u_1,u_4\}=\{0,\ 0\}$. The output systematic bits are $x_B=\{x_2,\ x_3\}=\{u_2,\ u_3\}$. The output parity check bits are $x_{B^C}=\{x_1,\ x_4\}=\{u_2+u_3,\ 0\}$ (x_4 would be punctured in practical systems). It is seen that the original information bits $u_A=\{u_2,u_3\}$ are visible in the output bits of the encoder. Therefore, to convert a non-systematic polar to a systematic polar code.

The above encoding method does not change the generation matrix. Hence the computation complexity and decoding algorithm of systematic polar code are the same as non-systematic polar code.

As seen in Figures 3.44 and 3.45, when the mother code length $N=256$, code rate=1/2, using AWGN channel and considering BER=10^{-5}, systematic polar outperforms non-systematic polar by 0.25 dB. However, their BLER performance is identical [94]. It is also verified in [95] that BER performance of systematic polar code is better than that of non-systematic polar.

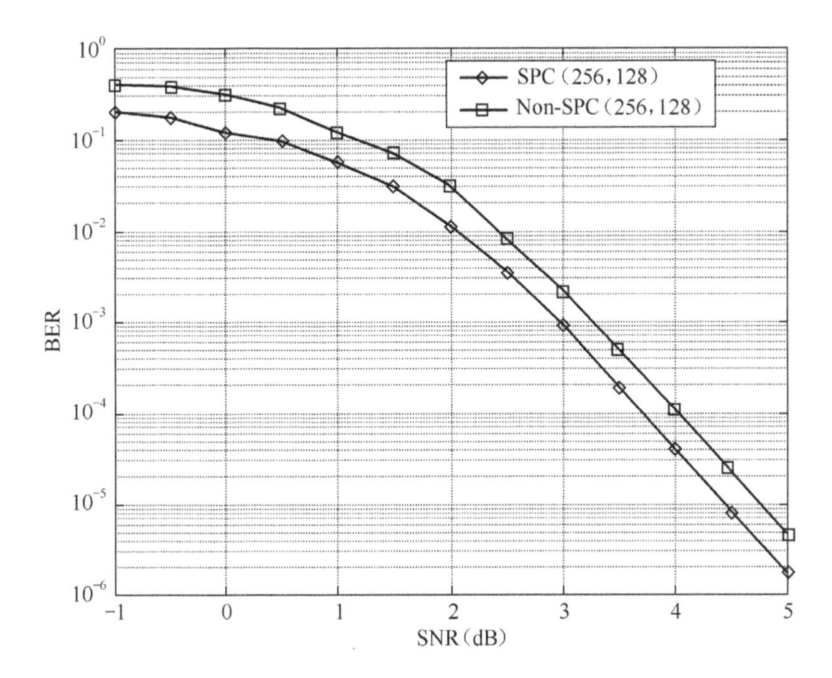

FIGURE 3.44 BER performance of the systematic polar vs. non-systematic polar code [94].

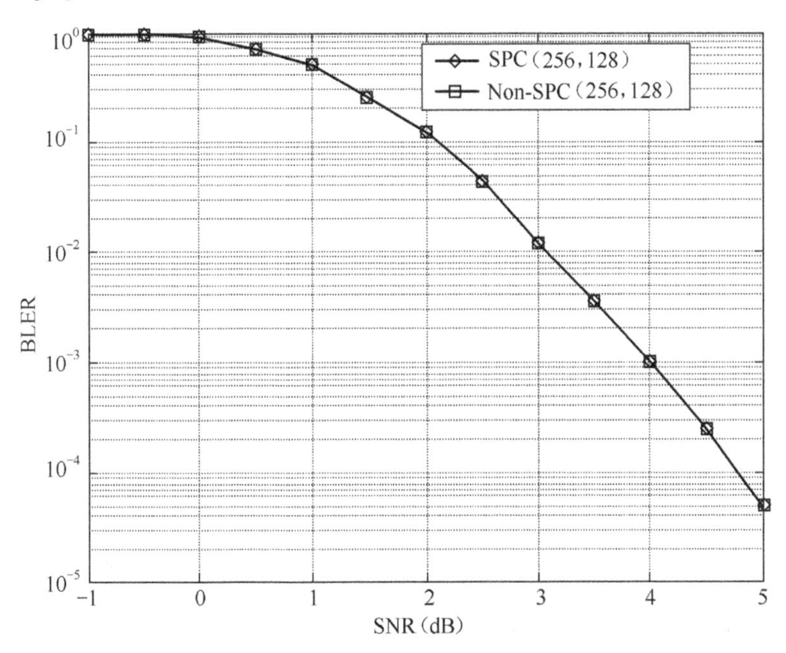

FIGURE 3.45 BLER performance of systematic polar vs. non-systematic polar code [94].

In summary, the systematic polar code can be considered as a special case of non-systematic polar code. It can be obtained by selecting appropriate bits as frozen bits (or alternatively to select appropriate bit locations for information bits).

3.12 2D POLAR CODE

A two-dimensional (2D) polar code [96] was proposed by Arikan in 2009. "2D" emphasizes that the information bits should be arranged into a regular two-dimensional array form, instead of the conventional single-column form. In the following, we use an example to explain how a 2D polar code is constructed. Note that the concatenated code mentioned in Section 4.2.5 of [38] can also be considered as 2D polar code.

Consider $G_8 = \begin{bmatrix} 1,0,0,0,0,0,0,0 \\ 1,0,0,0,1,0,0,0 \\ 1,0,1,0,0,0,0,0 \\ 1,0,1,0,1,0,1,0 \\ 1,1,0,0,0,0,0,0 \\ 1,1,0,0,1,1,0,0 \\ 1,1,1,1,0,0,0,0 \\ 1,1,1,1,1,1,1,1 \end{bmatrix}$. Based on channel reliability (ref. Section 3.6), pick the fourth, sixth, seventh, and eighth rows to form the sub-matrix $G_P(8,4) = \begin{bmatrix} 1,0,1,0,1,0,1,0 \\ 1,1,0,0,1,1,0,0 \\ 1,1,1,1,0,0,0,0 \\ 1,1,1,1,1,1,1,1 \end{bmatrix}$. If setting u_1, u_2, u_3, and u_5 as the frozen bits, then we get $x_1^8 = (u_1, u_2, u_3, \ldots\ldots, u_8) \bullet G_8 = (u_4, u_6, u_7, u_8) \bullet G_P(8,4)$.

One strategy for realizing two-dimensional polar codes is to select an array of information bits $U = (u_{i,j})$ so that each row of U contains the allowed bits of the specific polar code P(N, K) and each column is a codeword of a linear block code. For example, considering a 4×8 matrix U. Each row of U contains the allowable information of P(8, 4), and each column is a length-4 codeword of Table 3.2 where the index (N, K, d) denotes

TABLE 3.2 Column Codewords Used to Form 4×8 Array U

Index	Types of Codeword
1	(4, 0, ∞)
2	(4, 0, ∞)
3	(4, 0, ∞)
4	(4, 1, 4)
5	(4, 0, ∞)
6	(4, 3, 2)
7	(4, 3, 2)
8	(4, 4, 1)

the codeword length *N*, dimension *K* and Hamming distance *d*. For example, (4, 0, ∞) refers to codewords containing all 0. (4, 1, 4) includes 0000 and 1111. An allowable information array $U = \begin{bmatrix} 0,0,0,1,0,1,0,1 \\ 0,0,0,1,0,0,1,1 \\ 0,0,0,1,0,1,0,0 \\ 0,0,0,1,0,0,1,1 \end{bmatrix}$

After polar encoding of the allowable information array, we get $X = U \bullet G_8 = \begin{bmatrix} 1,0,0,1,1,0,0,1 \\ 1,0,1,0,0,1,0,1 \\ 0,1,1,0,0,1,1,0 \\ 1,0,1,0,0,1,0,1 \end{bmatrix}$. Until now, the encoding of 2D polar is completed. Then the 32-bit codeword is to be transmitted.

The decoding of 2D polar is based on the structure of U and carried out in a row–column interlaced fashion on the received data Y. Because the first three columns are all 0's, the decoder would automatically set them to 0. The fourth column of U contains the actual information. The decoder would estimate the values of $\tilde{u}_{i,4}, 1 \le i \le 4$ row by row (using SC for polar). Then use the ML algorithm to decode this column. Next, the fifth column of U is all 0's. Hence, the decoder sets them to 0. Then decode the sixth, seventh, and eighth columns.

Simulation results [96], as illustrated in Figure 3.46, show that 2D polar of 4×256 outperforms 1D polar of 1×256 and exhibits similar performance as longer 1D polar (1×1024).

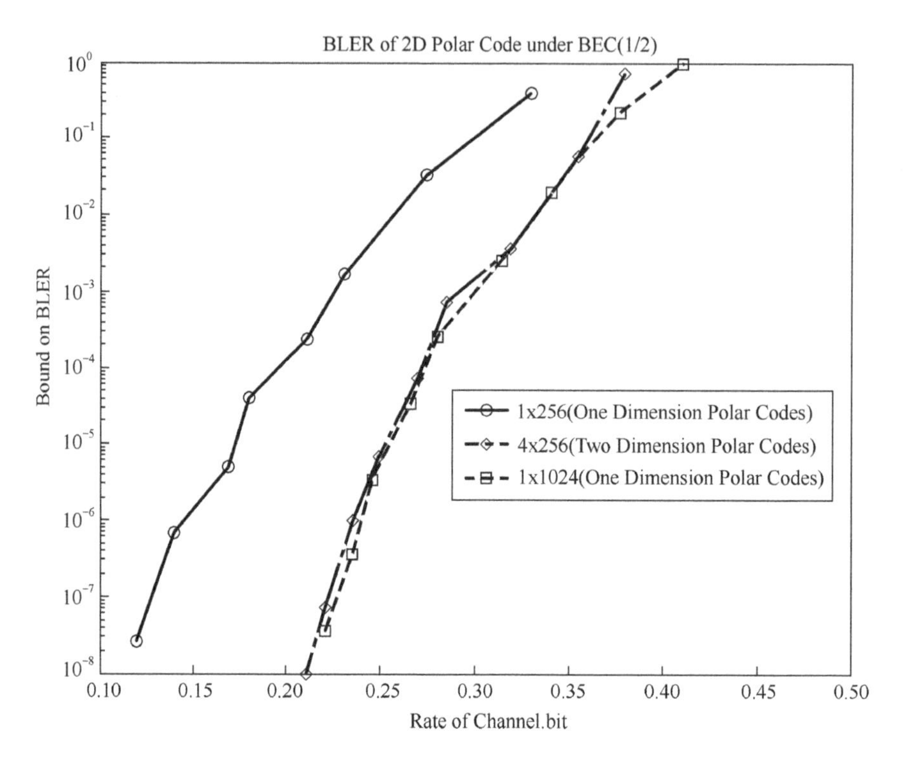

FIGURE 3.46 Performance comparison between 2D polar and 1D polar [96].

3.13 DECODING ALGORITHMS FOR POLAR CODES

Similar to other coding schemes, for short-to-medium block length, the performance of the polar code without CRC aid is not competitive [38]. However, with CRC, polar code performs quite well, especially when using the CRC-aided successive cancelation list (CA-SC-L) algorithm for decoding. When the list depth is large, the polar code performance would approach that of ML decoding [13].

3.13.1 SC Algorithm

A detailed description of the SC algorithm can be found in [1,34]. During SC decoding, when making the decision on the i-th bit, we need to calculate the transfer probability $W_N^{(i)}(y_1^N, \hat{u}_1^{i-1} | \hat{u}_i)$ of the channel $W_N^{(i)}$. Given the received sequence y_1^N, the a-posteriori probability (APP) of the partial decoding sequence $\hat{u}_1^i$ is

$$P_N^{(i)}(\hat{u}_1^i | y_1^N) = \frac{W_N^{(i)}(y_1^N, \hat{u}_1^{i-1} | \hat{u}_i) \bullet \Pr(\hat{u}_i)}{\Pr(y_1^N)} \tag{3.29}$$

When the information is equally probable between {0, 1}, e.g., $\Pr(\hat{u}_i = 0) = \Pr(\hat{u}_i = 1) = 0.5$, the probability of the received sequence being y_1^N is

$$\Pr(y_1^N) = \frac{1}{2^N} \bullet \sum_{u_1^N} W_N(y_1^N | u_1^N) \tag{3.30}$$

In addition, the APP can be recursively calculated as

$$P_N^{(2i-1)}(\hat{u}_1^{2i-1} | y_1^N) = \sum_{\hat{u}_1^{2i} \in \{0,1\}} P_{N/2}^{(i)}(\hat{u}_{1,0}^{2i} \oplus \hat{u}_{1,e}^{2i} | y_1^{N/2}) \bullet P_{N/2}^{(i)}(\hat{u}_{1,e}^{2i} | y_{1+N/2}^N) \tag{3.31}$$

$$P_N^{(2i)}(\hat{u}_1^{2i} | y_1^N) = P_{N/2}^{(i)}(\hat{u}_{1,0}^{2i} \oplus \hat{u}_{1,e}^{2i} | y_1^{N/2}) \bullet P_{N/2}^{(i)}(\hat{u}_{1,e}^{2i} | y_{1+N/2}^N) \tag{3.32}$$

When $N=1$, we have

$$P_1^{(1)}(\hat{u} | y) = \Pr(\hat{u} | y) = \frac{W(y|\hat{u})}{2 \bullet \Pr(y)} = \frac{W(y|\hat{u})}{W(y|0) + W(y|1)}.$$

Based on this posteriori probability, $\hat{u}_1^i$ can be determined by the following formula:

$$\hat{u}_i = \begin{cases} 0, \ if \ i \in A \text{ and } \dfrac{P_N^{(i)}(\hat{u}_1^{i-1}, u=0|y_1^N)}{P_N^{(i)}(\hat{u}_1^{i-1}, u=1|y_1^N)} \geq 1 \\ 0, \ if \ i \in A \text{ and } \dfrac{P_N^{(i)}(\hat{u}_1^{i-1}, u=0|y_1^N)}{P_N^{(i)}(\hat{u}_1^{i-1}, u=1|y_1^N)} < 1 \\ u_i, \ if \ i \in A^C \end{cases} \tag{3.33}$$

Taking the mother code length $N=4$ as an example, the SC algorithm can be represented as a tree, as illustrated in Figure.3.47. It is observed that the SC algorithm always proceeds along the path that has the largest posteriori probability. In the next section, we will see that sometimes SC would lead to the wrong path (resulting in a decoding error).

3.13.2 SC-L Algorithm

SC-L algorithm is built on top of the SC algorithm where L paths with high posteriori probabilities are preserved for the next level of decoding.

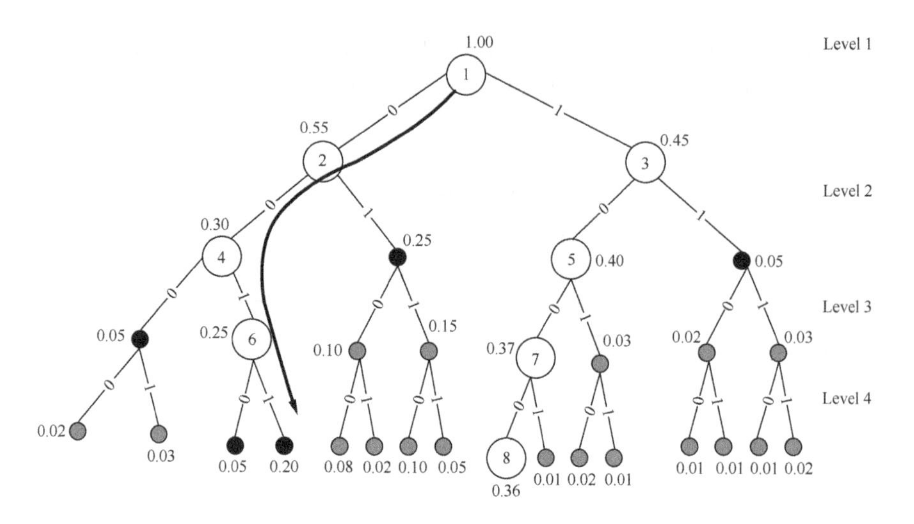

FIGURE 3.47 Tree structure of the SC algorithm [34].

In the end, the path with the largest posteriori probability is selected. Compared to the SC algorithm, the SC-L algorithm requires path storage and fallback operation. A more detailed description of SC-L can be found in [13,14].

When the list length $L=1$, the SC-L algorithm degenerates to the SC algorithm. When $L=\min(2^N, 2^{\text{Length(A)}})$, SC-L becomes the ML algorithm, where length(A) is the length of the information block. As Figure 3.48 shows, when $L=32$, the performance of SC-L algorithm is already quite close to that of ML at BLER=1%.

Still take mother code length $N=4$ as an example, SC-L algorithm can also be represented in a tree structure. As seen in Figure 3.49, assuming the list depth $L=2$, in the first layer, the left path {0, x, x, x} and the right path {1, x, x, x} are found according to the maximization of APP. These two paths are stored in a list with $L=2$ entries. In the second layer, the left path {0, 0, x, x} and the right path {1, 0, x, x} are found according to the maximization of APP. In the third layer, the left path {0, 0, 1, x} and the right path {1, 0, 0, x} are found according to the maximization of APP. In the fourth layer, the left path {0, 0, 1, 1} and the right path {1, 0, 0, 0} are found according to the maximization of APP. Since the APP of the right path is higher than that of the left path. SC-L would output the result of right path: {1, 0, 0, 0} and finish the decoding. The right path selected by SC-L is in fact the path according to ML.

In addition, successive cancelation with stack (SC-S) algorithm and successive cancelation with hybrid stack+list (SC-H) algorithm are studied in [34]. Their performance is very close to that of ML algorithm.

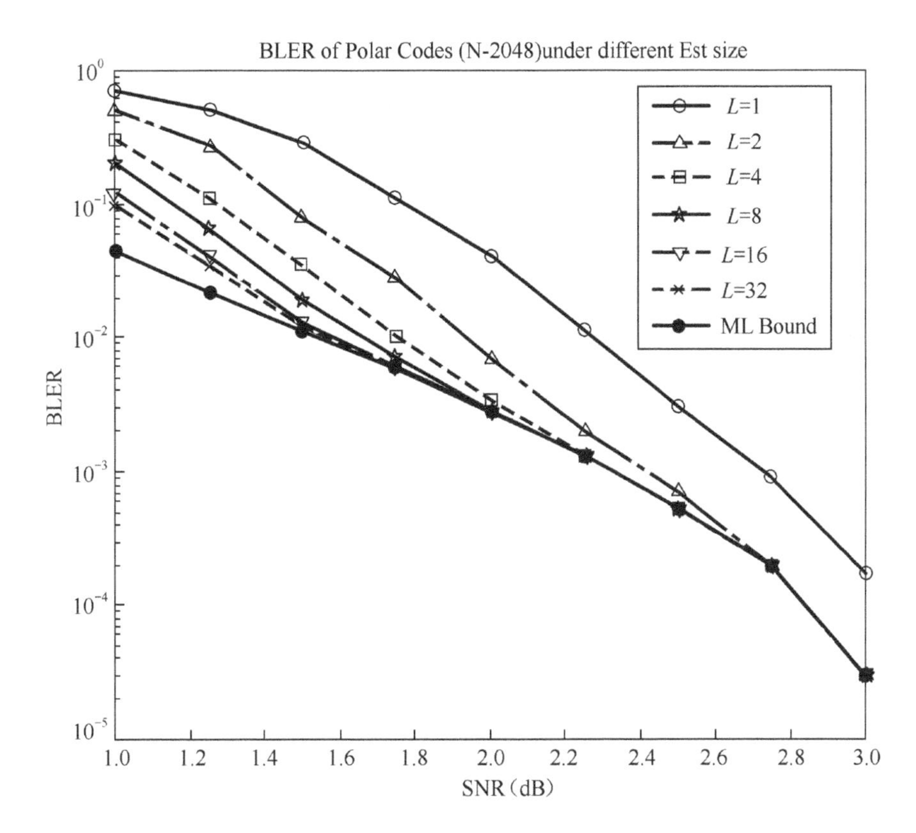

FIGURE 3.48 Performance of SC-L with different list depths [13].

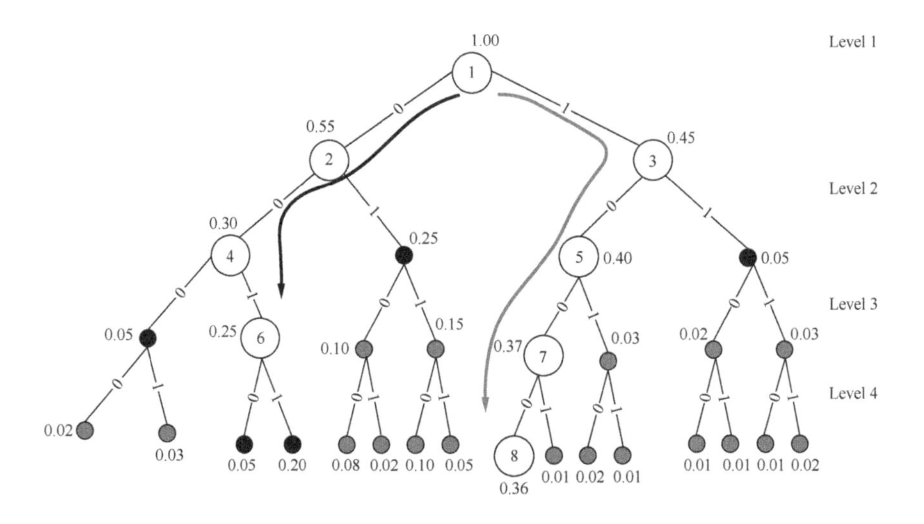

FIGURE 3.49 SC-L decoding algorithm in a tree structure [34].

3.13.3 Statistic Ordering-Based Decoding Algorithm

The statistic ordering-based algorithm [36] has the following steps:

Step 1. the decoder arranges the received data $y = u \bullet G + n$ in the descending order, based on their belief. That is, $z = \lambda(y)$ where $z = \{z_1, z_2, z_3, \ldots\ldots, z_N\}$, $\lambda(\bullet)$ is the permutation operation (e.g., a permutation matrix). After this operation, we get $|z_1| \geq |z_2| \geq |z_3| \geq \ldots\ldots \geq |z_N|$.

Step 2. apply the above operation $\lambda(\bullet)$ to the polar code generation matrix and get $S = \lambda(G) = [g_1, g_2, g_3, \ldots\ldots, g_N]$, where $g_1, g_2, g_3, \ldots\ldots, g_N$ are N column vectors of matrix S. N is the mother code length of polar code.

Step 3. starting from the left-most column of matrix S, search for K columns that have the least correlation (e.g., having the most Hamming distance) with the first column, and put them as the first K columns of matrix T. K is the information block length before polar encoding. The rest (N–K) columns of matrix S will be placed to (K+1)-th to N-th columns of matrix T (note that it is not clear from [36] whether ordering is performed for these (N–K) columns). This operation can be represented as another permutation function $T = \phi(S)$. Hence, $T = \phi(\lambda(G))$.

Step 4. perform Step 3 for the output of Step 1 and get the vector $v = \phi(z) = \phi(\lambda(y))$

Step 5. perform Gaussian elimination on matrix T and get a matrix P where its first K columns/rows constitute a $K \times K$ identity matrix

Step 6. to perform a hard decision on the first K elements in vector v and get the vector $a = \{a_1, a_2, a_3, \ldots\ldots, a_K\}$

Step 7. use matrix P to encode the output of Step 6 and get the codeword $c = a \bullet P$. Note that (N–K) should be added at the beginning of the vector a.

Step 8. calculate $d = \lambda^{-1}(\phi^{-1}(c))$ and perform hard decision decoding on d to get $e = HD(d)$, where $\lambda^{-1}(\bullet)$ is the inverse operation of $\lambda(\bullet)$. $\phi^{-1}(\bullet)$ is the inverse operation of $\phi(\bullet)$. $HD(\bullet)$ denotes hard decision.

In the absence of noise, we would get $e = u$, by going through the above steps.

In order to improve the decoder's performance, the L-th order statistic ordering decoder was proposed in [36]. That is, perform bit flipping one-by-one on (up to) L bits in vector a obtained from Step 6 to get a new codeword c, and calculate the Euclidean distance between the modulation sequence of the codeword c and the received vector v. Then use the codeword c with the minimum Euclidean distance to compute d, and perform the hard decision decoding to get the decoded bits.

In [36] CRC is also combined with the above algorithm. Simulation results show that at a high code rate, e.g., R=3/4, the CRC-aided L-th order statistic ordering algorithm outperforms the CRC-aided SC list algorithm by about 0.1 dB. At a lower rate, e.g., 1/4, CRC-aided L-th order statistic ordering algorithm underperforms CA-SCL-32.

3.13.4 Belief Propagation (BP) Algorithm

Arikan mentioned in [1] that the polar code generation matrix G can be represented by a factor graph so that BP may be used for the decoding. In the following, we use $N=8$ polar code to describe how to apply BP for polar code decoding [39,40].

The factor graph of the G8 polar code is shown in Figure 3.50 which contains $N \bullet (1+\log_2 N)=32$ factors, e.g., a total of 32 nodes, involving decoding

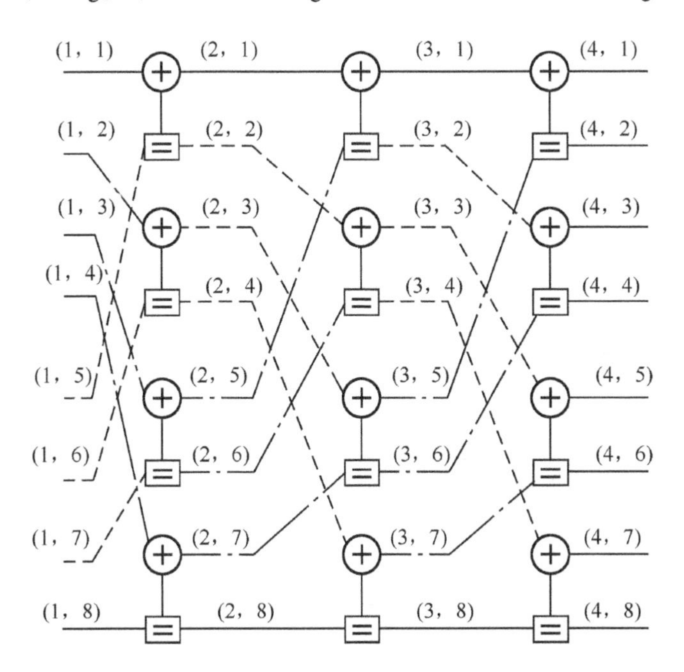

FIGURE 3.50 Factor graph of G8 polar code.

computation. Each node is represented with a notation (i, j), where i denotes the level number, e.g., $1 \le i \le 1+\log_2 N$. The first level means the information bits to be encoded. The last level means the information bits after the encoding. j means the index of a node in the current level $1 \le j \le N$.

The above factor graph can be decomposed into the computation between $2\times2=4$ nodes, or basic computation blocks as shown in Figure 3.51, where $R_{i,j}$ represents the message from the current node to the node on the right. $L_{i,j}$ represents the message from the current node to the node on the left. In the first level, we have

$$L_{1,j} = \begin{cases} 0, & \text{if } j \in A \\ \infty, & \text{if } j \in A^C \end{cases} \tag{3.34}$$

In the last level, we have

$$R_{1+\log_2 N, j} = \log\left(\frac{P(y_j \mid x_j = 0)}{P(y_j \mid x_j = 1)}\right) \tag{3.35}$$

BP decoding algorithm is implemented via message passing between nodes as illustrated in Figure 3.51. The information on each node can be calculated based on the neighboring nodes, as represented in the following equations:

$$L_{i,j} = g(L_{i+1,2j-1}, (L_{i+1,2j} + R_{i,j+N/2})) \tag{3.36}$$

$$L_{i,j+N/2} = g(R_{i,j}, L_{i+1,2j-1}) + L_{i+1,2j} \tag{3.37}$$

$$R_{i+1,2j-1} = g(R_{i,j}, (L_{i+1,2j} + R_{i,j+N/2})) \tag{3.38}$$

$$R_{i+1,2j} = g(R_{i,j}, L_{i+1,2j-1}) + R_{i,j+N/2} \tag{3.39}$$

where $g(x, y) = \ln(1 + xy/(x+y))$.

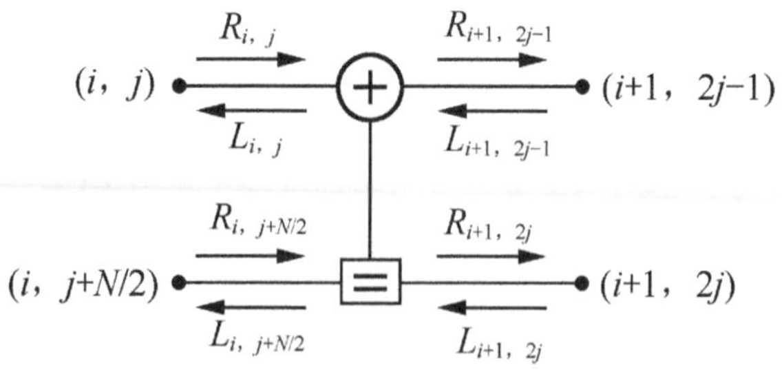

FIGURE 3.51 Basic computation block of BP algorithm for the polar code.

Using the above equations to iteratively perform the update of each node forward (from the left to the right) and backward (from the right to the left). The iteration process would terminate once the CRC check passes or it reaches the maximum number of iterations.

According to [40], with N=512, code rate=1/2, 60 iterations, BER=5*10–5 and AWGN channel, BP algorithm outperforms SC algorithm by about 0.3 dB.

3.13.5 Parallel Decoding for Polar Code

The SC algorithm has a relatively long delay, especially when the mother code length is long, e.g., N=220. This would limit the application of polar code. To reduce the decoding latency, parallel decoding has been studied, which basically has two types: a full-parallel decoding algorithm based on BP [41,42,99] and multi-SC parallel decoding based on the recursive structure of polar code [43,113]. The latter is the focus of this section.

As in [43], we denote $u_1^N = \{u_1, u_2, u_3, \ldots\ldots, u_{N-1}, u_N\}$ as a block of N information bits. This is made up of two sections: $u_1^{N/2} = \{u_1, u_2, u_3, \ldots\ldots, u_{N/2-1}, u_{N/2}\}$ in the first half and $u_{1+N/2}^N = \{u_{1+N/2}, u_{2+N/2}, u_{3+N/2}, \ldots\ldots, u_{N-1}, u_N\}$ in the latter half. $G_N = F^{\otimes \log_2(N)}$ is the generation matrix of polar code. $x_1^N = \{x_1, x_2, x_3, \ldots\ldots, x_{N-1}, x_N\}$ are the encoded bits. Hence,

$$x_1^N = u_1^N \bullet G_N = u_1^N \bullet F^{\otimes \log_2(N)} = u_1^N \bullet \begin{bmatrix} F^{\otimes \log_2(N/2)}, & 0 \\ F^{\otimes \log_2(N/2)}, & F^{\otimes \log_2(N/2)} \end{bmatrix} \quad (3.40)$$

$$x_1^N = \left[x_1^{N/2}, x_{1+N/2}^N\right] = [a_1^{N/2} \bullet F^{\otimes \log_2(N/2)}, b_1^{N/2} \bullet F^{\otimes \log_2(N/2)}] \quad (3.41)$$

where $a_1^{N/2} = u_1^{N/2} \oplus u_{1+N/2}^N$, $b_1^{N/2} = u_{1+N/2}^N$.

This means that for a polar code of mother code length N, it can be decomposed into two polar codes each of mother code length $N/2$ and then decoded in parallel. It should be noted that here $a_1^{N/2}$ and $b_1^{N/2}$ are correlated.

When the data $y_1^N = \{y_1, y_2, y_3, \ldots\ldots, y_{N-1}, y_N\}$ is received, the first half $y_1^{N/2} = \{y_1, y_2, y_3, \ldots\ldots, y_{N/2-1}, y_{N/2}\}$ is sent to the first SC decoder, and we get $a_1^{N/2}$. The latter half $y_{1+N/2}^N = \{y_{1+N/2}, y_{2+N/2}, y_{3+N/2}, \ldots\ldots, y_{N-1}, y_N\}$ is sent to the second SC decoder, and we get $b_1^{N/2}$.

When calculating LLR of each bit, the processing of these two SC decoders is exactly the same as the conventional SC decoder, e.g.,

$$L_{N/2}^{(i)}(y_1^{N/2}, a_1^{\hat{i}-1}) = \log \frac{W_{N/2}^{(i)}(y_1^{N/2}, a_1^{\hat{i}-1} \mid a_i = 0)}{W_{N/2}^{(i)}(y_1^{N/2}, a_1^{\hat{i}-1} \mid a_i = 1)} \quad (3.42)$$

$$L_{N/2}^{(i)}(y_{1+N/2}^{N}, b_1^{\hat{i}-1}) = \log \frac{W_{N/2}^{(i)}(y_{1+N/2}^{N}, b_1^{\hat{i}-1} \mid b_i = 0)}{W_{N/2}^{(i)}(y_{1+N/2}^{N}, b_1^{\hat{i}-1} \mid b_i = 1)} \quad (3.43)$$

The decision of a_i and b_i can be either independently or jointly carried out. After that, we get $u_1^{N/2} = \{a_i \oplus b_i\}$, $u_{1+N/2}^{N} = \{b_i\}$.

Similarly, a polar code with mother code length N can also be decomposed into four polar codes each of mother code length $N/4$ and decoded in parallel, or decomposed into eight polar codes each of mother code length $N/8$ and decoded in parallel, and so on.

According to [43], when using the above parallel decoding algorithm, the BLER performance is quite similar to the conventional SC algorithm.

3.14 COMPLEXITY, THROUGHPUT AND DECODING LATENCY

Complexity refers to the number of computations and amount of memory needed to implement a channel coding scheme, which includes encoder's complexity and decoder's complexity. Normally, encoder's complexity is lower than decoder's. In addition, certain balance should be maintained between complexity and performance, and between complexity and throughput. This section discusses complexity and throughput. Polar code performance is discussed in the next section.

3.14.1 Computation Complexity

Arikan pointed out in [1,44] that with the SC decoding algorithm, the complexities of polar encoder and polar decoder are in the order of $O(N \bullet \log(N))$. This is because, for a polar code of mother code length N, there are log2(N) number of layers. During the encoding, $(\log_2 N) \bullet N/2$ times of modulo-2 summations are needed. During the decoding, N* log2(N)/2 check node operations and N* log2(N)/2 variable node operations are needed.

From another angle, denoting C_N as the complexity to decode a polar code of mother code length N, we can decompose the length-N Polar into two polar each of length $N/2$, then [44]

$$C_N = 2 \bullet C_{N/2} + k \bullet N = N + k \bullet N \quad (3.44)$$

where k is a constant. The number of computations calculated in Eq. (3.44) contains some repeated counting. In fact, Arikan pointed out in [1] that by appropriate grouping and sharing of computed LLR, the complexity can be made as low as

$$C_N = N + N \bullet \log(N) = (1 + \log(N)) \bullet N \quad (3.45)$$

In the case of the SC-L algorithm, the decoder's complexity is $O(L \bullet N \bullet \log(N))$ [13] where L is the list depth. In [109], the complexity of SC-L is estimated as $L*N*\log(N) + L*N + 2*L*K*\log(2*L)$, where K is the size of the information block to be encoded. When the SC-S algorithm is used, the decoder's complexity is also $O(L \bullet N \bullet \log(N))$, but it is SNR-dependent [34], e.g., decreases as SNR increases.

In [97] decoding complexities of several algorithms are compared, as Figure 3.52 shows. Here the length of the code block (e.g., N) is 1024, and the code rate is 1/2. It is seen that for polar code, the complexities of

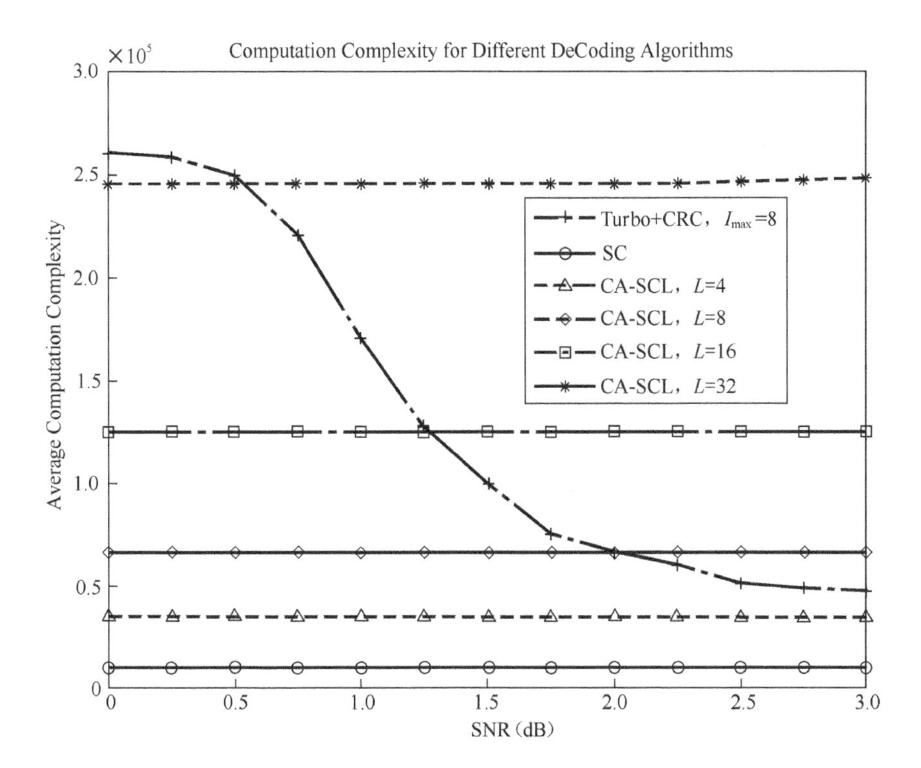

FIGURE 3.52 Decoding complexity comparison between polar code and turbo codes [97].

decoding algorithms being considered are independent of SNR. According to Section 3.13.2, when the list depth $L=8$, the performance of CA-SCL would be quite close to that of ML decoding. Hence, when estimating the complexity of CA-SCL, we usually assume $L=8$ as the baseline. When the SNR is less than 2 dB, the complexity of the SC-L algorithm of $L = 8$ is lower than that of the turbo codes.

Computation complexity is compared in [104] for short block length. It is shown that the complexity of PC-polar is much lower than that of RM and similar to that of Golay code. In [105], it is shown that at BLER$=10^{-4}$, the computation complexity of polar code is comparable to that of tail-biting convolutional codes.

3.14.2 Memory Complexity

For a polar code with a mother code of length N, there are $\log_2(N)$ layers. In each layer, N LLR values need to be stored. To find the path that corresponds to the originally transmitted bits, total $N^* \log_2(N)$ LLR values should be stored. Hence, for the SC algorithm, the memory complexity is $O(N \bullet \log(N))$ [34]. In [13], it was proposed to store the data of variable nodes in fixed nodes, so that the memory complexity can be reduced to $O(N)$ for SC decoding.

For the SC-L algorithm, since each list requires a set of posteriori probability values, the memory complexity is $O(L \bullet N \bullet \log(N))$ [34]. In [13] it was claimed that the memory complexity of the SC-L algorithm can be reduced to $O(L^*N)$.

The memory complexity of the SC-S algorithm is $O(D^*N)$ [34] where D is the depth of the stack. The memory complexity of the SC-H algorithm is roughly $O(L^*N^*N)$ or lower [34].

In Section 3.3.1 of [105], it is shown that with $L=8$, the memory complexity of polar is a little higher than that of tail-biting convolutional codes.

Overall speaking, the requirement for memory is relatively low for Polar codes.

3.14.3 Throughput

In [98], it is shown that by using the proposed method, the throughput of polar decoder can reach 4.4 Gbps, or even 6.4 Gbps. Based on the BP algorithm, a decoding throughput of 4.68 Gbps can be achieved [99] for the mother code length $N=1024$. The decoding method in [100] can achieve a throughput of 12 Gbps, with 100 MHz bandwidth, 256 QAM, eight antennas, and code rate$=0.95$.

Currently, polar code is used for physical control channel and NR-PBCH where the number of information bits is at most 500 or 1000 bits, whose throughput requirement is very low.

3.14.4 Decoding Latency

According to [98], the decoding latency for a code block of $N=1024$ and $R=1/2$ is 2ns * 56 = 112 ns. It was reported in [100] that the decoding latency for a code block of $N=512$ and $R=5/6$ is about 0.54 μs (e.g., 540 ns). It was shown in [101] that for a very large data block, polar code decoding can be finished in 16 μs. For an information bit size of 80 bits, the decoding can be completed in 1.7 μs [102]. For a 40-bit information block, it is shown in [103] that polar code decoding can be finished in 0.22 μs ($N=512$, $L=8$)

Decoding latency is compared in Section 4.3.1 of [105] where the latency of polar code ($L=8$) for 44 blind decodings is 0.47 μs, a little more than the decoding latency of tail-biting convolutional codes which is 0.22 μs.

In 5G NR, when 15 kHz subcarrier spacing is used, the OFDM symbol duration (including cyclic prefix) is 71 μs. When the subcarrier spacing is increased to 480 kHz, one OFDM symbol duration would be as short as 2.2 μs. So it is important that the decoding latency of polar can be kept small in order to support the self-contained subframe structure of 5G NR.

In summary, the complexity, throughput, and decoding latency of polar code are good enough to meet the requirements for 5G NR.

3.15 PERFORMANCE OF POLAR CODES

3.15.1 Minimum Hamming Distance

Hamming distance refers to the number of different bits between two code blocks. For instance, if the code block $A=\{1, 0, 0, 1\}$ and the code block $B=\{0, 1, 0, 1\}$, their Hamming distance is $d=2$. Minimum Hamming distance is the smallest distance of all possible codewords after encoding. Hamming distance represents the capability of a channel coding scheme to differentiate two codewords. In general, the greater the minimum Hamming distance is, the better the performance of the coding scheme. Table 3.3 provides the minimum Hamming distances of several channel coding schemes.

3.15.2 Block Error Rate

BLER of polar code can be computed [1,34]. As shown in Figure 3.53, the results of theoretical computations and simulations are quite close. According to [1][34], BLER can be computed in the following way. Denote

TABLE 3.3 Minimum Hamming Distance of Golay Code, RM Code, and Polar Code [104]

N	Type	K=3	K=4	K=5	K=6	K=7	K=8	K=9	K=10	K=11	K=12
20	Golay	10	10	8	8	8	8	7	6	5	4
	RM	8	8	8	8	6	6	6	6	4	N/A
	PC-Polar	11	9	8	8	6	6	6	6	4	4
24	Golay	12	12	10	10	9	8	8	8	8	8
	RM	10	9	9	9	7	7	6	6	4	N/A
	PC-Polar	8	8	8	8	8	8	8	6	4	4
32	Golay	17	16	16	13	12	11	11	10	9	9
	RM	16	16	16	16	12	12	12	12	10	N/A
	PC-Polar	18	16	8	16	12	12	12	12	8	N/A

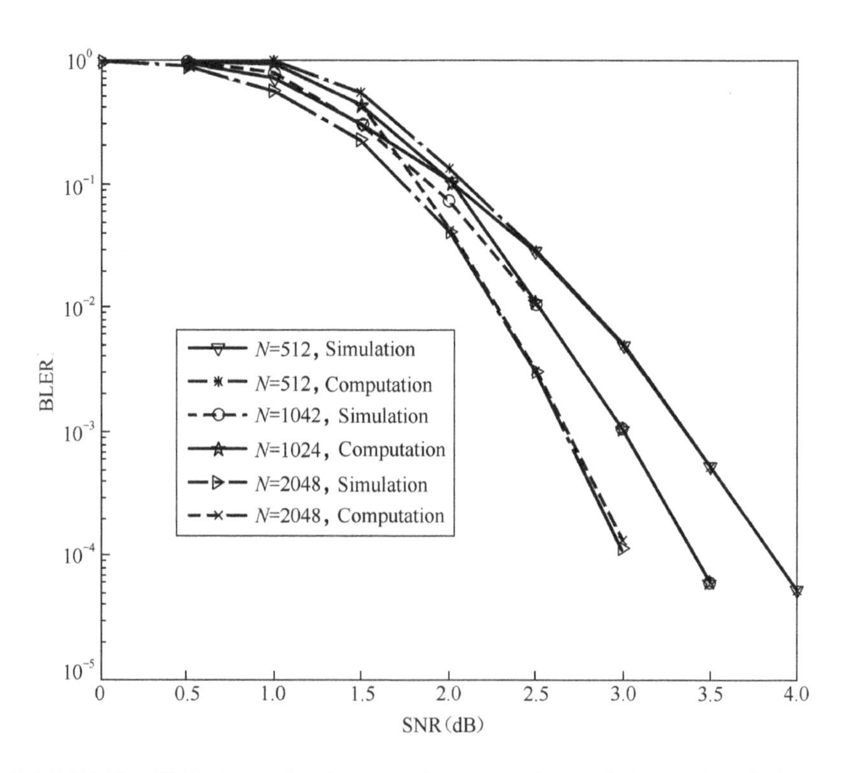

FIGURE 3.53 BLER from the theoretical computation and simulations [34].

the event of block error as $\varepsilon = \bigcup_{i \in A} B_i$ where A is the information set, i is the layer index during the decoding, B_i is the event of first-time erroneous decision which is defined as

$$B_i = \{(u_1^N, y_1^N) \in X^N \times Y^N : u_1^{i-1} = \hat{U}_1^{i-1}(u_1^N, y_1^N), u_i \neq \hat{U}_i(u_1^N, y_1^N)\} \quad (3.46)$$

It is noticed that $B_i = \{(u_1^N, y_1^N) \in X^N \times Y^N : u_1^{i-1} = \hat{U}_1^{i-1}(u_1^N, y_1^N), u_i \neq h_i(y_1^N, \hat{U}_1^{i-1}(u_1^N, y_1^N))\} \subset \varepsilon_i$ where $\varepsilon_i = \{(u_1^N, y_1^N) \in X^N \times Y^N : W_N^{(i-1)}(y_1^N, u_1^{i-1} | u_i) \leq W_N^{(i-1)}(y_1^N, u_1^{i-1} | u_i \oplus 1)\}$.

Then we get $\varepsilon \subset \bigcup_{i \in A} \varepsilon_i$ and $P(\varepsilon) \leq \sum_{i \in A} P(\varepsilon_i)$.

It is also noticed that $P(\varepsilon_i) = \sum_{u_1^N, y_1^N} \frac{1}{2^N} W_N(y_1^N | u_1^N) 1_{\varepsilon_i}(u_1^N, y_1^N) \leq \sum_{u_1^N, y_1^N} \frac{1}{2^N} W_N(y_1^N | u_1^N) \sqrt{\frac{W_N^{(i)}(y_1^N, u_1^{i-1} | u_i \oplus 1)}{W_N^{(i)}(y_1^N, u_1^{i-1} | u_i)}} = Z(W_N^{(i)})$

Then the upper bound of BLER is obtained:

$$P(\varepsilon) \leq \sum_{i \in A} Z(W_N^{(i)}) \tag{3.47}$$

where $Z(W_N^{(i)})$ is the Bhattacharyya parameter in Section 3.3.1.

3.15.3 False Alarm Rate

FAR is also an important parameter to evaluate the performance of a channel coding scheme. In the field of channel coding, false alarm refers to the situation when the receiver thinks that the decoding is successful while the decoded bits are different from the original information bits transmitted, or there is nothing transmitted to the receiver. In the case of PDCCH, multiple blind decoding is needed to detect whether any DCI has been transmitted to the receiver. It is possible that the DCIs not intended for this receiver are also detected, or there is no DCI transmitted from the base station. Such a situation can happen even if there are CRC bits. Hence, FAR is especially important for PDCCH.

In 4G LTE, the requirement for FAR is $2^{-\text{CRC\_Length}}$ where the CRC of PDCCH has 16 bits. Hence, the FAR requirement is about $2^{-16} \approx 1.5 \times 10^{-5}$. In 5G NR, the FAR requirement should be at least similar to that of LTE.

For the CA-polar code, the FAR can be calculated as FAR=$2^{\wedge(\log 2(L) - \text{CRC\_Length})}$ where L is the list depth. Hence, if 19-bit CRC is used and the list depth $L = 8$, the FAR is about 1.5×10^{-5}, as shown in Figure 3.54.

3.15.4 Performance Comparison with Other Codes

In the current 5G NR standards, polar code is mainly used for short code blocks. Hence, our focus in this section is the performance for extremely short and short block lengths.

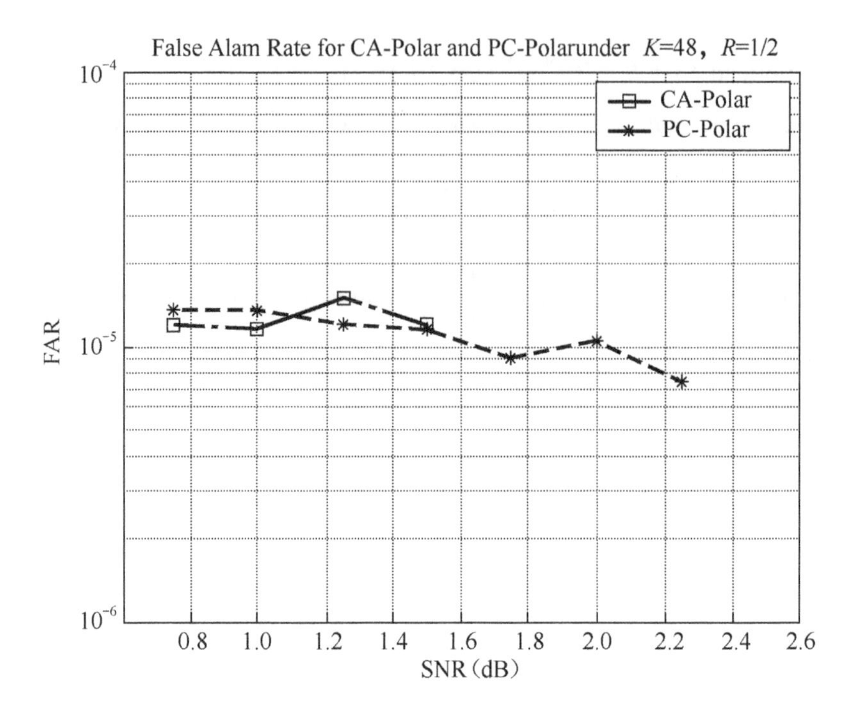

FIGURE 3.54 Simulation results of FAR [106].

3.15.4.1 *Extremely Short Block Length ($K \leq 12$ without CRC)*

As illustrated in Figure 3.55, the performance difference between polar code, RM code, and Golay code is quite small for extremely short block length. Since the minimum Hamming distance of polar code is relatively small, its performance at a high code rate (e.g., $K=12$, corresponding to code rate=0.6, 0.5, 0.375, respectively) is a little inferior to RM code and Golay code.

3.15.4.2 *Short Block Length ($12 \leq K < 200$)*

It is observed from Figure 3.56 that the performance of polar code is about 1 dB better than that of dual-RM code used by LTE. Polar code outperforms Golay code by about 0.6 dB. The reason is that for information block length equal to or larger than 13 bits, dual-RM code and dual-Golay code require segmentation of the original information block into two sub-blocks of similar length, each to be encoded by the (24, O) code. By contrast, no segmentation is required for polar code, e.g., (48, O) can be directly used.

It is seen in Figure 3.57 that polar code outperforms tail-biting convolutional code (TBCC) by about 0.6~1.5 dB.

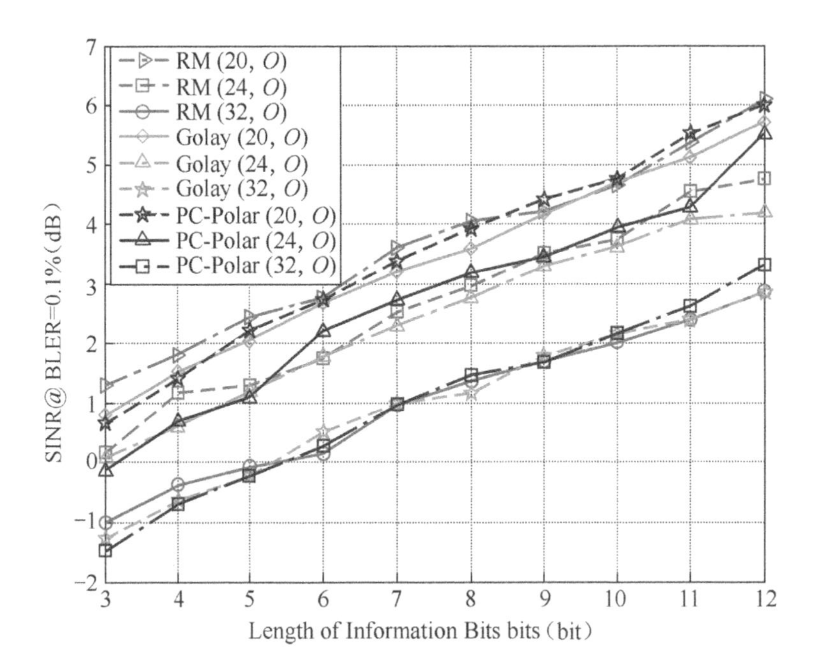

FIGURE 3.55 Performance comparison for polar code, RM code, and Golay code for extremely short block length [107].

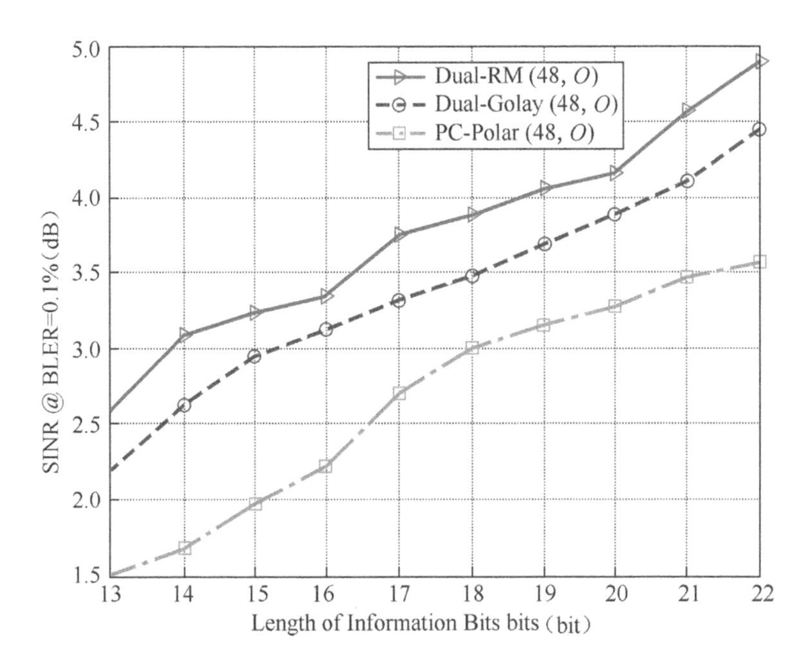

FIGURE 3.56 Performance comparison between polar code, dual-RM code, and dual-Golay code [107].

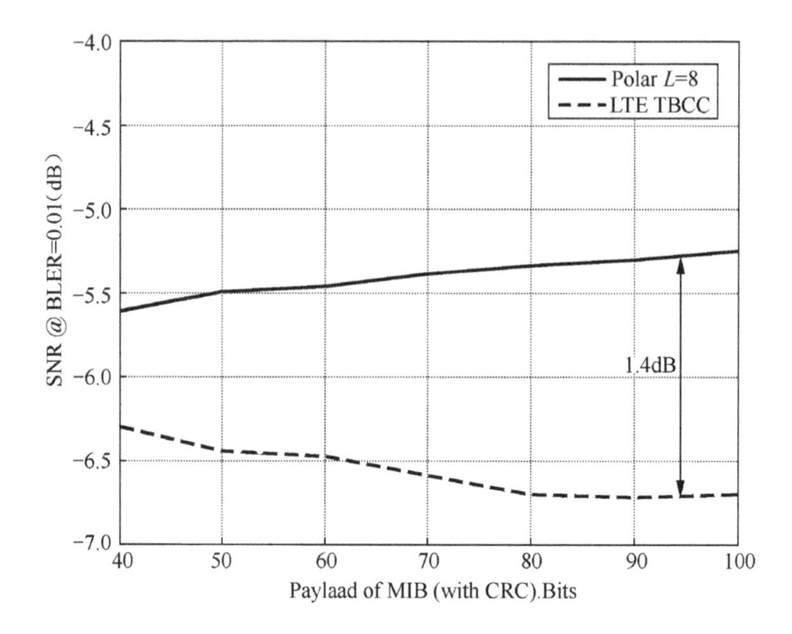

FIGURE 3.57 Performance comparison between polar code and TBCC [108].

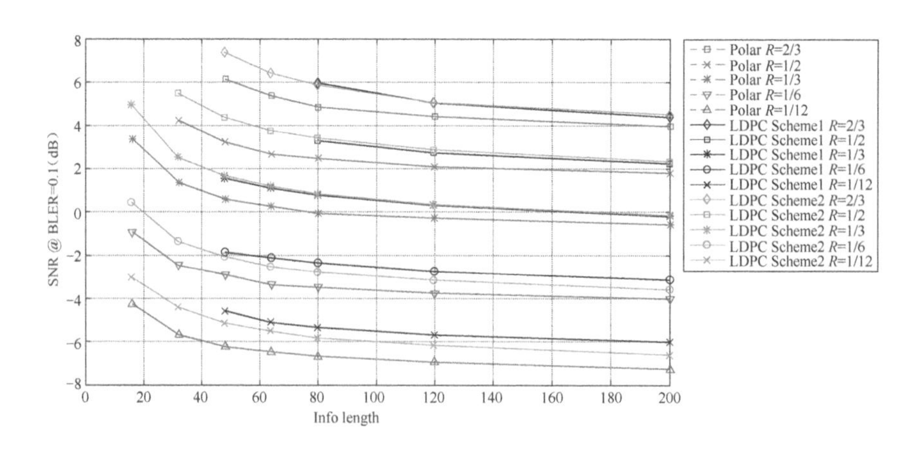

FIGURE 3.58 Performance comparison between polar code and LDPC for short block length [108].

It is seen in Figure 3.58 that under certain simulation settings [108], for a block length less than 200 bits, polar code outperforms LDPC by 0.2~0.5 dB.

3.15.4.3 Medium Block Length (200 ≤ K < 1000)

As seen in Figure 3.59, for $K=400$ bits and BLER = 1%, polar code outperforms turbo codes by about 0.6 dB.

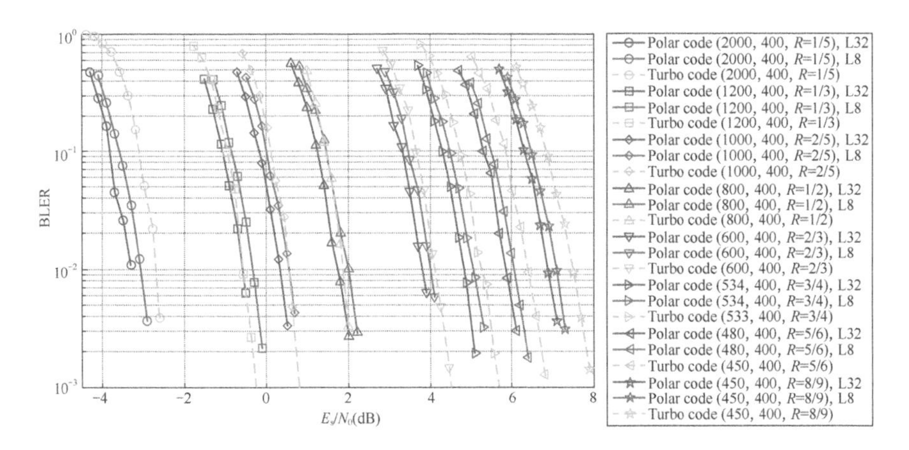

FIGURE 3.59 Performance comparison between polar code and turbo code for $K=400$ [109].

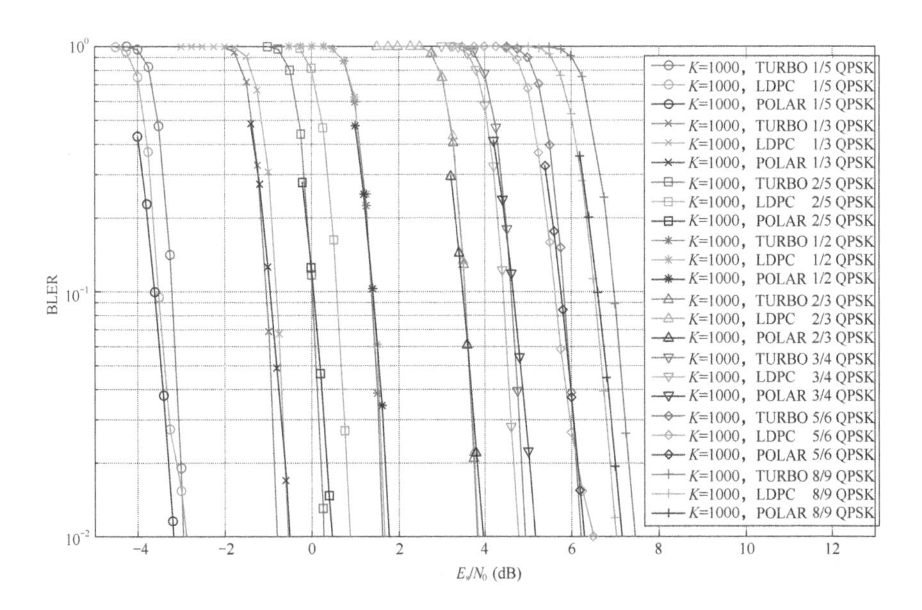

FIGURE 3.60 Performance comparison between turbo codes, polar code, and LDPC codes for $K=1000$ bits [110].

3.15.4.4 Long Block Length ($K \geq 1000$)

As seen in Figure 3.60, when $K=1000$ bits, QPSK modulation, AWGN channel, and BLER = 1%, the performances of turbo codes, polar code and LDPC code are quite comparable, with a difference less than 0.5 dB.

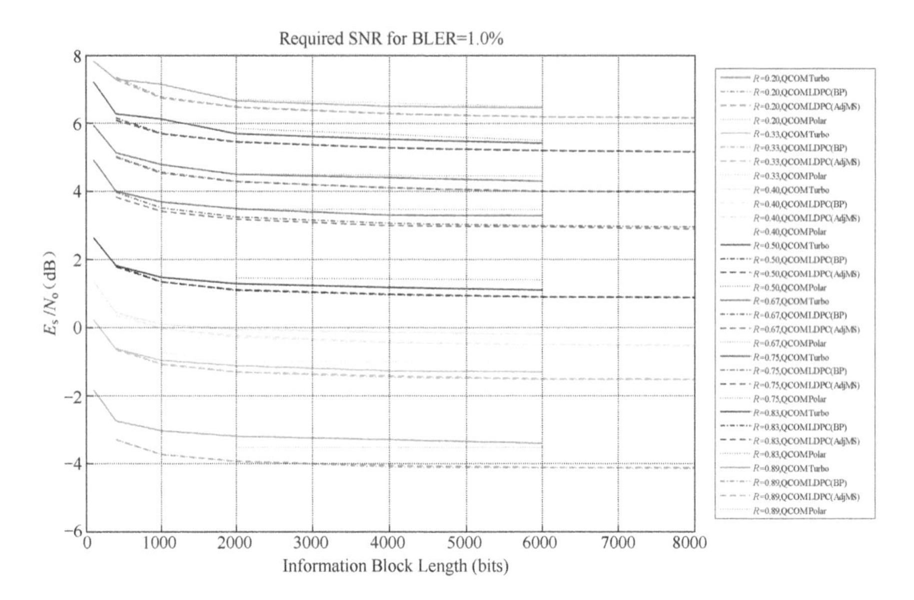

FIGURE 3.61 Performance comparison between polar code, turbo codes, and LDPC code for long block length [111].

As seen from Figure 3.61, when $K \geq 1000$ and R=1/5, polar code is inferior to LDPC code by about 0.6 dB but outperforms turbo codes by about 0.2 dB. Their difference is smaller for higher code rates.

More simulation results can be found in [111]. The overall trend is similar to the above-mentioned one. Due to the performance benefit of polar code for short or medium block length, as well as the reasonable complexity, throughput, and latency of its decoding, polar code was adopted by 3GPP as the channel coding scheme for 5G NR physical control channels (e.g., control information) and NR-PBCH [30,33].

3.16 POLAR CODE IN 3GPP SPECIFICATION

In this section, we describe how the various designs of polar code mentioned above are reflected in the 3GPP specification (Table 3.4).

TABLE 3.4 Polar Code Design Reflected in 3GPP Specification

3GPP Specification (TS38.2120)	Comment
6.3.1.2.1 UCI encoded by polar code If the payload size $A \geq 12$, code block segmentation and CRC attachment is performed according to Section 5.2.1. If $A \geq 360$ and $E \geq 1088$, $I_{seg} = 1$; … If $12 \leq A \leq 19$, the parity bits $p_{r0}, p_{r1}, p_{r2}, ..., p_{r(L-1)}$ in Section 5.2.1 are computed by setting L to 6 bits and … If $A \geq 20$, the parity bits $p_{r0}, p_{r1}, p_{r2}, ..., p_{r(L-1)}$ in Section 5.2.1 are computed by setting L to 11 bits and … **5.2 Code block segmentation and code block CRC attachment** **5.2.1 Polar coding** The input bit sequence to the code block segmentation is… if $I_{seg} = 1$ $C = 2$; … for $i = 0$ to $A' - A - 1$ $a'_i = 0$; end for … The sequence $c_{r0}, c_{r1}, c_{r2}, c_{r3}, ..., c_{r(A'/C-1)}$ is used to calculate the CRC parity bits $p_{r0}, p_{r1}, p_{r2}, ..., p_{r(L-1)}$ according to section 5.1 …	Segmentation of information block, to refer Section 3.10 Segmentation is applied only when UCI info block size A≥360 and codeword length≥1088 (after rate matching) 6-bit CRC for UCI length A=12~19; 11-bit CRC for UCI length A≥20 Up to two segments. If necessary, add a "0" to the front of 1st segment (as frozen bit). CRC appended to each segment

5.3 Channel coding
5.3.1 Polar coding
The bit sequence input for a given code block to channel coding is denoted by $c_0, c_1, c_2, c_3, ..., c_{K-1}$, where K is the number of bits to encode. After encoding the bits are denoted by $d_0, d_1, d_2, ..., d_{N-1}$, where $N = 2^n$ and the value of n is determined by the following:
Denote by E the rate matching output sequence length as given in Section 5.4.1;
If $E \leq (9/8) \cdot 2^{(\lceil \log_2 E \rceil - 1)}$ and $K/E < 9/16$
$n_1 = \lceil \log_2 E \rceil - 1$;
else
$n_1 = \lceil \log_2 E \rceil$;
end if
$R_{\min} = 1/8$;
$n_2 = \lceil \log_2 (K / R_{\min}) \rceil$;
$n = \max\{\min\{n_1, n_2, n_{\max}\}, n_{\min}\}$
where $n_{\min} = 5$.
	Polar code encoding, refer to Sections 3.5–3.8
	Mother code length is a power of 2
	Determination of mother code length, refer to Section 3.7
	If code rate $R=K/E$ is low and E is small, mother code length toward lower
	"9/8" is the β factor in rate matching. "9/16" is the code rate threshold for puncturing.
	$R_{\min}$ is a minimum code rate
	Max. mother code length is 1024 for uplink ($n_{\max}$=10) and 512 ($n_{\max}$ = 9) for downlink

(*Continued*)

TABLE 3.4 (CONTINUED)

<table>
<tr><th>3GPP Specification (TS38.2120)</th><th>Comment</th></tr>
<tr><td>

5.3.1.1 Interleaving

The bit sequence $c_0,c_1,c_2,c_3,...,c_{K-1}$ is interleaved into the bit sequence $c_0,c_1,c_2,c_3,...,c_{K-1}$ as follows:

$c'_k = c_{\Pi(k)}, k=0,1,...,K-1$

where the interleaving pattern $\Pi(k)$ is given by the following…

…where $\Pi_{IL}^{\max}(m)$ is given by Table 5.3.1-1 and $K_{IL}^{\max}=....$

Table 5.3.1-1: Interleaving pattern $\Pi_{IL}^{\max}(m)$

<table>
<tr><td>m</td><td>$\Pi_{IL}^{\max}(m)$</td><td>:</td><td>m</td><td>$\Pi_{IL}^{\max}(m)$</td></tr>
<tr><td>0</td><td>0</td><td>:</td><td>196</td><td>197</td></tr>
<tr><td>:</td><td>:</td><td>:</td><td>:</td><td>:</td></tr>
<tr><td>27</td><td>49</td><td>:</td><td>223</td><td>223</td></tr>
</table>

</td><td>

Pre-encoder interleaving

No pre-encoder interleaving for UCI ($I_{IL}=0$)

Pre-encoder interleaving for DCI and PBCH (distributed CRC, refer to Section 3.5.1.3)

$\Pi(k)$: pattern for pre-encoder interleaving

</td></tr>
<tr><td>

5.3.1.2 Polar encoding

The polar sequence $\mathbf{Q}_0^{N_{\max}-1}=\{Q_0^{N_{\max}},Q_1^{N_{\max}},...,Q_{N_{\max}-1}^{N_{\max}}\}$ is given in Table 5.3.1-2, where $0 \le Q_i^{N_{\max}} \le N_{\max}-1$ denotes a bit index before polar encoding for $i=0,1,...,N-1$ and $N_{\max}=1024$. The polar sequence $\mathbf{Q}_0^{N_{\max}-1}$ is in ascending order of reliability $W(Q_0^{N_{\max}})<W(Q_1^{N_{\max}})<...<W(Q_{N_{\max}-1}^{N_{\max}})$, where $W(Q_i^{N_{\max}})$ denotes the reliability of bit index $Q_i^{N_{\max}}$.

...

Denote $\mathbf{G}_N=(\mathbf{G}_2)^{\otimes n}$ as the n-th Kronecker power of matrix $\mathbf{G}_2$, where $\mathbf{G}_2=\begin{bmatrix} 1 & 0 \\ 1 & 1 \end{bmatrix}$.

…

Generate $\mathbf{u}=[u_0\ u_1\ u_2\ ...\ u_{N-1}]$ according to the following:

…

The output after encoding $\mathbf{d}=[d_0\ d_1\ d_2\ ...\ d_{N-1}]$ is obtained by $\mathbf{d}=\mathbf{u}\mathbf{G}_n$.

Table 5.3.1-2: Polar sequence $\mathbf{Q}_0^{N_{\max}-1}$ and its corresponding reliability $W(Q_i^{N_{\max}})$

<table>
<tr><td>$W(Q_i^{N_{\max}})$</td><td>$Q_i^{N_{\max}}$</td><td>:</td><td>$W(Q_i^{N_{\max}})$</td><td>$Q_i^{N_{\max}}$</td></tr>
<tr><td>0</td><td>0</td><td>:</td><td>896</td><td>966</td></tr>
<tr><td>:</td><td>:</td><td>:</td><td>:</td><td>:</td></tr>
<tr><td>127</td><td>274</td><td>:</td><td>1023</td><td>1023</td></tr>
</table>

</td><td>

Single nested polar code sequence (refer to Section 3.6)

Reliability $W(Q_i^{N_{\max}})$ represented in integer, the higher, the more reliable

To construct generation matrix G (refer to Section 3.5.2)

G2 is Arikan kernel

Encoding: x=u*G

Polar code sequence

$Q_i^{N_{\max}}$ is an index of sub-channel

</td></tr>
</table>

(*Continued*)

TABLE 3.4 (CONTINUED)

3GPP Specification (TS38.2120)	Comment
5.4.1 Rate matching for polar code The rate matching for polar code is defined per coded block and consists of sub-block interleaving, bit collection, and bit interleaving. … **5.4.1.1 Sub-block interleaving** The bits input to the sub-block interleaver are the coded bits $d_0, d_1, d_2, ..., d_{N-1}$. The coded bits $d_0, d_1, d_2, ..., d_{N-1}$ are divided into 32 sub-blocks. The bits output from the sub-block interleaver are denoted as $y_0, y_1, y_2, ..., y_{N-1}$, generated as follows: … where the sub-block interleaver pattern $P(i)$ is given in Table 5.4.1.1-1. **Table 5.4.1.1-1: Sub-block interleaver pattern $P(i)$**	Rate matching, refer to Section 3.7 Interleaving at the sub-block level Interleaving over 32 sub-blocks Interleaving pattern at the sub-block level
6.3.1.3.1 UCI encoded by polar code Information bits are delivered to the channel coding block. They are denoted by $c_{r0}, c_{r1}, c_{r2}, c_{r3}, ..., c_{r(K_r-1)}$, where r is the code block number, and K_r is the number of bits in code block number r. The total number of code blocks is denoted by C and each code block is individually encoded by the following: If $18 \le K_r \le 25$, the information bits are encoded via polar coding according to Section 5.3.1, by setting $n_{\max} = 10$, $I_{IL} = 0$, $n_{PC} = 3$, $n_{PC}^{wm} = 1$ if $E_r - K_r + 3 > 192$ and $n_{PC}^{wm} = 0$ if $E_r - K_r + 3 \le 192$, where E_r is the rate matching output sequence length as given in Section 6.3.1.4.1. If $K_r > 30$, the information bits are encoded via polar coding according to Section 5.3.1, by setting $n_{\max} = 10$, $I_{IL} = 0$, $n_{PC} = 0$, and $n_{PC}^{wm} = 0$. After encoding the bits are denoted by $d_{r0}, d_{r1}, d_{r2}, d_{r3}, ..., d_{r(N_r-1)}$, where N_r is the number of coded bits in code block number r. **6.3.1.4 Rate matching** **6.3.1.4.1 UCI encoded by polar code** … Rate matching is performed according to Section 5.4.1 by setting $I_{BIL} = 1$ and …	Polar code encoding for UCI 6-bit CRC appended to 12~19 bit UCI →18~25 bits Apply polar code encoding, 3-bit PC (use PC-CA-polar code). "$I_{IL} = 0$" means no pre-encoder interleaving (e.g., no distributed CRC) 11-bit CRC appended to 20-bit UCI → 31 bits or more. Use CA-polar code Max. mother code length=2^{10}=1024 ($n_{\max}$=10) "I_{BIL}=1" means post-encoder interleaving (triangular matrix) is applied.

(Continued)

TABLE 3.4 (CONTINUED)

3GPP Specification (TS38.2120)	Comment
7 Downlink transport channels and control information **7.1 Broadcast channel** **7.1.4 Channel coding** Information bits are delivered to the channel coding block. They are denoted by $c_0, c_1, c_2, c_3, \ldots, c_{K-1}$, where K is the number of bits, and they are encoded via polar coding according to Section 5.3.1, by setting $n_{\max} = 9$, $I_{IL} = 1$, $n_{PC} = 0$, and $n_{PC}^{wm} = 0$. **7.1.5 Rate matching** … Rate matching is performed according to Section 5.4.1 by setting $I_{BIL} = 0$.	PBCH uses distributed CRC ($I_{IL} = 1$. CA-Polar code. There is pre-interleaving G(j) in Section 7.1.1 (not shown here) Max mother code length $= 2^9 = 512$ ($n_{\max} = 9$) Use distributed CRC ($I_{IL} = 1$) CA-polar code. "$I_{BIL} = 0$" means no post-encoder interleaving (no triangular matrix). No PC bits.
7.3 Downlink control information **7.3.3 Channel coding** Information bits are delivered to the channel coding block. They are denoted by $c_0, c_1, c_2, c_3, \ldots, c_{K-1}$, where K is the number of bits, and they are encoded via polar coding according to Section 5.3.1, by setting $n_{\max} = 9$, $I_{IL} = 1$, $n_{PC} = 0$, and $n_{PC}^{wm} = 0$. **7.3.4 Rate matching** … Rate matching is performed according to Section 5.4.1 by setting $I_{BIL} = 0$. …	Polar code encoding for DCI Max. mother code length $= 2^9 = 512$ ($n_{\max} = 9$) Use distributed CRC ($I_{IL} = 1$) CA-polar code No post-encoder interleaving for DCI ($I_{BIL} = 0$)

3.17 MERITS, SHORTCOMINGS, AND FUTURE TRENDS OF POLAR CODES

The reason that polar code can be adopted by the highly selective 3GPP standards is due to its merits listed below [56]:

- So far, polar code is the only channel coding scheme that can reach Shannon's limit [34,38]. All other channel codes can only approach instead of reach Shannon's limit.
- Polar code has a solid information theory foundation.
- No error floor. At a high SNR, the BLER of polar code is upper-bound by $2^{-\sqrt{N}+O(\sqrt{N})} \approx 2^{-\sqrt{N}}$.
- Relatively low complexity of encoding and decoding.
- Very fine granularity of code rate adjustment (down to one-bit level).

- Recursive property of polar code can be used for implementation (polar code with mother code length N can be implemented by two polar codes each with mother code length $N/2$).

Certainly, polar code is not perfect. It has a few shortcomings [56] listed below:

- Hamming distance of polar code is relatively small, which may affect its performance, especially for short block length. However, this may be mitigated to some extent by selecting appropriate frozen bits
- SC decoding has a relatively long latency, which can be mitigated by parallel decoding [41–43,99,113]

In August 2016, some research institutes proposed turbo code 2.0 (with tail-biting, new puncturing pattern, new interleaver patterns, etc.) [114]. Future polar codes would also have version 2.0. The following may be potential directions:

- Polar code kernels [15,36,115] that are different from the Arikan kernel. High-order polar code kernels may expedite the speed of polarization but would also lead to multiple polarized values (no longer noiseless and pure noise channels) [36]
- Space-time polar code [34], to apply polar code to each codeword in MIMO
- Adding outer code (apart from CRC). For instance, RS code [116] to further improve system performance
- Low-complexity parallel decoding [41–43,99,113]
- Systematic polar code like Golay code [104], e.g., partial or entire systematic bits as part of codeword
- Multi-dimensional polar code [96], which requires extensive computer simulation to search for good code

REFERENCES

[1] E. Arikan, "Channel polarization: A method for constructing capacity achieving codes for symmetric binary-input memoryless channels," *IEEE Trans. Inf. Theory*, vol. 55, July 2009, pp. 3051–3073.

[2] US201161556862P. Methods AND Systems for decoding Polar codes. 2011.11.08.
[3] E. Arikan, "On the origin of polar coding," *IEEE J. Sel. Commun.*, vol. 34, no. 2, 2015, pp. 209–223.
[4] E. Arıkan, "Sequential decoding for multiple access channels," Tech. Rep. LIDS-TH-1517, Lab. Inf. Dec. Syst., M.I.T., 1985.
[5] E. Arıkan, "An upper bound on the cut-off rate of sequential decoding," *IEEE Trans. Inf. Theory*, vol. 34, Jan. 1988, pp. 55–63.
[6] E. Arıkan, "Channel combining and splitting for cut-off rate improvement," *IEEE Trans. Inf. Theory*, vol. 52, Feb. 2006, pp. 628–639.
[7] J. L. Massey, "Capacity, cut-off rate, and coding for a direct-detection optical channel," *IEEE Trans. Commun.*, vol. COM-29, no. 11, Nov. 1981, pp. 1615–1621.
[8] N. Hussami, "Performance of polar codes for channel and source coding," *Proceedings of the IEEE International Symposium on Information Theory*, Jul, 2009, pp. 1488–1492.
[9] R. Mori, "Performance of polar codes with the construction using density evolution," *IEEE Commun. Lett.*, vol. 13, no. 7, July 2009, pp. 519–521.
[10] A. Eslami, "On bit error rate performance of polar codes in finite regime," *Communication, Control, and Computing (Allerton), 2010 48th Annual Allerton Conference*, 2011, pp. 188–194.
[11] E. Şaşoğlu, "Polar codes for the two-user binary-input multiple-access channel," *2010 IEEE Information Theory Workshop on Information Theory (ITW 2010, Cairo)*, 2010, pp. 1–5.
[12] H. Mahdavifa and A. Vardy, "Achieving the secrecy capacity of wiretap channels using Polar codes," *Information Theory Proceedings (ISIT)*, 2010, pp. 913–917.
[13] I. Tal and A. Vardy, "List decoding of polar codes," *Information Theory Proceedings (ISIT), 2011 IEEE International Symposium on Information Theory*, 2011, pp. 1–5.
[14] C. Leroux, I. Tal, and A. Vardy, "Hardware architectures for successive cancellation decoding of polar codes," *Acoustics, Speech and Signal Processing (ICASSP)*, 2011, pp. 1665–1668.
[15] V. Miloslavskaya, "Design of binary polar codes with arbitrary kernel," *2012 IEEE Information Theory Workshop*, 2012, pp. 119–123.
[16] H. Si, "Polar coding for fading channels," *2013 IEEE Information Theory Workshop (ITW)*, 2013, Sept 2013, pp. 1–5.
[17] P. Giard, "Fast software polar decoders," *2014 IEEE International Conference on Acoustics, Speech and Signal Processing (ICASSP)*, 2014, pp. 7555–7559.
[18] V. Miloslavskaya, "Shortened polar codes," *IEEE Trans. Inf. Theory*, vol. 61, no. 9, Sept. 2015, pp. 4852–4865.
[19] 3GPP, RP-160671, New SID Proposal: Study on New Radio Access Technology, NTT DOCOMO, RAN#71, March, 2016.
[20] 3GPP, R1-1610659, Evaluation on Channel coding candidates for eMBB control channel, ZTE, RAN1 #86b, October, 2016.

[21] 3GPP, R1-1608863, Evaluation of channel coding schemes for control channel, Huawei, RAN1#86b, October, 2016.
[22] 3GPP, R1-1610419, UE Considerations on Coding Combination for NR Data Channels, MediaTek, RAN1#86b, October, 2016.
[23] 3GPP, R1-1613078, Evaluation on channel coding candidates for eMBB control channel, ZTE, RAN1#87, November, 2016.
[24] 3GPP, R1-1611114, Selection of eMBB Coding Scheme for Short Block length, ZTE, RAN1#87, November, 2016.
[25] 3GPP, R1-1611256, Performance evaluation of channel codes for small block sizes, Huawei, RAN1#87, November, 2016.
[26] 3GPP, R1-1613343, Comparison of coding candidates for eMBB data channel of short codeblock length, MediaTek, RAN1#87, November, 2016.
[27] 3GPP, R1-1608867, Considerations on performance and spectral efficiency Huawei, RAN1#86b, October, 2016.
[28] 3GPP, R1-1611259, Design Aspects of Polar and LDPC codes for NR, Huawei, RAN1#87, November, 2016.
[29] 3GPP, R1-1613061, Comparison of coding candidates for DL control channels and extended applications, MediaTek, RAN1#87, November, 2016.
[30] 3GPP, Final_Minutes_report_RAN1#87, Nov. 2016. http://www.3gpp.org/ftp/tsg_ran/WG1_RL1/TSGR1_87/Report/Final_Minutes_report_RAN1%2387_v100.zip.
[31] 3GPP, R1-1719520, Remaining details of Polar coding, ZTE, RAN1#91, November, 2017.
[32] 3GPP, Final Report of 3GPP TSG RAN WG1 #AH1_NR, January 2017. http://www.3gpp.org/ftp/tsg_ran/WG1_RL1/TSGR1_AH/NR_AH_1701/Report/Final_Minutes_report_RAN1%23AH1_NR_v100.zip.
[33] 3GPP, Final_Minutes_report_RAN1#89_v100, Aug 2017. http://www.3gpp.org/ftp/tsg_ran/WG1_RL1/TSGR1_89/Report/Final_Minutes_report_RAN1%2389_v100.zip
[34] K. Chen, *Theoretical study and practical solutions for Polar codes*, Ph. D thesis, Beijing University of Post & Telecommunications, March 2014.
[35] K. Chen, K. Niu, J Lin, "A reduced-complexity successive cancellation list decoding of Polar codes," *IEEE Vehicular Technology Conference (VTC Spring)*, 2013, pp. 1–5.
[36] D. Wu, *Construction of Polar code and decoding algorithm study*, Ph. D thesis, Xidian University, April 2016.
[37] D. Wu, Y. Li. "Construction and block error rate analysis of polar codes over AWGN channel based on Gaussian approximation. *IEEE Commun. Lett.*, 2014, 18 (7), pp. 1099–1102.
[38] J. Wang, *Encoding and decoding algorithm study for Polar codes*, Master Thesis, Harbin Institute of Technology, June 2013.
[39] T. Lu, *Study and simulation of Polar encoding and decoding*, Master Thesis, Nanjing Science & Technology University, March 2013.
[40] G. Chen, *Encoding and decoding for Polar codes*, Master Thesis, Nanjing Science & Technology University, Feb. 2014.

[41] E. Arikan, "Polar codes: A pipelined implementation," *Presented at "4th International Symposium on Broadband Communication (ISBC 2010)" July 11–14*, 2010, Melaka, Malaysia.
[42] S.M. Abbas, Y. Fan, J. Chen and C.Y. Tsui, "Low complexity belief propagation polar code decoder," *2015 IEEE Workshop on Signal Processing Systems (SiPS)*, 2015, pp. 1–6.
[43] B. Li, "Parallel Decoders of Polar Codes," Computer Science, 4 Sep, 2013.
[44] E. Arikan, *Polar coding tutorial*, Simons Institute UC Berkeley, Berkeley, California, USA, Jan. 15, 2015.
[45] L. Zhang, *Study on decoding algorithms for Polar code and their applications*, Ph. D Thesis, Zhejiang University, April 2016.
[46] 3GPP, R1-1709178, FRANK polar construction for NR control channel and performance comparison, Qualcomm, RAN1#89, May, 2017.
[47] E. Arikan, "On the rate of channel polarization," *IEEE International Conference on Symposium on Information Theory*, 2009, pp. 1493–1495.
[48] 3GPP, Draft_Minutes_report_RAN1#90_v010. Prague, Czech Rep, August 2017. http://www.3gpp.org/ftp/tsg_ran/WG1_RL1/TSGR1_90/Report/Draft_Minutes_report_RAN1%2390_v010.zip.
[49] 3GPP, Draft_Minutes_report_RAN1#91_v020, 2017.12.08, http://www.3gpp.org/ftp/tsg_ran/WG1_RL1/TSGR1_91/Report/Draft_Minutes_report_RAN1%2391_v020.zip.
[50] 3GPP, R1-1704248, On channel coding for very small control block lengths, Huawei, WG1#88bis, April 2017.
[51] 3GPP, R1-1611254, Details of the Polar code design, Huawei, RAN1#87, November, 2016.
[52] 3GPP, R1-1700088, Summary of polar code design for control channels, Huawei, RAN1 Ad-Hoc Meeting, USA, 16th -20th January 2017.
[53] 3GPP, R1-1714377, Distributed CRC Polar code construction, Nokia, RAN1#90, August 2017.
[54] 3GPP, R1-1712167, Distributed CRC for Polar code construction, Huawei, RAN1#90, August 2017.
[55] 3GPP, R1-1700242, Polar codes design for eMBB control channel, CATT, RAN1#AH_NR, January 2017.
[56] E. Arikan, "Challenges and some new directions in channel coding," *J. Commun. Netw.*, vol. 17, no. 4, 2015, pp. 328–338.
[57] 3GPP, R1-1713707, PBCH coding design for reduced measurement complexity, MediaTek, RAN1#90, August 2017.
[58] 3GPP, R1-1716223, NR PBCH coding design, MediaTek, RAN1#Ad-Hoc#3, September 2017.
[59] 3GPP, R1-1713705, Polar rate-matching design and performance, MediaTek, WG1#90, August 2017.
[60] CN 201710036000.0, Beihang University, A method of construction of Polar code with variable length 2017.01.17.
[61] 3GPP, Final_Minutes_report_RAN1#AH1_NR_v100. http://www.3gpp.org/ftp/tsg_ran/WG1_RL1/TSGR1_AH/NR_AH_1701/Report/Final_Minutes_report_RAN1%23AH1_NR_v100.zip.

[62] R. Mori, and T. Tanaka, "Performance of polar codes with the construction using density evolution", *IEEE Comm. Lett.*, July 2009, vol. 13, no. 7, pp. 519–521.
[63] I. Tal, and A. Vardy, "How to construct polar codes," *IEEE Trans. Info. Theory*, vol. 59, no. 10, Oct. 2013, pp. 6562–6582.
[64] P. Trifonov, "Efficient design and decoding of polar codes," *IEEE Trans. Comm.*, vol. 60, no. 11, Nov 2012, pp. 3221–3227.
[65] 3GPP, R1-1611254, Details of the Polar code design, Huawei, RAN1#87, November, 2016.
[66] S. ten Brink, "Design of low-density parity-check codes for modulation and detection," *IEEE Trans. Commun.*, vol.52, no.4, April 2004, pp. 670–678.
[67] 3GPP, R1-1711218, Sequence construction of Polar codes for control channel, Qualcomm, RAN1#NR Ad-Hoc#2, June 2017.
[68] 3GPP, R1-1710749, Design of combined-and-nested polar code sequences, Samsung, RAN1#NR Ad-Hoc#2, June 2017.
[69] 3GPP, R1-1708051, Design of a Nested polar code sequences, Samsung, RAN1#89, May 2017.
[70] 3GPP, R1-1705425, Design of a Nested Sequence for Polar Codes, Samsung, RAN1#88bis, April 2017.
[71] CN 201710660483.1, ZTE, A method and apparatus of sequence generation, data decoding. 2017.08.04.
[72] 3GPP, R1-1713234, Performance evaluation of sequence design for Polar codes, ZTE, RAN1#90, August, 2017.
[73] 3GPP, R1-1712168, Sequence for Polar code, Huawei, RAN1#90, August 2017.
[74] 3GPP, R1-1712174, Summary of email discussion [NRAH2-11] Polar code sequence, Huawei, RAN1#90, August 2017.
[75] 3GPP, Draft_Minutes_report_RAN1#AH_NR2_v010. http://www.3gpp.org/ftp/tsg_ran/WG1_RL1/TSGR1_AH/NR_AH_1706/Report/Draft_Minutes_report_RAN1%23AH_NR2_v010.zip.
[76] 3GPP, R1-1714793, Information sequence design for Polar codes, Ericsson, RAN1#90, August 2017.
[77] 3GPP, R1-1705084, Theoretical analysis of the sequence generation, Huawei, RAN1#88bis, April 2017.
[78] 3GPP, R1-1715000, Way Forward on rate-matching for Polar Code, MediaTek, RAN1#90, August 2017.
[79] 3GPP, R1-1708649, Interleaver design for Polar codes, Qualcomm, RAN1#89, May, 2017.
[80] 3GPP, R1-1714691, Channel interleaver for Polar codes, Ericsson, RAN1#90, August 2017.
[81] 3GPP, R1-167871, Examination of NR coding candidates for low-rate applications, MediaTek, RAN1 #86, August, 2016.
[82] 3GPP, R1-167871, Polar code design and rate matching, Huawei, RAN1 #86, August, 2016.
[83] 3GPP, R1-1704385, Rate matching of polar codes for eMBB, ZTE, RAN1#88b, April, 2017.

[84] 3GPP, R1-1704317, Rate Matching Schemes for Polar Codes, Ericsson, RAN1#88b, April, 2017.
[85] 3GPP, R1-1707183, Polar codes construction and rate matching scheme, ZTE, RAN1 #89, May, 2017.
[86] 3GPP, R1-1710750, Design of unified rate-matching for polar codes, Samsung, RAN1 NR Ad Hoc #2, June, 2017.
[87] 3GPP, R1-1711702, Rate matching for polar codes, Huawei, RAN1 NR Ad Hoc #2, June, 2017.
[88] 3GPP, R1-1714939, Rate matching scheme for Polar codes, ZTE, RAN1#90, August, 2017.
[89] 3GPP, R1-1714381, Implicit timing indication for PBCH, Nokia, RAN1#90, August, 2017.
[90] 3GPP, R1-1710002, Support of implicit soft combining for PBCH by Polar code construction, Huawei, RAN1# NR Ad-Hoc#2, June, 2017.
[91] 3GPP, Draft_Minutes_report_RAN1#90b, 9th–13th October, 2017.
[92] 3GPP, R1-1704384, Further Consideration on Polar codes with maximum mother code, ZTE, RAN1 #88b April, 2017.
[93] 3GPP, R1-1713237, Segmentation of Polar codes for large UCI, ZTE, RAN1#90, August 2017.
[94] E. Arikan, "Systematic polar coding," *IEEE Comm. Lett.*, vol. 15, no. 8, Aug. 2011, pp. 860–862.
[95] S. Zhao, "Puncturing of systematic polar code based on decoding reliability," *J. South-East Univ. (Nat. Sci.)*, vol. 47, no. 1, Jan. 2017, pp. 23–27.
[96] E. Arikan, "Two-dimensional polar coding," *Tenth International Symposium on Coding Theory and Applications (ISCTA'09)*, July 13–17, 2009, Ambleside, UK.
[97] K. Niu, K. Chen, "CRC-aided decoding of polar codes," *IEEE Comm. Lett.*, vol. 16, no. 10, Oct. 2012, pp. 1668–1671.
[98] B. Yuan, "Algorithm and VLSI architecture for Polar codes decoder," A Dissertation submitted to the faculty of the graduate school of the University of Minnesota, Dissertations & Theses - Gradworks, 2015.
[99] Y. S. Park, "A 4.68Gb/s belief propagation polar decoder with bit-splitting register file," *2014 Symposium on VLSI Circuits Digest of Technical Papers*, June 2014, pp. 1–2.
[100] P. Giard, "A Multi-Gbps unrolled hardware list decoder for a systematic Polar code," *Conference on Signals, Systems & Computers*, 2017, pp. 1194–1198.
[101] 3GPP, R1-1608865, Design aspects of Polar code and LDPC for NR, Huawei, RAN1#86bis, Oct, 2016.
[102] 3GPP, R1-1611081, Final Report of 3GPP TSG RAN WG1 #86bis v1.0.0, RAN1#87, November 2016.
[103] 3GPP. R1-1718505, Channel Coding for URLLC, Tsofun, RAN1#90bis, 9th–13th, October 2017.
[104] 3GPP, R1-1705636, Evaluation of the coding schemes for very small block length, Qualcomm, RAN1#88b, April 2017.

[105] 3GPP, R1-1613086, Control Channel Complexity Considerations, Qualcomm, RAN1#87, 14th–18th November 2016.
[106] 3GPP, R1-1704772, Design Aspects of Polar code, Intel, RAN1#86bis, October 2016.
[107] 3GPP, R1-1704386, Considerations on channel coding for very small block length, ZTE, RAN1#88b, April 2017.
[108] 3GPP, R1-1704249, Channel coding for PBCH, Huawei, RAN1#88b, April 2017.
[109] 3GPP, R1-1610060, Evaluation of polar codes for eMBB, NTT, RAN1#86bis, October 2016.
[110] 3GPP, R1-1609583, Update on eMBB coding performance, Nokia, RAN1#86bis, October 2016.
[111] 3GPP, R1-1610423, Summary of channel coding simulation data sharing, InterDigital, RAN1#86bis, October 2016.
[112] 3GPP, TS38.212, NR Multiplexing and channel coding (Release 15), http://www.3gpp.org/ftp/Specs/archive/38_series/38.212/.
[113] C. Leroux, "A semi-parallel successive-cancellation decoder for polar codes," *IEEE Trans. Signal Process.*, vol. 61, no. 2, 2013, pp. 289–299.
[114] 3GPP, R1-167413, Enhanced Turbo codes for NR: Implementation details, Orange and Institut Mines-Telecom, RAN1#86, August 2016.
[115] V. Miloslavskaya, "Performance of binary polar codes with high-dimensional kernel," *Proceedings of International Workshop on Algebraic and Combinatorial Coding Theory*, 2012, pp. 263–268.
[116] Y. Wang, "An improved concatenation scheme of polar codes with reed–solomon codes," *IEEE Commun. Lett.*, vol. 21, no. 3, 2017, pp. 468–471.

CHAPTER 4

Convolutional Codes

Jin Xu and Yifei Yuan

Convolutional codes are a type of channel coding scheme with a long history. It is "old" in the sense that since its birth in 1955, convolutional codes have been widely used in mobile communications, satellite communications, deep-space communications, etc. It is "modern" because convolutional codes recently found their application in the physical traffic channel of narrowband IoT (NB-IoT) standards. In the following, we will show how convolutional codes work, where they are used and what new techniques are.

4.1 BASICS OF CONVOLUTIONAL CODES

4.1.1 Principle of Convolutional Codes and Decoding Algorithms

The convolutional code was first proposed by Prof. Peter Elias of the MIT in 1955 [1], as an alternative to block codes. As implied by its name, convolutional code has many similarities with convolution filters in discrete-time signal processing, except that convolutional codes are normally defined in finite fields, and most often in the binary field, e.g., GF(2). This means that the input, the output, and the tap coefficients of convolutional codes are mostly binary, as illustrated in Figure 4.1.

The redundant bits in Figure 4.1 are generated using the polynomials $[1+D^2, 1+D+D^2]$, where D is the delay operator, similar to Z^{-1}. By using the delay operator, the time-domain convolution can be converted into multiplications in transform domain D or represented by Eq. (4.1):

$$H(D)=\frac{1+D^2}{1+D+D^2} \tag{4.1}$$

DOI: 10.1201/9781003336174-4

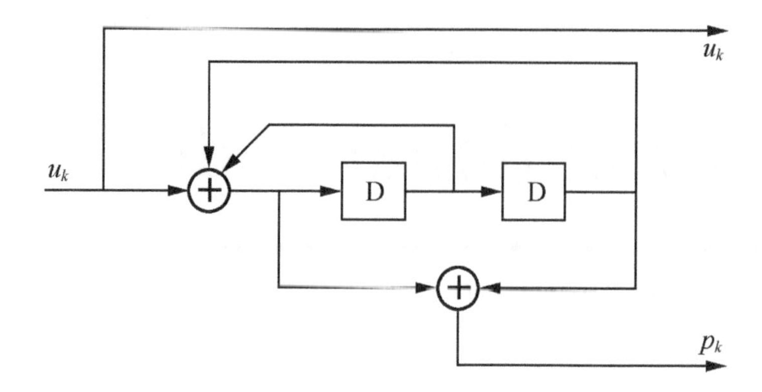

FIGURE 4.1 Illustration of convolutional encoder.

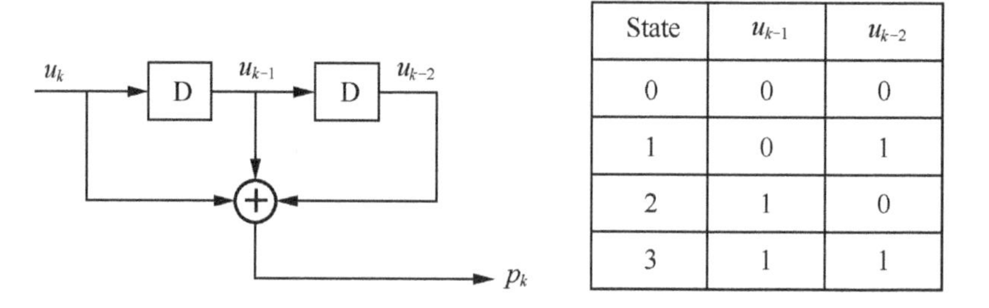

State	u_{k-1}	u_{k-2}
0	0	0
1	0	1
2	1	0
3	1	1

FIGURE 4.2 Illustration of non-recursive convolutional encoder.

There are two types of convolutional codes: non-recursive and recursive. The example in Figure 4.1 is recursive. A recursive convolutional code resembles an infinite impulse response (IIR) filter with a feedback loop. A simple form of recursive convolution is the accumulator which has been widely used. Before the advent of turbo codes [2], non-recursive convolutional codes were more popular and once set the record in approaching Shannon's limit. However, during the study on turbo codes, people began to realize the importance of recursive convolutional codes in the course of decoding iterations.

A non-recursive convolutional code resembles a finite impulse response (FIR) filter without a feedback loop. As the example shown in Figure 4.2, the non-recursive convolutional code has a constraint length of 3 and memory lengths of 2 and 4 states. Generation formula of non-recursive convolutional codes can be written in simple binary forms. For the example in Figure 4.2, the binary form of the generator is $g = (111)_2$. When the

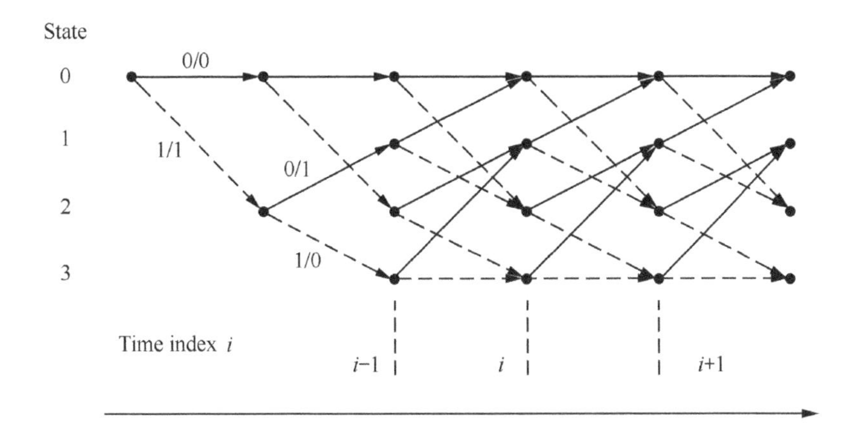

FIGURE 4.3 A trellis for the four-state convolutional encoder.

constraint length is relatively long, the octal form is more often used. For instance, $(110100101)_2$ can be succinctly written as $g=(645)_8$.

The performance of a convolutional code is largely determined by its free distance d_{free} and the structure type (systematic, non-systematic, recursive, non-recursive). The free distance can be calculated from the generator polynomials (note: exact calculation of d_{free} is difficult [3]). For instance, for a convolutional code (n, k) that has δ number of delay unit D, its free distance can be estimated as [4]

$$d_{free} \leq (n-k) \bullet (1 + floor(\delta / k)) + \delta + 1 \tag{4.2}$$

After the free distance is estimated, computer simulations can be carried out to find generator polynomials whose free distance is close to the estimated value [5]. The determination of generator polynomials is also discussed in [6].

Convolutional codes can also be represented by a trellis, as illustrated in Figure 4.3, which corresponds to the generator polynomials in Figure 4.2. In this trellis, each branch represents that the convolutional encoder transits from one state to another state (maybe the same state). i represents the current instant. If the input bit is 0, the branch would be a solid line. If the input bit is 1, the branch would be a dashed line. The notation "a/b" beside the branch represents input bit "a" and output bit, "b", respectively.

The optimal decoding algorithm for convolutional codes is the maximum likelihood (ML)-based Viterbi algorithm [3,7] which was first proposed by Andrew Viterbi in 1967. For convolutional codes, the criterion of

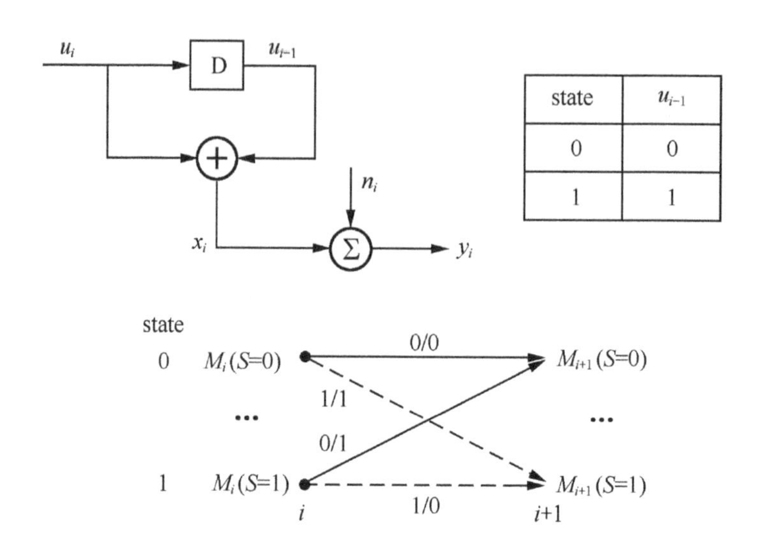

FIGURE 4.4 A two-state convolutional code and its trellis.

ML is reflected in the most probable path in the trellis. If to search blindly, even for binary convolutional codes, we need to exhaustively search over 2^N times (N is the length of the encoded sequence). When the sequence length is long, such an exhaustive search is definitely not feasible. The Viterbi algorithm is a type of dynamic programming method whose computation complexity is proportional to the number of states, rather than the sequence length. In the following, we describe the basic operation of the Viterbi algorithm. For simplicity without losing the generality, a two-state non-recursive convolutional code is used here as an example, as illustrated in Figure 4.4. Only the redundancy bit x_i is considered. The nose is represented as n_i, which is modeled as additive white Gaussian noise (AWGN) with zero mean and a variance of σ^2. Let us specifically look at the transition from the time instant i to (i+1). Here, there are four branches, each representing the input/output bit (u_i/x_i) of the convolutional encoder at the time instant i. The metric $M_i(s)$ contains the accumulated information from observations y_0 to y_i. The metric $M_{i+1}(s)$ represents the update of $M_i(s)$, based on the new observation y_{i+1}. The updating formula is

$$
\begin{aligned}
M_{i+1}(S=0) &= \min[M_i(S=0)+(y_i-1)/(2\delta^2)] \\
M_{i+1}(S=1) &= \min[M_i(S=0)+(y_i-0)/(2\delta^2)]
\end{aligned}
\tag{4.3}
$$

This updating processing is usually called "pruning" because when two branches point to the same node, e.g., the state of the convolutional encoder

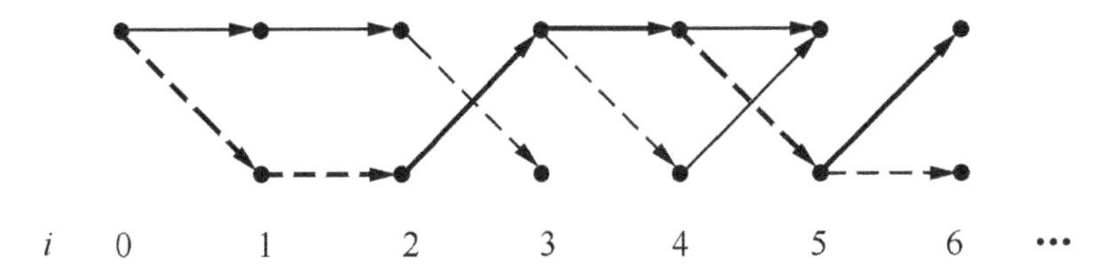

FIGURE 4.5 Survival path in a trellis of a two-state convolutional code.

at $(i+1)$ time instant, they would update the metric differently. Among these two updates, only the branch resulting in a smaller metric would be kept, while the other branch would be pruned. Discarded branches are also called competing branches. All the surviving branches form a path. As illustrated in Figure 4.5, the thick bold lines represent the survival path starting from the initial state of 0 of the convolutional encoder and going through a number of Viterbi decoding. The survival path is the optimal path in the sense of ML.

In theory, only when the entire sequence has been processed can the optimal survival path be obtained. In practical decoding, due to the limited memory size, the path tracing can only be done within a limited size of window. Generally speaking, the searching window length should be at least 4~5 times the constraint length of the convolutional code; otherwise, the decoding may not converge, e.g., unable to achieve the purpose of ML decoding [3]. In summary, the Viterbi algorithm basically has the following four steps:

1. To initialize all the metrics $M_0(s)$, except for tail-biting convolutional codes or other very specific convolutional codes. Normally, the initial states are all set to 0, e.g., $M_0(S=0)=1$, and $M(S \neq 0)=0$
2. Using Eq. (4.3) to update the metrics which involve "sum" and "comparison" operation
3. To slide the searching window and store the survival path, as seen in the thick bold lines in Figure 4.5
4. To trace back along the survival path to the start of the searching window and determine the input bit at this time instant

Complexity of the Viterbi algorithm increases exponentially with the constraint length of the convolutional codes. For classical convolutional codes,

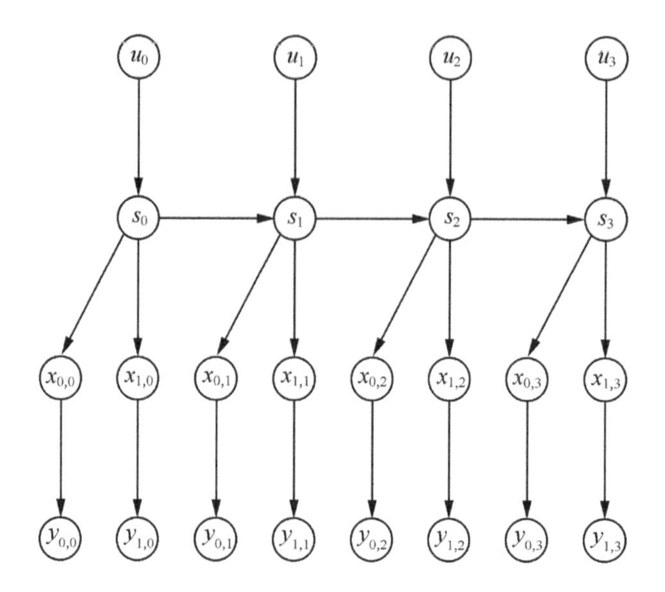

FIGURE 4.6 Illustration of factor graph for convolutional codes.

the constraint lengths are typically between 7 and 9. While convolutional codes have fairly good performance, the complexity is still an issue for long constraint lengths. There are some sub-optimal decoding algorithms such as the Fano algorithm, whose complexity is lower. After turbo codes were invented, the focus of convolutional codes has been shifted away from finding long constraint lengths to approach Shannon's limit. New study directions include list decoding [8,9], iteration-based sum-product algorithm (SPA) [10], and a new construction like (u|u+v) [11].

Similar to the principle of message passing discussed in Chapter 2 for low-density parity check (LDPC), the decoding operation of convolutional codes can also be described by a factor graph, as illustrated in Figure 4.6.

Its differences from the factor graph of LDPC are as follows:

- The factor node of convolutional codes reflects the state of a convolutional encoder (finite machine), rather than a value. Convolutional encoder has memory, and its state is related to many adjacent variables. The relationships are quite complex and so far there is no closed-form representation
- Factor graph of convolutional codes has a very definite direction for message passing, e.g., mono-direction from the beginning state to the end state.

4.1.2 Basic Performance

Compared to many block codes, convolutional codes have performance benefits. However, its performance analysis is in general quite difficult. Instead, performance bounds are often used. For the AWGN channel, if the Viterbi algorithm is used, the upper bound (or union bound) of a block error rate of convolutional codes can be represented as Eq. (4.4) [3]:

$$P(E) \le \left(\frac{L}{2}\right) \sum_{d=d_f}^{\infty} a(d) z^d \tag{4.4}$$

where L is the length of the code block, d_f is the free distance of the convolutional code. $a(d)$ is the number of paths whose distance to the all-zero sequence is d. $z(.)$ is a function of signal-to-noise ratio (SNR) γ and can be approximated as follows:

$$z(\gamma) = e^{-\gamma} \tag{4.5}$$

Taking the Universal Mobile Telecommunication System (UMTS) standards as an example, its convolutional code has the mother code rate of 1/3. The generator polynomials can be written in octal form as $G_0=557$, $G_1=663$, and $G_2=711$. It is not difficult to calculate the weight distribution of the codeword sequence, represented as the transfer function like

$$T(D) = 5D^{18} + 7D^{20} + 36D^{22} + \ldots \tag{4.6}$$

Equation (4.6) shows that in the trellis of UMTS convolutional codes, there are five paths whose Hamming distance is 18. There are seven paths whose Hamming distance is 20. There are 36 paths whose Hamming distance is 22, and so on. Considering that these convolutional codes are well designed, $a(d)$ would not increase with d exponentially. Hence, we may focus only on the first three terms. The upper bound of block error rate (BLER) vs. SNR can be represented as

$$P_e(\gamma) \le \frac{L}{2\beta}\left[5e^{-18\gamma} + 7e^{-20\gamma} + 36e^{-22\gamma}\right] \tag{4.7}$$

The constant β in Eq. (4.7) can be slightly adjusted to match the property of practical codes.

The above performance analysis of convolutional codes can only provide a relatively loose upper bound because paths are not totally independent. For instance, in Eq. (4.6), between the five paths of Hamming distance 18, and between the paths of different Hamming distances, there are certain correlations. Hence, strictly speaking, the error probability cannot be directly added, e.g., certain cross-terms (e.g., correlated part) should be removed. Unfortunately, such work is usually very difficult, rendering the absence of exact formulae for the relationship between BLER and SNR.

For certain convolutional codes with short constraint lengths, if the channel is binary symmetric (BSC) with a cross-probability of p, the transfer matrix of the corresponding Markov process can be analytically represented. In the following example, consider a simple accumulator $[1, 1+D]$ with a code rate of 1/2. The Hamming distance is 2, and there are five states in the Markov machine: (2, 0), (1, 0), (0, 0), (0, 1), and (0, 2). The transfer probabilities between these states are illustrated in Figure 4.7.

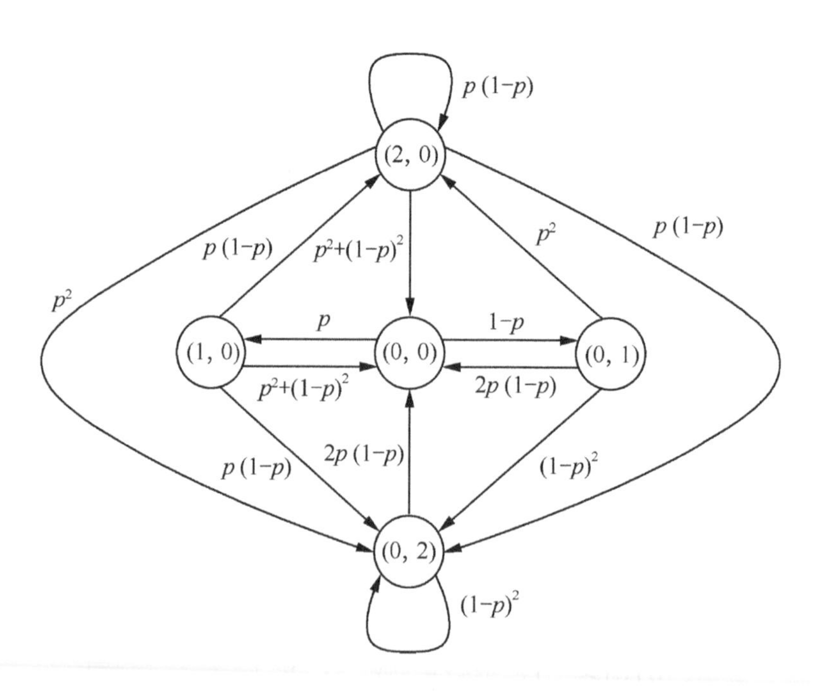

FIGURE 4.7 Markov machine for a convolutional code (accumulator, $[1, 1+D]$) of binary symmetric channel (BSC).

The transfer probabilities can also be written in matrix form:

$$T=\begin{pmatrix} (1-p)^2 & 0 & 2p(1-p) & 0 & p^2 \\ (1-p)^2 & 0 & 2p(1-p) & 0 & p^2 \\ 0 & 1-p & 0 & p & 0 \\ p(1-p) & 0 & p^2+(1-p)^2 & 0 & p(1-p) \\ p(1-p) & 0 & p^2+(1-p)^2 & 0 & p(1-p) \end{pmatrix} \tag{4.8}$$

Let us use vector $\pi=(\pi_0,\pi_1,\ldots,\pi_{M-1})'$ to represent the probabilities of Markov machine being in state 0, …, state M–1. After the state machine enters a steady state, we have

$$\pi=T'\pi \tag{4.9}$$

and $\pi_0+\pi_1+\ldots+\pi_{M-1}=1$. Hence, the probabilities of these five states are

$$\pi=\frac{1}{1+3p^2-2p^3}\begin{pmatrix} 1-4p+8p^2-7p^3+2p^4 \\ 2p-5p^2+5p^3-2p^4 \\ 2p-3p^2+2p^3 \\ 2p^2-3p^3+2p^4 \\ p^2+p^3+2p^4 \end{pmatrix} \tag{4.10}$$

With Eq. (4.10), we can calculate the error probabilities. Note that in the Viterbi algorithm, sometimes two branches would have the same metric. In this situation, coin flipping can be used to select one of the branches as the winner. The exact equation for error probability is

$$P_e=\frac{p^2\left(14-23p+16p^2+2p^3-16p^4+8p^5\right)}{\left(1+3p^2-2p^3\right)\left(2-p+4p^2-4p^3\right)} \tag{4.11}$$

The above exact formula can be approximated using Taylor series expansion, like

$$P_e=7p^2-8p^3-29p^4+64p^5+47p^6+O\left(p^7\right) \tag{4.12}$$

For this convolutional code, the free distance is 3. The union bound of error probability is $P_e \le K \cdot p^{3/2}$. Hence, compared to Eq. (4.4), even if using Eq. (4.12) as an approximation, the representation is much more fined than the union bound.

4.1.3 Decoding Complexity and Throughput Analysis

The basic Viterbi algorithm is a successive processing. Its complexity is proportional to the number of states. Normally, the number of states of a convolutional encoder is an exponential of its constraint length. The number of states also determines the memory size. Hence, the constraint lengths of commonly used convolutional codes are no greater than 7. For some special applications such as deep-space communications, the constraint length can go up to 15. After turbo codes were introduced, it is found that increasing the constraint length of convolutional codes is a less efficient way compared to using turbo codes. In fact, the constituent convolutional codes inside turbo codes can be very simple.

As mentioned before, using a time-domain sliding window can reduce the need for memory. In order to improve the throughput of decoding, an entire code block can be segmented into smaller sub-blocks where the Viterbi algorithm can be carried out in parallel [12], as illustrated in Figure 4.8. The length of each sub-block should be at least five times of constraint length.

For the parallel decoding in Figure 4.8, the decoding latency can be calculated as

$$Latency_{[\mu s]} = \frac{I \cdot \left(\left\lceil \frac{L}{P} \right\rceil + 2 \cdot D \right)}{f_{c[MHz]}} \tag{4.13}$$

where I is the number of rounds. P is the number of parallel processing. L is the block length of information bits (without CRC). D is the overlapped portion. f_c is the clock frequency for processing.

4.1.4 Tail-Biting Convolutional Code (TBCC)

To simplify the decoding, classic convolutional decoder needs to know the initial state of the shift registers of convolutional encoder and the state of the shift registers when the last bit enters the encoder. If the decoder does not have the above information, it has to perform 2^m blind trials to figure out the initial state and another 2^m blind trials for the ending state where m is the number of shift registers. $m+1$ is the constraint length.

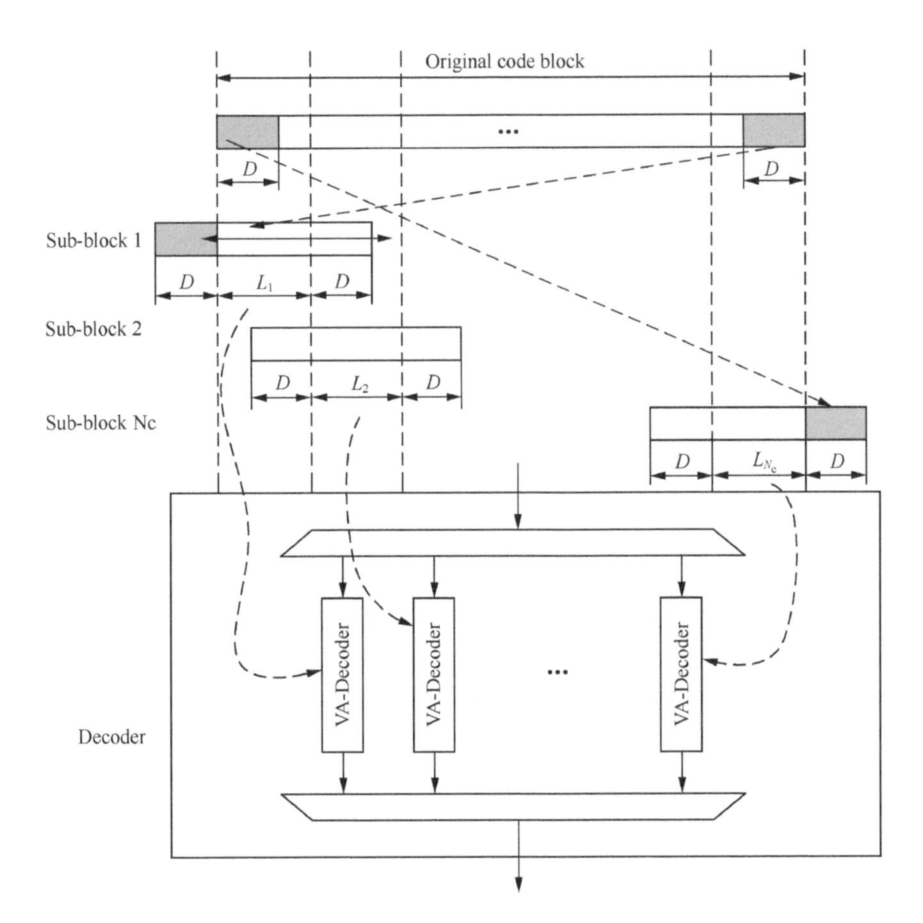

FIGURE 4.8 Parallel decoding for convolutional codes, with a certain overlap between sub-blocks.

It is relatively trivial to set the initial state. However, the ending state depends on the last m bits to be encoded. During each encoding, m bits may vary, resulting in 2^m possible ending states for the decoder to guess. When the number of shift registers is relatively large, e.g., $m=6$, the complexity of blind trials would be high. To reduce this complexity, in classic convolutional encoders, the ending state is made the same as the initial state. For instance, first set the initial state to be 0, and then append m terminating "0"s to the end of the original information block so that the ending state of a non-recursive convolutional encoder is 0.

Consider a non-recursive convolutional code of constraint length=7 (e.g., $m=6$) and code rate $R=1/3$. Assuming the length of original

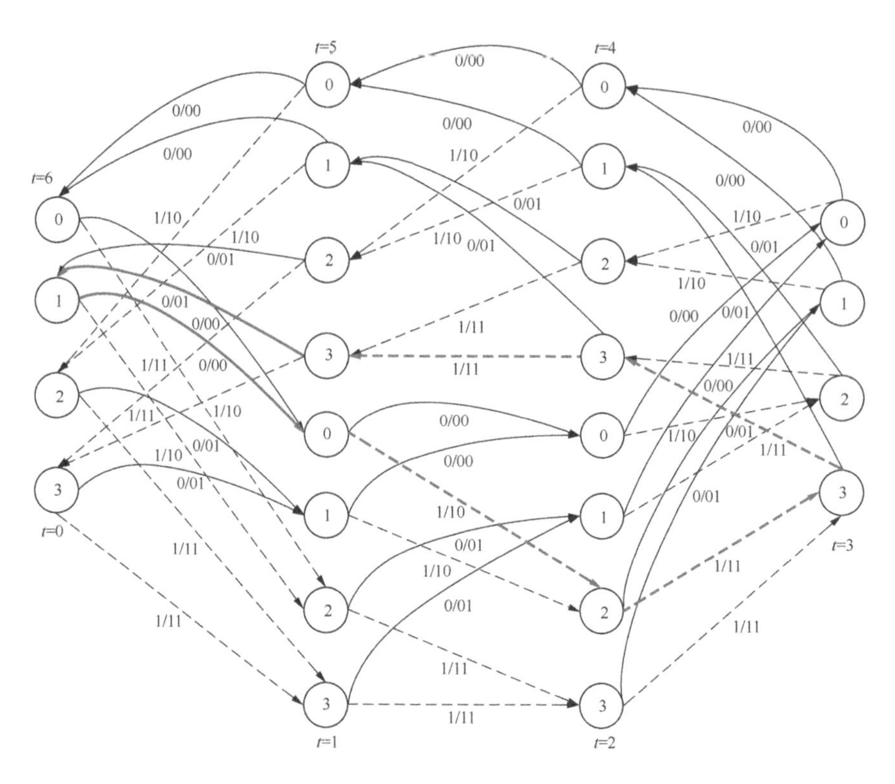

FIGURE 4.9 State transition of four-state TBCC ($m=2$).

information bits $k=40$ bits, if no terminating bits "0" are to be added, the code block size becomes $N=k/R=40/(1/3)=120$ bits after encoding. If m terminating bits "0" are added, the code block size would be increased to $N=(k+m)/R=(40+6)/(1/3)=138$ bits, which is 138/120–100%=15% overhead. This would degrade the performance where $m/(m+k)$ is called code rate loss.

When the constraint length is relatively long and the information block size is relatively short, the code rate loss due to the overhead of terminating bits would be very significant. To eliminate this rate loss, TBCC was proposed [3,13,14]. Corresponding to Figure 4.1, the trellis of a TBCC of $m=2$ (4 states) is illustrated in Figure 4.9. Assuming that the initial state of TBCC is 1, after going through the following states, 0→2→3→3→3, the TBCC circulates back to the original state 1 (highlighted in bold in Figure 4.9), resembling a snake that bites its tail. According to [3], the trellis of TBCC is strictly symmetric, e.g., it always circulates the ending state and thus forms a cyclic structure.

Practically, there are two ways for encoding of TBCC:

- For the initial values of m shift registers $\{D_1, D_2, D_3, \ldots\ldots, D_m\}$, put them in reverse order to the end of the original information block $\{B_{k-m+1}, B_{k-m+2}, B_{k-m+3}, \ldots\ldots, B_{k-1}, B_k\}$. That is: $D_m = B_{k-m+1}$, $D_{m-1} = B_{k-m+2}, D_{m-2} = B_{k-m+3}, \ldots\ldots, D_1 = B_k$. D_1 is the shift register closest to the input. D_m is the shift register closest to the output. B_k is the last bit in the original information block. After this setting, encoding can be started.
- Fill in the last m bits $\{B_{k-m+1}, B_{k-m+2}, B_{k-m+3}, \ldots\ldots, B_{k-1}, B_k\}$ in the natural order to the convolutional encoder, but discard the output and finish the initialization of the shift registers. Here, B_k is the last bit in the original information block. Encoding can be started after this initialization.

There are two categories of decoding algorithms [15] for TBCC.

- Optimal decoding algorithm
- Sub-optimal decoding algorithm

For the optimal decoding algorithm, the decoding has the following steps:

Step 1. pick an arbitrary initial state.

Step 2. use ML decoding (e.g., Viterbi) to find an optimal path, with the ending state can be arbitrary.

Step 3. check whether the initial state is the same as the ending state. If yes, terminate. If not, go to Step 4.

Step 4. use the ending state obtained in Step 2 as the initial state. If this state has been tried, then go to Step 1. If not tried it before, then go to Step 2.

Apparently, because the receiver does know in advance the initial state of the encoder, the decoding complexity of TBCC is higher than that of the classic convolutional codes, e.g., some blind decoding is needed.

For sub-optimal decoding, there are mainly three types [16]:

- Based on Circular Viterbi Algorithm (CVA) decoding

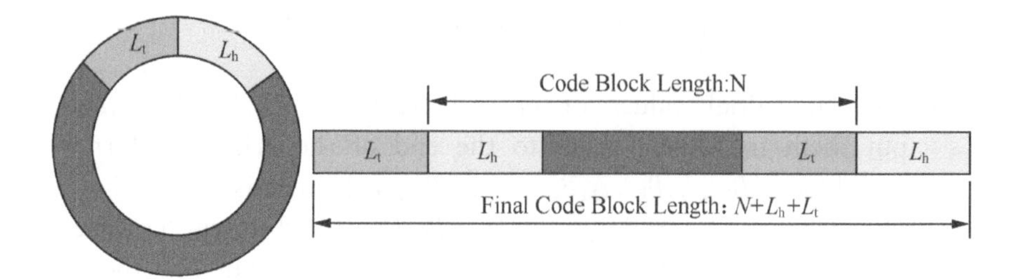

FIGURE 4.10 Illustration of code block reconstruction based on circular buffer [15].

- Based on Viterbi algorithm plus heuristic approach, e.g., Viterbi-Heuristic (VH)
- Based on Bidirectional Efficient Algorithm for Searching code Trees (BEAST)

In [15], a decoding method is proposed as shown in Figure 4.10 which is based on the reconstruction of code block in the circular buffer. The operations are as follows. To copy the front bits in a code block (Lh part in Figure 4.10) and put them to the end. Copy the last part of the code block (the segments denoted "Lt" in Figure 4.10) and put them to the beginning. The final information block has the size of N+Lh+Lt. The merit of this method is that after the code block reconstruction, the regular Viterbi algorithm can be used for decoding. As shown in Figure 4.11, compared to regular convolutional code (with tails), TBCC of $R=1/3$ and 64 states ($m=6$) can provide 0.1~0.6 dB performance gain [17].

4.2 APPLICATION OF CONVOLUTIONAL CODES IN MOBILE COMMUNICATIONS

Convolutional codes have been used extensively, for instance, dedicated control channel for CDMA2000 [18], traffic channels in wideband code division multiple access (WCDMA) [19], and physical downlink control channel (PDCCH) for long-term evolution (LTE) [20]. In the following, we will discuss the latter two applications.

4.2.1 Convolutional Codes in 3G UMTS (WCDMA)

The constraint length of convolutional codes in UMTS is 9, then there are 256 states. Code rates of 1/2 and 1/3 are supported, corresponding to the generator polynomials: $G_0=(561)_8$, $G_1=(753)_8$, and $G_0=(557)_8$, $G_1=(663)_8$, and $G_2=(711)_8$, respectively, also shown in Figure 4.12. Note that these

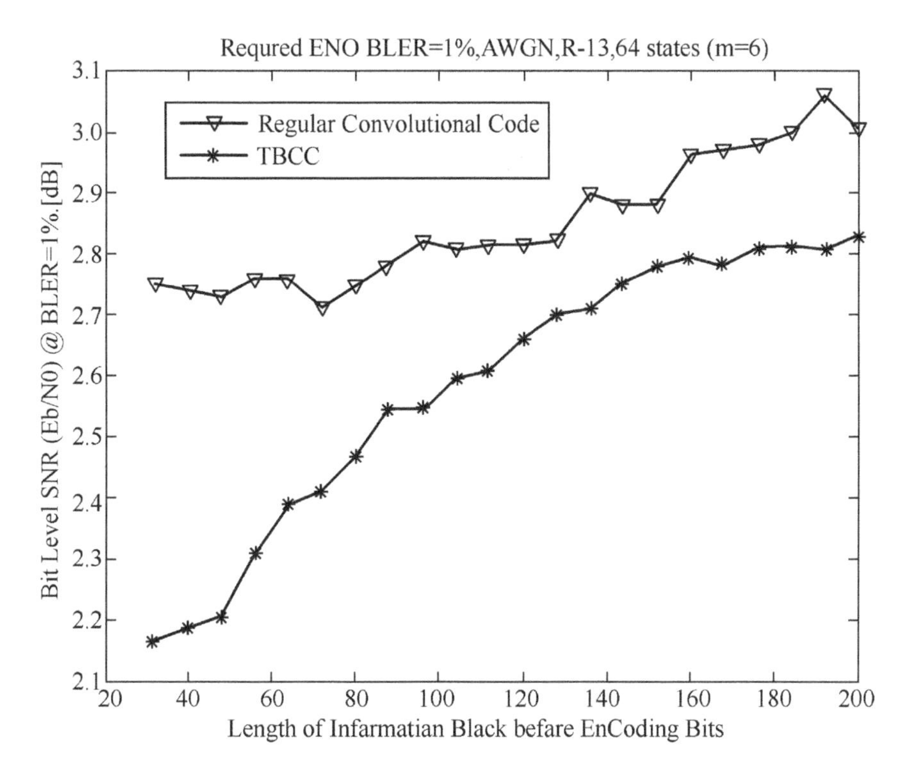

FIGURE 4.11 Performance comparison between TBCC and regular convolutional code [17].

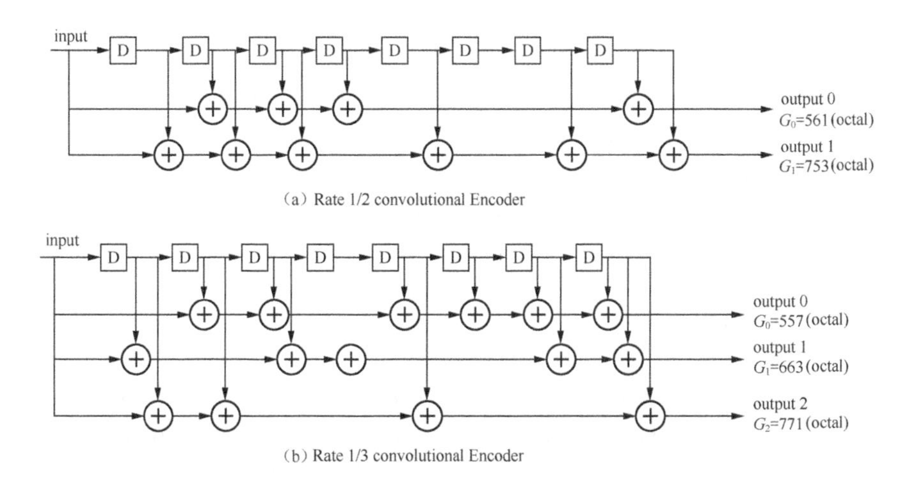

FIGURE 4.12 Rate 1/2 and 1/3 convolutional codes for UMTS [19].

TABLE 4.1 Distribution of Hamming Distance for Rate 1/3 of UMTS

	[557, 663, 711]	
d	A_d	C_d
18	5	11
20	7	32
22	36	195
24	85	564
26	204	1473
28	636	5129
30	1927	17434
32	5416	54092
34	15769	171117
36	45763	539486
38	131319	1667179
40	380947	5187615
42	1100932	16003037
44	3173395	49013235
46	9186269	150271658

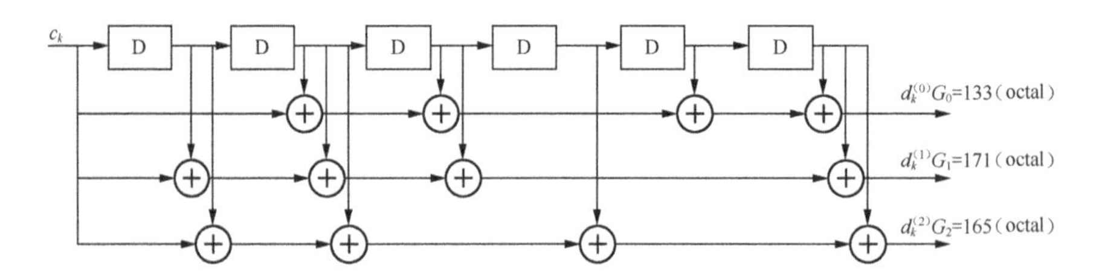

FIGURE 4.13 Rate 1/2 and 1/3 convolutional codes for LTE.

two sets of polynomials do not have nested properties and cannot be used together when doing the rate matching. The distribution of Hamming distance is listed in Table 4.1, where A_d is the number of minimum Hamming distances. C_d is the number of non-zero bits.

4.2.2 Convolutional Codes in LTE

LTE adopts TBCC for the PDCCH and narrowband physical downlink shared channel (N-PDSCH). The constraint length is 7, e.g., 64 states of the TBCC. Both 1/2 and 1/3 code rates are supported with the generator polynomials as $G_0=(133)_8$, $G_1=(171)_8$, and $G_2=(165)_8$, as illustrated in Figure 4.13. Note that the polynomials of these two rates have nested property, e.g., the rate can easily be matched from 1/2 to 1/3.

4.3 ENHANCEMENTS OF CONVOLUTIONAL CODES

Since its birth, convolutional codes have not gone through significant changes. Nevertheless, there are a few small improvements as discussed below.

4.3.1 Supporting Multiple Redundancy Versions

Circular buffer is used for rate matching of LTE's convolutional codes. In LTE, convolutional codes are mainly used for physical control channels without hybrid automatic retransmission request (HARQ) retransmissions. Hence, no redundancy version (RV) needs to be defined for convolutional codes. However, when convolutional codes are used to downlink the physical shared channel of NB-IoT, HARQ retransmission should be supported. Hence, it deserves consideration whether multiple redundancy versions should be specified for convolutional codes. Due to various reasons, this feature, as illustrated in Figure 4.14, has not yet been standardized for NB-IoT. In Figure 4.14, roughly, k0≈N_TB/2, meaning that the shift is about half of buffer size (N_TB).

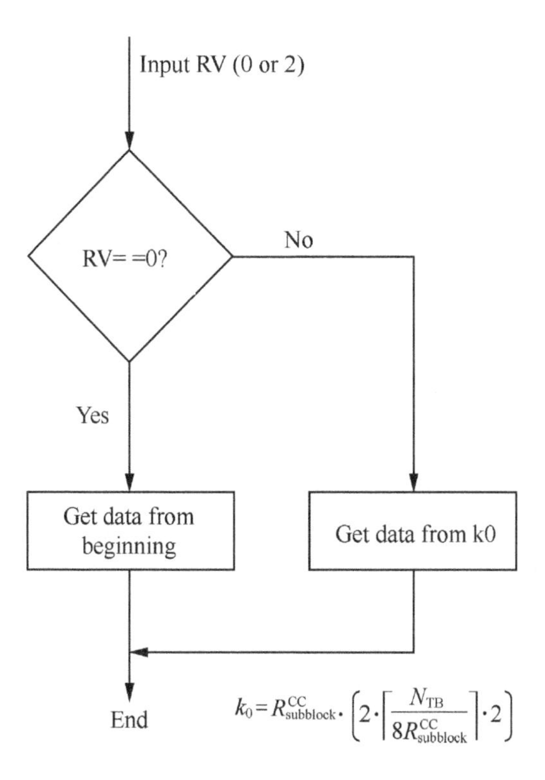

FIGURE 4.14 Convolutional codes supporting multiple redundancy versions.

TABLE 4.2 SNR Gain (dB) Over Single RV for AWGN Channel

		RH		
	TBS	[0 2 3 1]	[0 2 0 2]	[0 0 0 0]
$N_{RU}=1$	16	0	0	0
	40	0	0	0
	56	0	0	0
	104	0.3	0.22	0
	120	0.47	0.38	0
	136	0.64	0.59	0
	144	0.76	0.69	0
$N_{RU}=6$	152	0	0	0
	408	0.1	0.06	0
	712	0.23	0.13	0
	936	0.59	0.48	0

TABLE 4.3 SNR Gain (dB) Over Single RV for Fading Channel

		TBS		
	RV	[0 2 3 1]	[0 2 0 2]	[0 0 0 0]
$N_{RU}=1$	104	3.4	2.28	0
	120	3.61	3	0
	136	5.94	4.85	0
	144	7.32	6.46	0
$N_{RU}=6$	408	1.37	1.07	0
	712	3.33	3.14	0
	936	11.68	10.84	0

By introducing multiple redundancy versions, the performance of convolutional codes can be improved during HARQ retransmissions, which is summarized in Tables 4.2 and 4.3.

4.3.2 Supporting Lower Code Rate

The minimum code rate of convolutional codes in UMTS and LTE is 1/3. To effectively support very low SNR transmissions, a lower code rate needs to be designed. One way is to keep the constraint length to be the same as that of LTE, but add more nested generator polynomials, as shown in Table 4.4 where the first three rows are the polynomials already adopted in LTE. Columns 3~5 correspond to the minimum Hamming distance and the number of paths with this distance. As n increases, the code rate keeps reducing and the minimum Hamming distance continuously grows.

TABLE 4.4 Generator Polynomials for Constraint Length=7 [21]

n	Polynomial	d_f	A_d	C_d
1	133	-	-	-
2	171	10	11	36
3	165	15	3	7
4	117	20	2	3
5	135	25	1	1
6	157	30	1	2
7	135	36	4	8
8	123	40	1	1
9	173	46	3	6
10	135	51	2	4
11	171	56	2	3
12	135	61	1	1

TABLE 4.5 Hamming Distance Distribution of the Enhanced Convolutional Code (Constraint Length=9)

	[561, 753, 715]	
d	A_d	C_d
18	4	11
20	13	49
22	28	136
24	81	496
26	235	1652
28	646	5122
30	1889	16388
32	5608	53870
34	16007	167364
36	46407	529226
38	133484	1642089
40	386574	5104883
42	1113762	15714070
44	3220414	48331307
46	9297602	147890952

4.3.3 Further Optimized Polynomials

For the convolutional codes of UMTS with constraint length=9, their generator polynomials can be further optimized. For instance, by using $G_0=(561)_8$, $G_1=(753)_8$, and $G_2=(715)_8$, the number of paths with minimum Hamming distance is reduced from 5 to 4, which is beneficial to improve

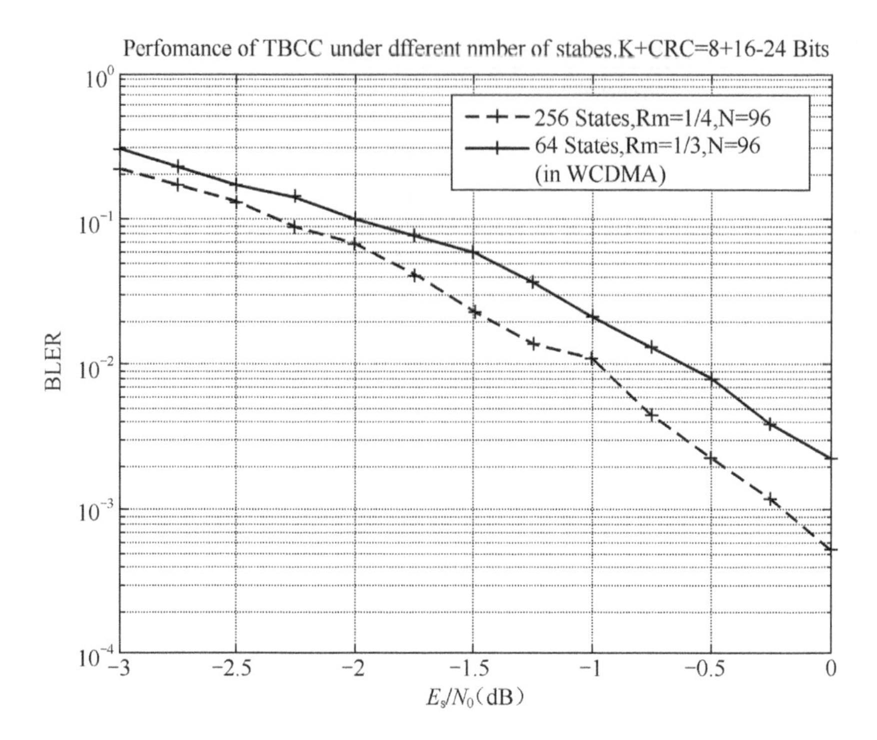

FIGURE 4.15 Performance of enhanced convolutional code compared to UMTS' convolutional code [21].

the performance at low SNR and low rate. Even though the number of paths for Hamming distance = 20 is increased, this would not cause a significant negative effect under typical SNR operating ranges. It is seen from Figure 4.15 [21] that at BLER = 1%, the enhanced convolutional code outperforms convolutional codes in WCDMA by about 0.3 dB.

4.3.4 CRC-Aided List Decoding

In the conventional Viterbi algorithm, normally only one optimal path would be decided. However, sometimes, the optimal path may be erroneous, whereas the sub-optimal path would be the correct one. In light of this, the list Viterbi algorithm (LVA) was proposed [8,9,22–25]. During list Viterbi decoding, a CRC check can be employed to select the correct path from the list, to improve the performance.

The process of CRC-aided list Viterbi decoding is depicted in Figure 4.16 [24]. The key step of LVA is to generate L candidates for the outcome of decoding (L paths). Here the convolutional code can be either conventional or tail-biting.

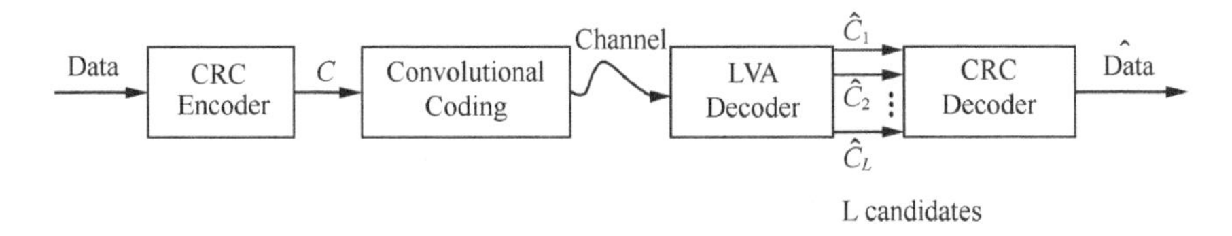

FIGURE 4.16 Process of CRC-aided list Viterbi decoding [24].

In order to generate L decoding paths, two methods are proposed in [9,25]: to simultaneously generate L decoding paths via parallel list Viterbi algorithm (PLVA), and one-by-one to generate L decoding paths via successive list Viterbi algorithm (SLVA). In SLVA, based on the first l optimal paths of ($1 \le l \le L-1$, L is list size), to find the $(l+1)$-th optimal path. The way to find the $(l+1)$-th optimal path is to compute the difference between the two sequences (paths), the path with a smaller difference would be the $(l+1)$-th optimal path [22,25]. Since PLVA requires computation and storage of metrics of L decoding paths, the complexity of PLVA is L times of the conventional Viterbi algorithm. In SLVA, only when all previous l decoding is erroneous, it would generate the path of $(l+1)$. Hence, its complexity is variable (e.g., related to information block length, constraint length, SNR, etc.) In practical engineering, L should not be set too large.

SLVA has two stages of processing: initialization and pathfinding. Initialization includes:

- To set the number of optimal paths already found to be $l=1$
- To use the Viterbi algorithm to get the optimal state path, as well as the metrics for each time instant, for each state
- To construct the matrix for the optimal state paths (to store L optimal paths), the vector of merging nodes for the next candidate path, to store the matrix of metric differences along L optimal paths (initialized to infinity), and the matrix of metrics of each optimal path

Pathfinding includes:

- **Step 1.** to compute the metric-difference long l-th optimal path and store it in the corresponding position.
- **Step 2.** to find the minimum in the first l-th row of the matrix storing metric differences, and store the time instantly in the merging

node vector for the next candidate path. To calculate the accumulated metric that passes this time instant.

- **Step 3.** to select the path with the largest accumulated metric among l candidate paths obtained from Step 2 and consider this path to be (l+1)-th optimal path. Then, store the merging instantly to the merging node vector for the next candidate path.
- **Step 4.** to remove the merging node corresponding to the candidate path (e.g., to set the metric difference of optimal path to be infinite).

According to [9,25], the performance of SLVA and PLVA are almost the same. As illustrated in Figure 4.17, at FER (e.g., BLER)=1%, with L=8, SLVA provides about 0.8 dB gain over the conventional Viterbi algorithm. The test results in [9] show that for convolutional code of polynomial $[561; 753]_8$ with k=148 bits, CRC=16 bits, m=8, L=2, and BLER=1%, PLVA outperforms VA by about 0.4 dB.

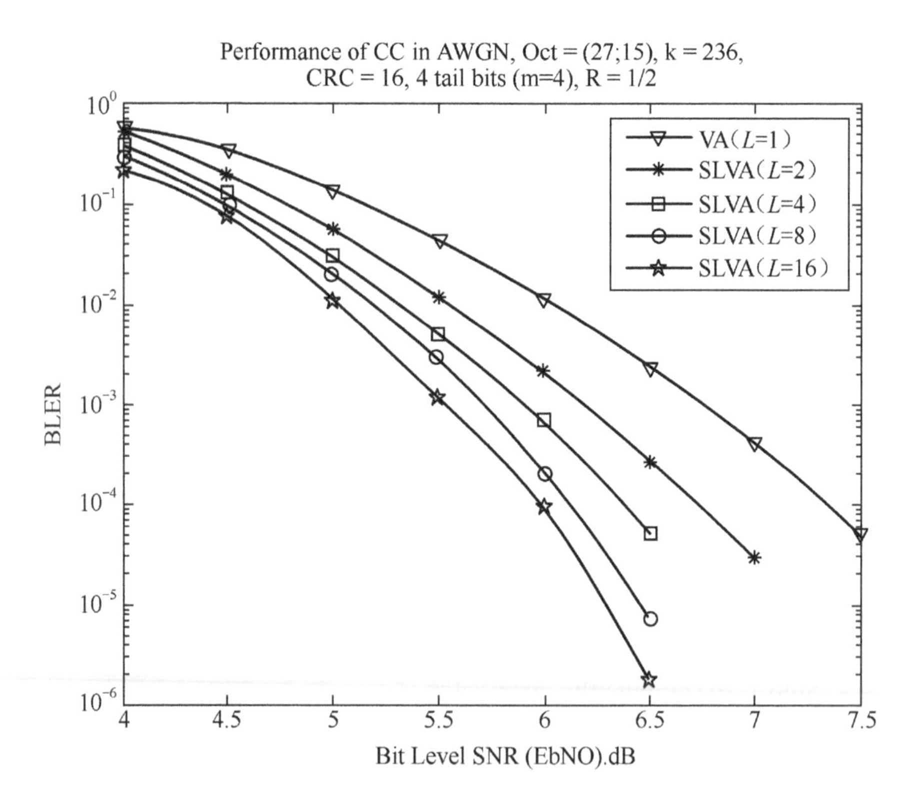

FIGURE 4.17 Performance of SLVA in the AWGN channel [25].

In [26], TBCC with per code block CRC was proposed to aid list decoding and improve the decoder's performance. Its basic idea is that at each stage of decoding, CRC is used to select the decoding path and prune the branches not satisfying CRC. Its complexity for the 1024 list is comparable to the 16 list of LVA. In general, by limiting the decoding complexity of the CRC-aided list Viterbi, the performance can be improved to some extent.

According to [27], the complexity of PLVA is roughly proportional to L. For practical systems, to balance the complexity and the performance, PLVA of L=4 or 8 is suitable. In addition, as L increases, the mis-detection probability would increase due to the increase of CRC checking. Under the same SNR, the mis-detection rate of CRC in PLVA (L=4) is about 2~3 times of CRC mis-detection rate of the Viterbi algorithm.

Essentially, the mis-detection in [27] is equivalent to false alarm rate (FAR), because LVA would anyway suggest a path, regardless of whether CRC passes or fails. If CRC passes, however, if the data (including CRC) is wrong, then this is a false alarm, e.g., FAR of LVA is about $\log_2(L)$ of FAR of the Viterbi algorithm. That is $FAR=2^{\log2(L)-CRC\_LENGTH}$.

In summary, after many years of development, even with some shortcomings (e.g., not suitable for long block length), convolutional codes can still be improved in their suitable use scenarios.

REFERENCES

[1] P. Elias, "Coding for noisy channels," *IRE Conv. Rec.*, vol. 6, 1955, pp. 37–47.

[2] C. Berrou, et al., "Near Shannon limit error-correcting coding and decoding: Turbo Codes," *Proc. IEEE Intl. Conf. Communication (ICC 93)*, May 1993.

[3] S. Lin, *Error control codes* (in Chinese, translated by J. Yan), 2nd Edition, Mechanical Industry Press, Beijing, China, June 2007.

[4] R. Hutchinson, "Convolutional codes with maximum distance profile," *Syst. Control Lett.*, vol. 54, no. 1, 2005, pp. 53–63.

[5] M. Cedervall, R. Johannesson, "A fast algorithm for computing distance spectrum of convolutional codes," *IEEE Trans. Commun.*, vol. 35, no. 6, pp. 1146–1159.

[6] CN201210382553.9, Z. Wu, "Method of determining polynomial of compatible convolutional code and encoding/decoding algorithms," Beijing University of Press and Telecommunications, Oct. 2013.

[7] A. J. Viterbi, "Error bounds for convolution codes and an asymptotically optimal decoding algorithm," *IEEE Trans. Inf. Theory*, vol. 13, no. 2, 1967, pp. 260–269.

[8] N. Seshadri, "List Viterbi decoding algorithms with applications," *IEEE Trans. Comm.*, vol. 42, April 1994, pp. 313–323.

[9] Z. Wang, *Sub-optimal path algorithm for convolutional codes applied in 3G generation mobile communications*, Master Thesis, Xidian University, Jan. 2009.

[10] X. Li, *Decoding algorithm study for convolutional codes*, Master Thesis, Xidian University, Jan. 2010.

[11] G. La Guardia, "Convolutional codes: techniques of construction," *Comput. Appl. Math.*, vol. 35, no. 2, July 2016, pp. 501–517.

[12] C. Lin, "A tiling-scheme Viterbi Decoder in software defined radio for GPUs," *International Conference on Wireless Communications*, 2011, pp. 1–4.

[13] G. Solomon, "A connection between block and convolutional codes," *SIAM J. Appl. Math.*, vol. 37, no. 2, Oct. 1979, pp. 358–369.

[14] H. Ma, "On tail-biting convolutional codes," *IEEE Trans. Commun*, vol. 34, no. 2, 2003, pp. 104–111.

[15] J. Ma, "Design and simulation of a highly efficient decoder for tail-biting convolutional codes," *Electron. Devices Appl.*, vol. 4, no.7, 2010, pp. 61–63.

[16] X. Wang, "Decoding algorithm of tail-biting convolutional codes based on reliability ordering," *J. Electron. Inf.*, vol. 37, no. 7, July 2015. pp. 1575–1579.

[17] 3GPP, R1-071323, Performance of convolutional codes for the E-UTRA DL control channel, Motorola, RAN1#48bis, March 2006.

[18] 3GPP2, C.S0002-0 v1.0 Physical Layer Standard for cdma2000 Spread Spectrum Systems, Oct. 1999.

[19] 3GPP, TS25.212 V5.10.0-Multiplexing and channel coding (FDD) (Release 5), June 2005.

[20] 3GPP. TS36.212 V14.0.0-Multiplexing and channel coding (Release 14), Sept. 2016.

[21] 3GPP, R1-1608871, On TBCC generator polynomials, Ericsson, RAN1#86bis, Oct. 2016.

[22] T. Hashimoto, "A list-type reduced-constraint generalization of the Viterbi algorithm," *IEEE Trans. Inf. Theory*, vol. 33, no. 6, Nov. 1987, pp. 866–876.

[23] B. Chen, "List Viterbi algorithms for wireless systems," *IEEE 51st Vehicular Technology Conference Proceedings*, vol. 2, 2000, pp. 1016–1020.

[24] D. Petrovic, "List Viterbi decoding with continuous error detection for Magnetic Recording," *IEEE Global Telecommunications Conference*, 2001, pp. 3007–3011.

[25] F. Hao, "List Viterbi algorithm and its application," China Sci &Tech Paper Online, 2011, http://www.paper.edu.cn.

[26] 3GPP, R1-162213, TBCC: rate compatibility: high level design, Qualcomm, RAN1#84b, April 2016.

[27] Y. Wei, *Study on decoding algorithms for 3GPPUMTS and LTE systems*, Ph. D Thesis, Shanghai Jiaotong University, May 2013.

CHAPTER 5

Turbo Codes

Jin Xu and Yifei Yuan

Turbo codes are a type of high-performance channel coding scheme invented by Berrou et al. in 1993 [1] which can approach Shannon's limit as close as 0.5 dB. After more than 20 years of development, turbo codes are already widely used in mobile communications systems such as 3G WCDMA [2], and 4G long-term evolution (LTE) [3]. During the study phase of 5G NR [4,5], Turbo code 2.0 [6,7] was one of the candidate channel coding schemes. In this chapter, we start with the principle of turbo codes, followed by the discussion of turbo codes in 4G LTE. In the end, Turbo code 2.0 is described.

5.1 PRINCIPLE OF TURBO CODES

The invention of turbo codes in 1993 [1] opened a flood of theoretical and technological breakthroughs in channel coding areas and had a profound impact on theory and practical engineering.

- Significance in theoretical analysis: turbo codes further demonstrate that Shannon's limit can only be approached by random channel coding for long information blocks. Inspired by the construction of turbo codes, some researchers re-examined the low-density parity check (LDPC) code proposed by Gallager in the early 1960s [8,9]. Although LDPC appears to be a type of block code which is different from the constituent convolutional codes in turbo codes, the construction of LDPC code emphasizes low-density and long-distance check nodes in a large block in order to get the effect of random coding. In this sense, the principle of constructing LDPC codes is quite

DOI: 10.1201/9781003336174-5

similar to that of turbo codes. Based on this principle, researchers conducted design optimization of LDPC codes and obtained a series of high-performance LDPC codes, some of which can approach Shannon's limit very closely [10].

- Significance in practical engineering: iterative decoding can drastically reduce the complexity of decoding. Optimal decoding of large random codewords is a hard problem whose complexity would exponentially grow with the block length. By contrast, the complexity of iterative decoders is only related to the number of iterations. For each iteration, the computation burden is similar to conventional decoding algorithms. Certainly, the iterative process is nonlinear. If the operating point is not suitable, the iteration process would not converge. We will discuss this later.

Due to its excellent performance and feasibility in engineering, polar codes are widely used in 3G [2] and 4G [3] mobile communications.

5.1.1 Concatenated Codes Prior to Turbo Codes Era

As its name implies, a concatenated code refers to combining two or more channel codes in a serial or parallel fashion that can be connected by an interleaver. As early as 1965, Forney proposed serially concatenated codes. In the next ~10 years, serially concatenated codes were extensively used in outer-space communications. During that period, the most commonly used inner code is convolutional codes. The typical outer code is the Reed–Solomon code.

Figure 5.1 shows a channel coding scheme adopted in deep-space communications standard where the outer code is Reed–Solomon (255, 223) and the original polynomial is $p(x)=x^8+x^7+x^2+x+1$. The generator polynomial is: $g(x)=\prod_{j=112}^{143}\left(x-a^{11j}\right)$, where α is the root of the polynomial $p(x)$; The inner code is a convolutional code of R = 1/2, constraint length = 7, with polynomial as $g_1=1111001$, $g_2=1011011$. The overall code rate of the concatenated code is 0.437. The Reed–Solomon code is non-binary and can effectively correct the bursty errors due to the decoding failure of convolutional code. The block interleaver can also affect the performance to some extent.

For the inner and outer code structure shown in Figure 5.1, there are two approaches to further improve the performance: The first approach is

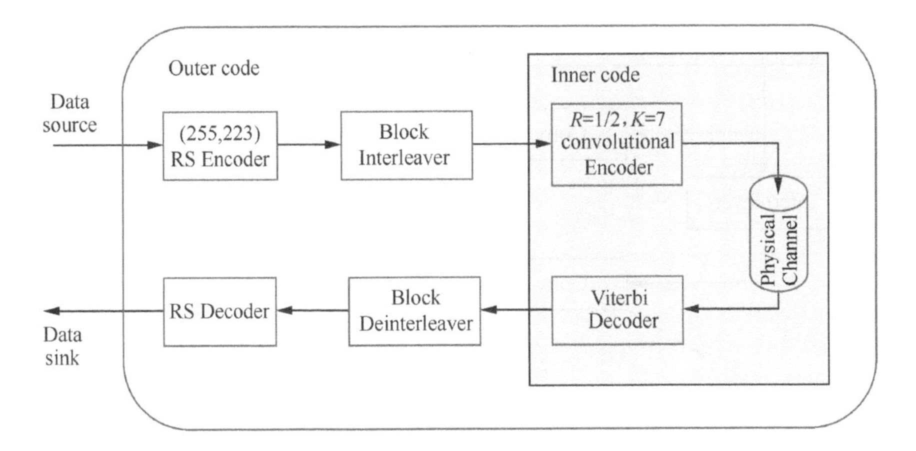

FIGURE 5.1 An example of concatenated code prior to the turbo codes era.

to increase the constraint length of the convolutional code. The gain from this approach may be limited and would significantly increase the decoding complexity; the second approach is to generate soft information which does not only contain the hard decisions but also the probability of right and wrong. Decoding algorithms that can provide soft information are Bahl-Cocke-Jelinek-Raviv (BCJR), soft-output Viterbi algorithm (SOVA), etc., to be discussed later. By following the second approach, the performance of Reed-Solomon (RS) and convolutional serially concatenated code was improved to some degree; however, there has been no significant breakthrough along this path.

5.1.2 Parallel Concatenated Convolutional Codes

Different from the concatenated codes in the early days, a classic turbo encoder consists of two constituent encoders that are parallel concatenated [11]. They are connected by an inner interleaver inside the turbo encoder, as illustrated in Figure 5.2. The data stream of systematic bits u_k passes through directly and reaches MUX (Multiplex) puncture. The second branch passes through the first constituent encoder to generate the redundant bits p_k^1. The third branch first passes through the inner interleaver and then goes through the second constituent encoder to generate the redundant bits p_k^2. These two streams of redundant bits are also input to MUX puncture. The mother code rate (without puncturing) is 1/3. It is noticed that both these two constituent codes are convolutional codes with the same generator polynomial (note that the generator polynomials

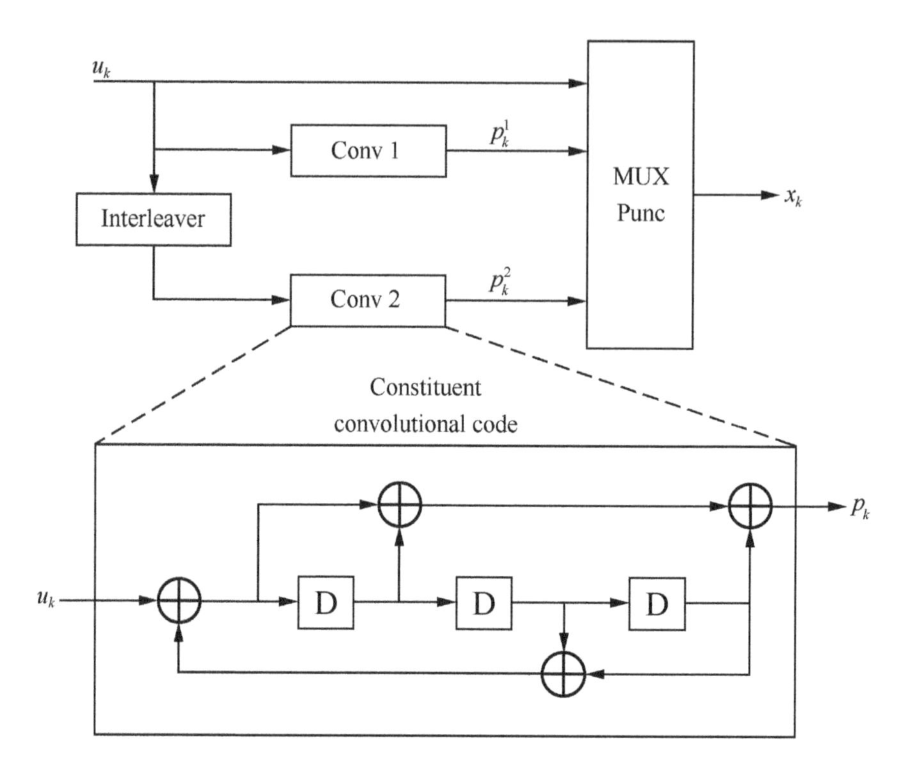

FIGURE 5.2 Turbo encoder for UMTS/LTE.

of two constituent codes may not necessarily be the same. In fact, sometimes better performance can be achieved by using different polynomials [12,13]). Different from the commonly used non-recurve convolutional codes, the constituent convolutional code for turbo codes is recursive, e.g., with self-feedback. This is one of the reasons for the excellent performance of turbo codes (the other reason is the inner interleaver). The MUX puncture is for rate matching purposes. Compared to previous concatenated codes, the ingenuity of turbo codes is the use of iterative decoding, which significantly releases the potential of soft information that can be refined continuously during the iterations.

In addition, other types of codes such as Bose-Chauduri-Hochquenghen (BCH) codes and Reed-Solomon (RM) codes can also be used as constituent codes for turbo codes. In [14], extended Hamming codes and Hadamard codes are studied for their use in turbo codes. The results show that with extended Hamming as the constituent code, the performance of turbo codes can approach Shannon's limit by about 1 dB.

5.1.3 Decoding Algorithms

Turbo decoders have two main characteristics: (1) operating on soft bits or soft information and (2) iterative decoding. They are reflected in the extrinsic information exchanged back and forth multiple times between the double engines in a turbo decoder. During this process, the extrinsic information is refined and the reliability (or belief) is increased. This is why the inventors chose "Turbo" as the name. As Figure 5.3 shows, **y** is the log-likelihood ratio (LLR) generated by the demodulator which can be considered as soft information of the observations. After de-puncture, three branches of LLR are obtained: $\mathbf{y}_s$, $\mathbf{y}_p^1$, and $\mathbf{y}_p^2$, corresponding to the systematic bits, the redundancy bits of the first constituent decoder (Decoder 1), and the redundancy bits of the second constituent decoder (Decoder 2), respectively. Based on $\mathbf{y}_s$, $\mathbf{y}_p^1$ as well as $\mathbf{ext}_{21}$ from Decoder 2 and Decoder 1 would calculate a posteriori probability (APP) of a bit taking value 0 or 1. This APP would be subtracted by the priori probability to get the "net" contribution to the increase of the reliability (or belief) by Decoder 1, denoted by $\mathbf{ext}_{12}$. Similarly, based on $\mathbf{y}_s$, $\mathbf{y}_p^2$ as well as $\mathbf{ext}_{12}$ from Decoder 1, Decoder 2 would calculate a posteriori probability (APP) of a bit taking value 0 or 1. This APP would be subtracted by the priori probability to get the "net" contribution to the increase of the reliability (or belief) by Decoder 1, denoted by $\mathbf{ext}_{21}$. After a certain number of iterations (e.g., 30), output the final reliability (belief) of each bit.

The structure of turbo codes can also be represented by a generalized factor graph. As shown in Figure 5.4, the factor graph for convolutional codes is already discussed in Chapter 4. Here we concatenate two constituent convolutional codes in parallel to get a code rate of 1/3. Note that there is an interleaver in between whose function is to scramble the bit positions to get another set of state sequences.

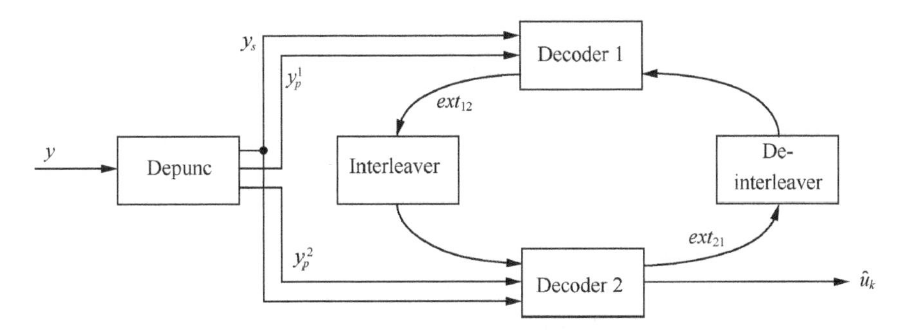

FIGURE 5.3 Information exchange flow in turbo decoder.

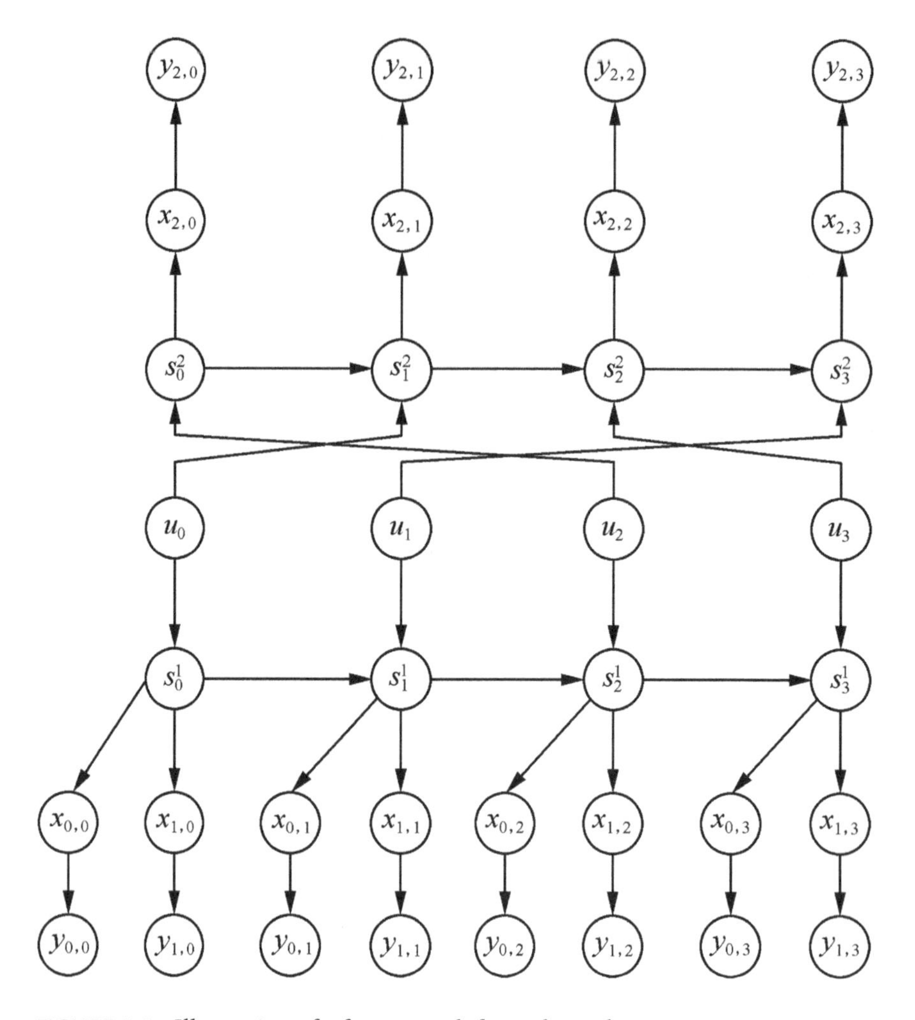

FIGURE 5.4 Illustration of a factor graph for turbo codes.

The optimal algorithm to compute a posteriori probability (APP) is the BCJR algorithm [15]. To simplify the discussion, the trellis of a simple convolutional code is shown in Figure 5.5. The finite machine of this convolutional code has two shift registers, e.g., four states. The solid line represents the state transition when the input bit is 0. The dashed line represents the state transition when the input bit is 1. The initial state of the finite machine is 0 (State 0, corresponding to the beginning of the block). The terminating/ending state is also 0, corresponding to the end of the block.

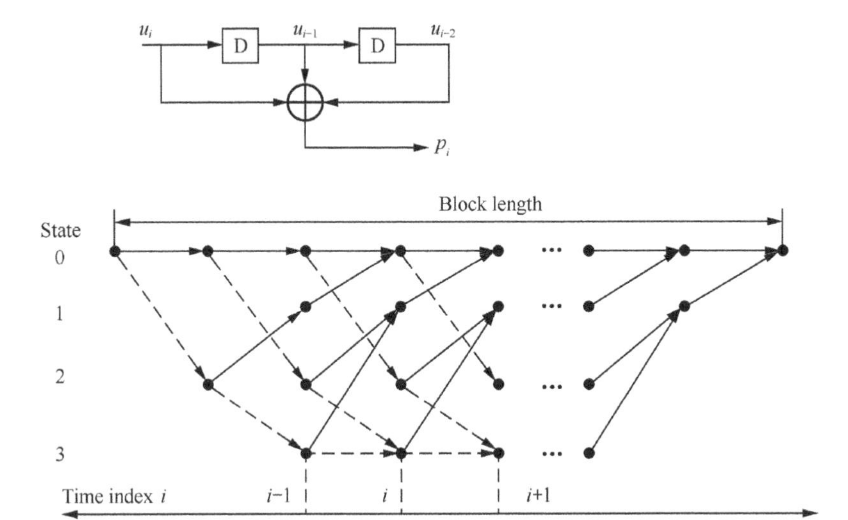

FIGURE 5.5 An example of a trellis for the BCJR algorithm.

For any bit, for instance, i-th information bit, its reliability (belief) can be measured by LLR which is

$$L(u_i)=\log\left(\frac{p(u_i=1|\mathbf{y})}{p(u_i=0|\mathbf{y})}\right) \tag{5.1}$$

It refers to that given a series of observations $\mathbf{y}$ (usually the output of the demodulator which is a code block contaminated by the noise and channel variation), the log-domain probability ratio whether i-th information bit is 1 or 0. Since the state of the encoder trellis depends only on the previous state and the current transfer probability, this is a Markov process so that Eq. (5.1) can be rewritten as

$$L(u_i)=\log\left(\frac{\sum_{S_1}\alpha_i(s')\gamma_i(s',s)\beta_i(s)}{\sum_{S_0}\alpha_i(s')\gamma_i(s',s)\beta_i(s)}\right) \tag{5.2}$$

where $\alpha_i(s')$ is the probability that the state of the encoder finite machine is s' at time instant i, given the observations from the beginning till the time instant i–1. $\beta_i(s')$ is the probability that the state of the encoder finite machine is s at time instant i+1i, given the observations from the end of the block till the time instant i+1. $\gamma_i(s',s)$ is the transition probability that the state of the encoder finite machine transfers from s' at time instant i to s at the time instant i+1, given the observation y_i. $\gamma_i(s',s)$ is partially determined by the priori probability. The BCJR algorithm can be basically summarized as first to compute, $\gamma_i(s',s)$ then to go forward from the beginning to the end of the code block to calculate $\alpha_i(s')$. Then move backward from the end to the beginning of the code block to calculate $\beta_i(s)$. Specifically, $\gamma_i(s',s)$ can be represented as

$$\gamma_i(s',s) = p(S_{i+1} = s \mid S_i = s') \tag{5.3}$$

The observation y_k corresponds to the finite machine's transition from state s' to s. Hence, Eq. (5.3) can be further written as

$$\gamma_i(s',s) = p(s \mid s')p(y_i \mid S_i = s', S_{i+1} = s) = p(u_i)p(y_i \mid u_i) \tag{5.4}$$

There are two terms on the right side of Eq. (5.4). The first term is the priori probability of the bit which is equal probability before the decoding. However, after the first iteration, this priori probability would keep increasing by taking in extrinsic information. The second term is the probability of getting the observation y_k if the encoder finite machine transfers from state s' to s. The second term depends on the trellis structure (which is deterministic) and the channel characteristics. For the additive white Gaussian noise (AWGN) channel with zero mean and variance of σ^2, the second term can be represented as

$$\begin{aligned} p(y_i \mid u_i) &\propto \exp\left[-\frac{\left(y_i^s - u_i\right)^2}{2\sigma^2} - \frac{\left(y_i^p - p_i\right)^2}{2\sigma^2}\right] \\ &= \exp\left[-\frac{\left(y_i^s\right)^2 + u_i^2 + \left(y_i^p\right)^2 + p_i^2}{2\sigma^2}\right] \cdot \exp\left[\frac{u_i y_i^s + p_i y_i^p}{\sigma^2}\right] \end{aligned} \tag{5.5}$$

Substituting Eq. (5.5) to Eq. (5.4), we get

$$\gamma_i(s',s) = B_i \exp\left[\frac{y_i^s(2u_i - 1) + p_i y_i^p}{\sigma^2}\right] \tag{5.6}$$

where p_i is the redundancy bit, determined by u_i and the encoder. B_i is a non-zero constant to be canceled in Eq. (5.2).

$\alpha_i(s)$ can be calculated by using Eq. (5.6) recursively, by summing over the entire set of previous states transitioning to the current state s, which is

$$\alpha_i(s) = \sum_{s' \in A} \alpha_{i-1}(s')\gamma_i(s',s) \tag{5.7}$$

The initial state of the encoder is 0. Hence,

$$\alpha_0(S=0)=1 \quad \text{and} \quad \alpha_0(S \neq 0)=0 \tag{5.8}$$

The recursive calculation of $\alpha_i(s)$ is carried out forwardly in the trellis.

Similarly, $\beta_i(s)$ can also be calculated using Eq. (5.6), by summing over the entire set of later states transitioning to the current state s, which is

$$\beta_{i-1}(s') = \sum_{s \in B} \beta_i(s)\gamma_i(s',s) \tag{5.9}$$

The state of the encoder for the last bit is 0. Hence,

$$\beta_N(S=0)=1 \quad \text{and} \quad \beta_N(S \neq 0)=0 \tag{5.10}$$

The recursive calculation of $\beta_i(s)$ is carried out backwardly in the trellis.

In summary, there are about four steps of BCJR:

- **Step 1.** using Eqs. (5.8) and (5.10) to initialize $\alpha_0(s)$ and $\beta_N(s)$
- **Step 2.** for each observation y_i, using Eq. (5.6) to compute $\gamma_i(s',s)$, using Eq. (5.7) to compute $\alpha_i(s)$. Store the values of γ and α for all the time instants and all the states.
- **Step 3.** using Eq. (5.9) recursively to calculate $\beta_i(s)$
- **Step 4.** multiply $\alpha_i(s)$, $\gamma_i(s',s)$ and $\beta_i(s)$ together. Based on Eq. (5.2), to sum over all the branches corresponding to $u_i=1$ or $u_i=0$

It is seen from the above that the complexity of BCJR is quite high. At each time instant, the total number of branches depends on the number of states of convolutional encoder which grows exponentially with the constraint length. The required memory size is proportional to the sequence length. In practical engineering implementations, the BCJR algorithm can be carried in parallel, e.g., to divide a relatively long code block into several sub-blocks which are also called processing windows. The forward and backward recursive calculations are carried out in parallel in each processing window. Normally, the size of each sub-block (processing window size) should be at least five times the constraint length.

Considering the complexity of the BCJR algorithm, some sub-optimal, but less complicated algorithms are of great interest in practical engineering. Among them is the SOVA. Different from the conventional Viterbi algorithm, SOAV can provide the soft information for each bit, in addition to finding an optimal path. In the computations of SOVA, Both the probability information of the survival path and the probability information of the competing path would be stored. Also, the priori probability can be incorporated to improve the posteriori probability. Figure 5.6 shows an example of SOVA.

Considering that at time instant $i=k$ (relative time $l=0$), there are two nodes with metrics $M_{i<k}^{(1)}$ and $M_{i<k}^{(2)}$, respectively. Here we use $i<k$ to emphasize that these two metrics are accumulated quantities. $\hat{u}_k(j_l)$ and $\hat{u}_k(j_l')$ denote the competing path j_l' at time instant (k+l), and the competing path j_δ' at time instant (k+δ). Both these two competing paths are with respect to the all-zero path j_δ. $M_{i<k+l}^{(1)}$ and $M_{i<k+l}^{(2)}$ are the two metrics at time instant l, corresponding to the survival path j_δ and a competing path j_l', respectively.

At time instant k, the joint probability of selecting path j and receiving sequence $y_{j\le k}$ can be written as

$$P(path=j, y_{j\le k}) \propto \exp[M_k(s)] \tag{5.11}$$

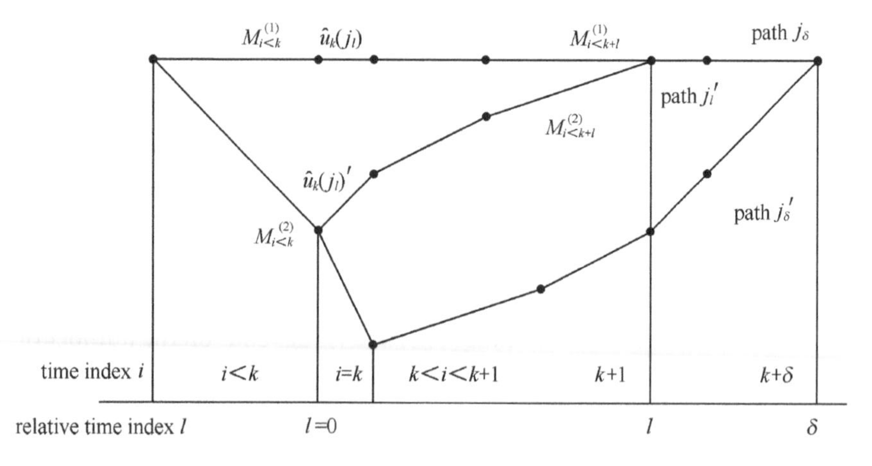

FIGURE 5.6 Illustration of SOVA algorithm.

Metric $M_k(s)$ depends on the states of the encoder along path j. As can be seen from Figure 5.6, SOVA tries to calculate the belief of bit u_k, δ time after Viterbi decoding. In the classic Viterbi algorithm, after the metric is updated at time instant k, the $path = j$ corresponding to the smaller metric in $M_k(s)$ would be selected as the survival path. At time instant $k+\delta$, we only need to select the survival path j_δ and discard the competing path j'_δ (terminating at the same trellis node). By contrast, in SOVA, extra computation is needed, e.g., the information of all the competing path j'_l should also be considered. Define the metric difference (accumulated till time instant $k+l$) between the survival path j_l and the competing path as j'_l.

$$\Delta_k^l = M_{k+1}(s^{(j_l)}) - M_{k+1}(s^{(j'_l)}) \geq 0 \tag{5.12}$$

The bigger Δ_k^l is, the more likely that the survival path is indeed j_l. The probability of selecting j_l as the survival path can be calculated as follows:

$$\begin{aligned} P_l(Correct) &= \frac{P(path = j_l, y_{j\leq k+l})}{P(path = j_l, y_{j\leq k+l}) + P(path = j'_l, y_{j\leq k+l})} \\ &= \frac{\exp[M_{k+1}(s^{j_l})]}{\exp[M_{k+1}(s^{j_l})] + \exp[M_{k+1}(s^{j'_l})]} \\ &= \frac{\exp(\Delta_k^l)}{1+\exp(\Delta_k^l)} \end{aligned} \tag{5.13}$$

Hence,

$$\Delta_k^l = \log \frac{P_l(Correct)}{1 - P_l(Correct)} \tag{5.14}$$

The LLR of the survival path is the metric difference in Eq. (5.12). Then, the LLR of bit u_k can be written as follows:

$$\begin{aligned} L(u_k) &= u_k \bullet \log \frac{\prod_{l=0}^{\delta}[1+\exp(\Delta_k^l)] + \prod_{l=0}^{\delta}[\exp(\Delta_k^l)-1]}{\prod_{l=0}^{\delta}[1+\exp(\Delta_k^l)] - \prod_{l=0}^{\delta}[\exp(\Delta_k^l)-1]} \\ &\approx u_k \bullet \min_{l=0,\ldots\ldots,\delta}(\Delta_k^l) \end{aligned} \tag{5.15}$$

For each trellis node along the survival path, trace back along the competing path to time instant k. if the decision on $\hat{u}_k(j_l')$ via the competing path is different from the decision on $\hat{u}_k(j_l)$ via the survival path, the LLR of bit u_k is updated. SOVA can include the priori information of $\hat{u}_k$. Hence Eq. (5.12) can be rewritten as

$$\Delta_k^l = M_{k+1}(s^{(j_l)}) - M_{k+1}(s^{(j'_l)}) + 0.5 \bullet L(u_k) \tag{5.16}$$

Here $L(u_k)$ is the priori information of bit $\hat{u}_k$, e.g., extrinsic information from the other constituent decoder.

The turbo decoding process itself is quite complicated, exhibiting strong nonlinear behavior. One noticeable characteristic is that the convergence rate is highly dependent on the signal-to-noise ratio (SNR) and code block length. The convergence rate directly determines the number of iterations needed in the receiver (e.g., decoder), which is of major concern in practical engineering. One effective tool to analyze the convergence behavior of iterative decoding (including the decoding algorithms for turbo codes and LDPC codes) is the Extrinsic Information Transfer (EXIT) chart [16,17]. Quite often there are two curves in an EXIT chart, representing the input/output response function of each constituent decoder, in the sense of belief of information bits. Figure 5.7 shows an example where the decoding starts from the input reliability in the bottom left corner. As the number of iterations increases, the belief of bits is growing by mapping between the two curves, and progressing toward the up-right corner.

5.1.4 Fundamental Performance

Since 1993, the understanding of turbo codes has been getting deeper and deeper. One of the breakthroughs in turbo code theory is the bound analysis [18,19], and the equivalence between turbo decoding and maximum likelihood (ML) decoding [20]. These finds are very useful in predicting the performance and designing turbo codes. Simply speaking, if the turbo interleaver is uniformly random, the bit error rate in the sense of ML can be estimated, assuming that the constituent channel encoders are recursive convolutional. It should be pointed out that the ML decoding for turbo codes is very complex which can rarely be used in practical systems. The iterative turbo decoding discussed earlier is a type of sub-optimal, but much simpler decoding algorithm. For the AWGN channel, the bit error rate in the sense of ML decoding can be approximated as follows:

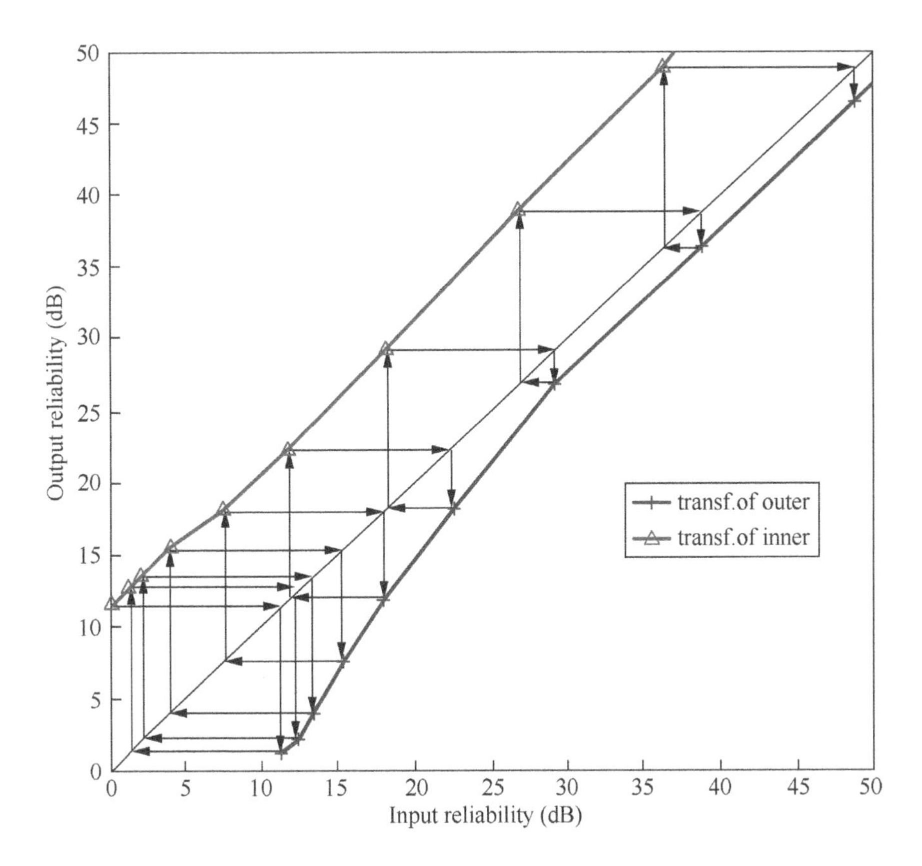

FIGURE 5.7 An example of an EXIT chart to illustrate the convergence behavior of iterative decoding.

$$P_b \approx \max_{w \geq 2}[\frac{w \bullet n_w}{N} \bullet Q(\sqrt{\frac{r \bullet d_{w,\min}^{TC} \bullet E_b}{N_0 / 2}})] \tag{5.17}$$

where w is the Hamming weight of the input (information) bit sequence, and n_w is the number of input bit sequences with weight w that corresponds to the minimum weight $d_{w,\min}^{TC}$ of the Turbo codeword. Both n_w and $d_{w,\min}^{TC}$ are related to the Turbo interleaver. $d_{w,\min}^{TC}$ is limited by the minimum weight $d_{w,\min}^{CC}$ of the constituent convolutional encoder. N is the length of Turbo interleaver, r is the code rate. E_b is the average energy per bit of the received signal, and $N_0/2$ is the double-sided power spectral density of AWGN. The function $Q(z)$ is defined as

$$Q(z) = \frac{1}{\sqrt{2\pi}} \int_{t=z}^{+\infty} \exp(-t^2 / 2) dt \tag{5.18}$$

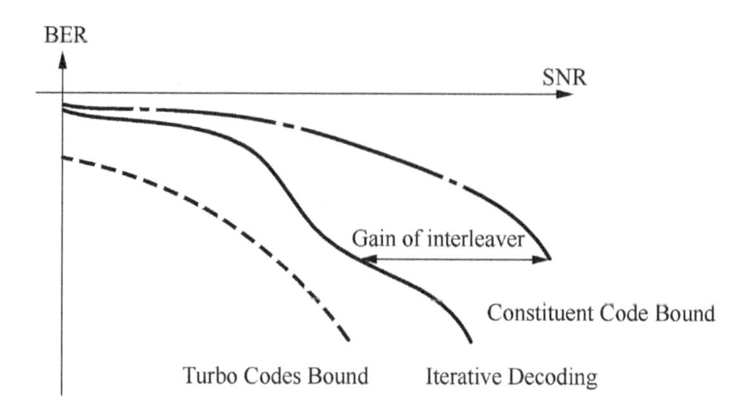

FIGURE 5.8 Illustration of the performance gain from the turbo interleaver.

It can be seen from Eq. (5.17) that the length of the turbo interleaver N has a big impact on the performance. The larger the N, the lower the bit error probability. Eq. (5.17) also shows the importance of recursive systematic convolutional (RSC) codes. For RSC, it is not hard to find $w \geq 2$, meaning that when at least two bits are in error, an error event would occur. However, for non-recursive convolutional codes, an error event would occur when one bit is in error. Low-weight codewords and their distributions can have a significant impact on the encoder's performance. In terms of Hamming distance, recursive convolutional codes do not have strong error correction capability. However, the diversity order is drastically increased after the randomization by the turbo interleaver, which helps to significantly improve the overall performance of turbo codes. The gain of the turbo interleaver can be illustrated in Figure 5.8.

There are three bit error rate (BER) vs. SNR curves in Figure 5.8. Among them, the performance bound of constituent convolutional codes can be represented as follows:

$$P_b = Q\left(\sqrt{\frac{r \bullet d_{w,\min}^{TC} \bullet E_b}{N_0 / 2}}\right) \tag{5.19}$$

The dashed line corresponds to the performance bound of turbo codes that can be represented by Eq. (5.17). The bold solid line is the performance of turbo codes with iterative decoding. Because of the term $1/N$ in Eq. (5.17), the performance bound of (in terms of BER) is much lower than that of constituent convolutional codes. The situation of iterative decoding depends on the SNR. At low SNR, the performance of iterative decoding is closer to that of constituent convolutional codes. As SNR increases, the BER of iterative decoding drops quickly over a narrow range of SNR, forming the so-called "waterfall" phenomenon. As the SNR continues growing, the BER of iterative decoding would gradually decrease and get close to that of the performance bound of turbo codes.

5.2 TURBO CODES IN LTE

In turbo codes for LTE [3], the most notable enhancement compared to Universal Mobile Telecommunication System (UMTS) turbo codes [2] is the use of quadratic permutation polynomial (QPP) inner interleaver.

5.2.1 Turbo Encoder of LTE

The encoder structure of LTE turbo codes is shown in Figure 5.9. Compared to the classic turbo codes as illustrated in Figure 5.2, there are two main differences. First, each constituent code uses its own tail bits to terminate the end state (as the branches connected to the lower switches in Figure 5.9). Second, there is no puncturing process here (in fact, it is moved to the rate

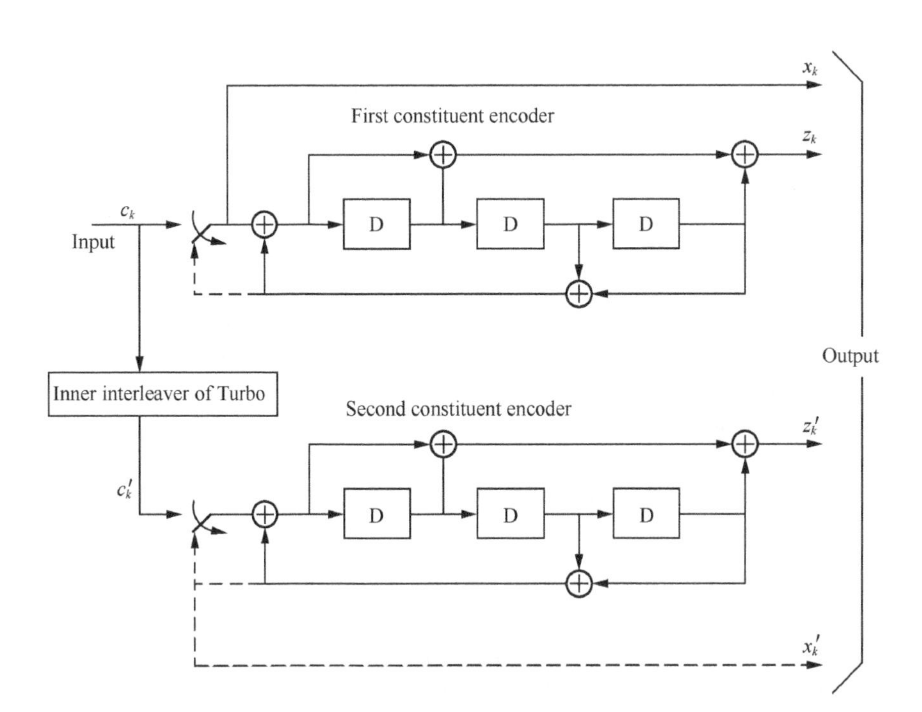

FIGURE 5.9 Encoder structure of LTE turbo codes [3].

matching module). The transfer function (generation matrix) of the eight-state constituent convolutional code can be represented as follows:

$$G(D)=[1,\frac{g_1(D)}{g_0(D)}] \tag{5.20}$$

where $g_0(D)=1+D^2+D^3$, $g_1(D)=1+D+D^3$..

As indicated in Section 5.1.2, by using different constituent convolutional codes, the performance of turbo codes would be improved [12,13]. Then why is the same polynomial used for the turbo codes in LTE? This is due to the complexity reduction consideration e.g. the same convolutional code can be reused by time sharing between the first parity bit stream and the second parity bit stream. Similarly, time sharing of the same convolutional code can also be implemented in the turbo decoder.

5.2.2 QPP Interleaver for LTE Turbo Codes

Normally, the performance of turbo codes does not quite rely on the optimization of each constituent code. Quite often, convolutional codes with simple polynomials can achieve good performance of turbo codes. Hence, a lot of research on turbo codes for practical applications focuses on the construction of turbo interleaver and efficient turbo decoding algorithms. From the standardization perspective, the design is mostly reflected in the selection of turbo interleaver, which constitute the major part of the standardization of turbo codes in 4G.

There are mainly two types of turbo interleavers: random interleavers and deterministic interleavers. The basic method is to use a random number generator. When the block length is very long, random interleavers can help to approach Shannon's limit. However, when the block length is short, pseudo-random interleavers would not be able to achieve the thorough shuffling of bits. One commonly used enhancement is the "S-Random" interleaver. Compared to a pure random interleaver, an S-random interleaver can ensure that for any pair of input points separated by S, the separation between the output points of this pair (e.g., after the interleaving) would not be less than S. By doing so, the bursty error in one constituent code would not propagate to the other constituent code, which can improve the performance of turbo codes when the block length is short. Figure 5.10 shows an example of an S-random interleaver.

Random interleavers usually require that both encoders and decoders maintain a memory to store the interleavers. For many practical

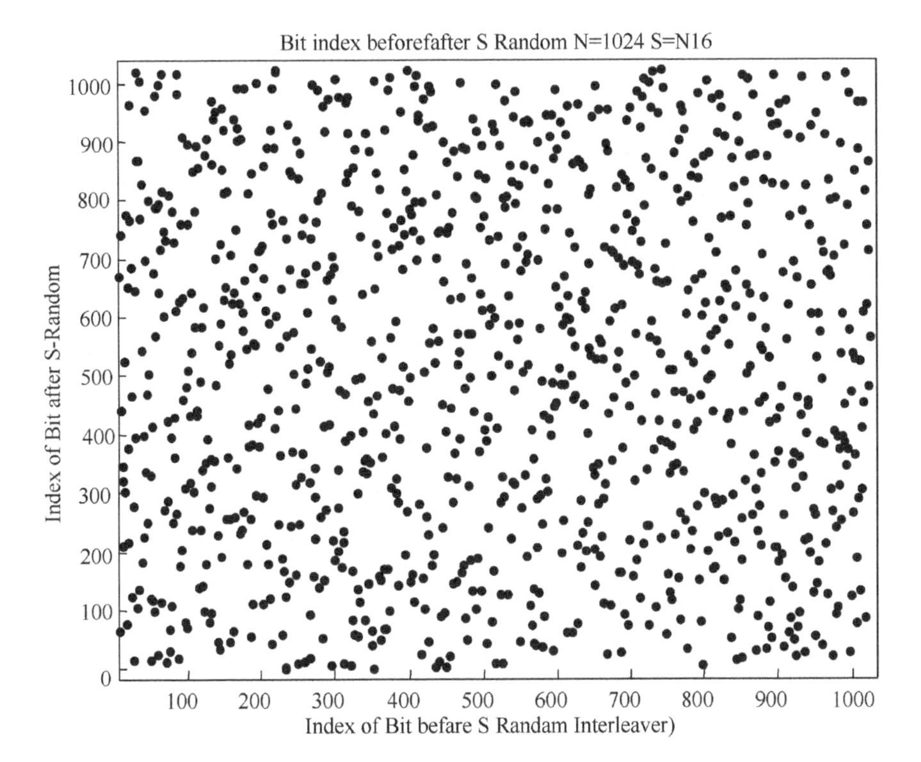

FIGURE 5.10 An example of a pattern of the S-random interleaver.

implementations, when the block length is very large, or multiple interleavers are needed, the memory size would be a big issue and can significantly increase the hardware cost. From this perspective, deterministic interleavers are more preferable, where the interleaving and de-interleaving process can be calculated, without the need to store the entire interleaver. Among the various kinds of deterministic interleavers, quadratic interleavers are more often used that is based on quadratic congruence algorithm. For a quadratic interleaver of size $N=2^n$, its basic form can be represented as [21]:

$$c_i = \text{mod}(\frac{k \bullet i \bullet (i+1)}{2}, N) \qquad ((5.21)$$

where the coefficient k is an odd number. The data in the position c_i would be changed to the position $\text{mod}(c_{i+1}, N)$ after the interleaving.

$$\pi_{QN}: c_i \mapsto \text{mod}(c_{i+1}, N), \quad \forall i \qquad (5.22)$$

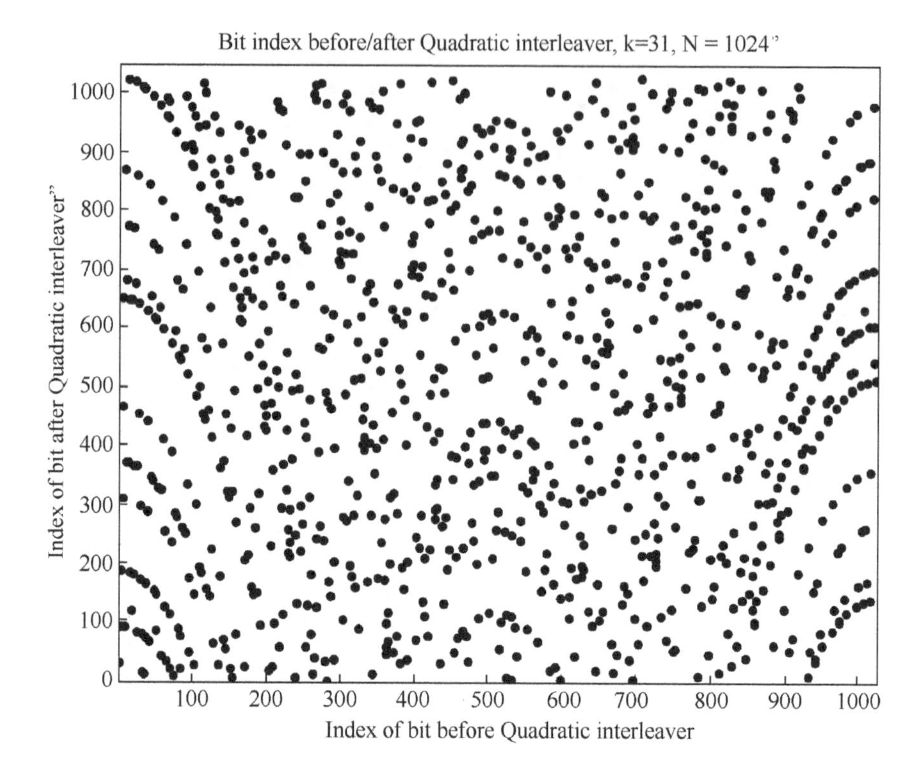

FIGURE 5.11 An example of a pattern of quadratic interleaver (noticeable parabolic patterns ↘ and ↗ seen on the left and the right).

Figure 5.11 shows an example of pattern of quadratic interleaver. Although the performance of deterministic quadratic interleaver is not as good as that of S-random interleaver, quadratic interleaver can reach the average performance of random interleavers [21].

Turbo interleaver in LTE is based on quadratic interleavers, which is called quadratic permutation polynomials (QPPs) interleaver [3]. Denote K as the length of information block, for i-th bit, the order after QPP interleaving becomes

$$\pi(i) = \mathrm{mod}((f_1 \bullet i + f_2 \bullet i^2), K) \tag{5.23}$$

Here the coefficients f_1 and f_2 are both smaller than K. The rather simple generation polynomial makes it easier to implement QPP interleaver in practical engineering. For instance, the QPP address can be calculated

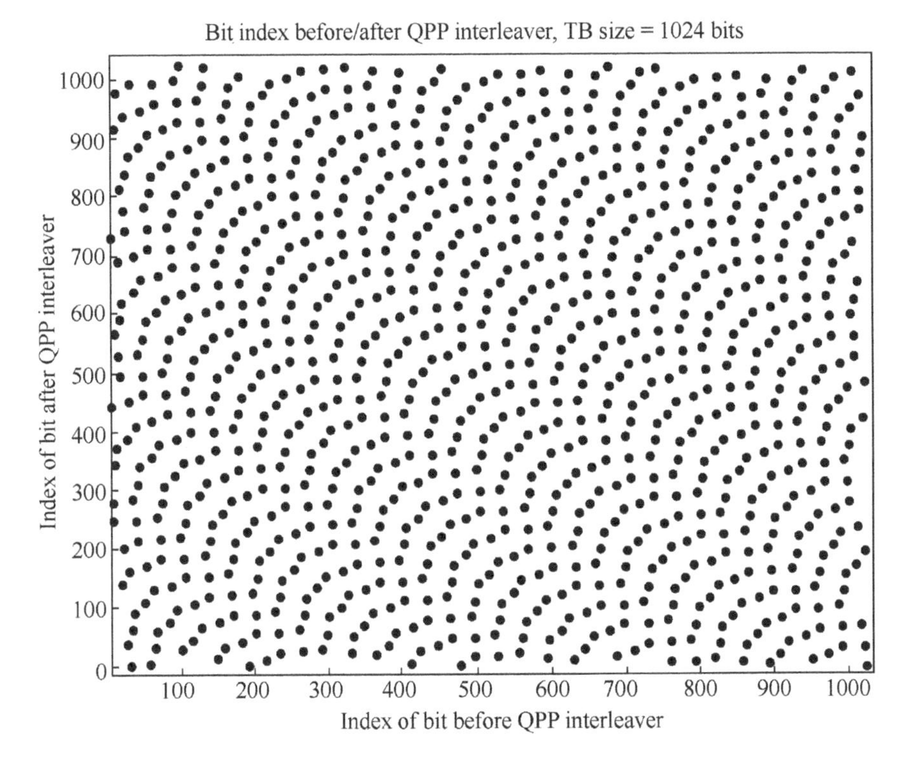

FIGURE 5.12 An example of a pattern of QPP interleaver (noticeable parabolic pattern ⌒).

recursively, without explicit computation of multiplication or modulo operation, like the following:

$$\begin{aligned} \Pi(x+1) &= f_1 \bullet (x+1) + f_2 \bullet (x+1)^2 && \text{mod} \quad K \\ &= \left(f_1 x + f_2 x^2\right) + \left(f_1 + f_2 + 2 f_2 x\right) && \text{mod} \quad K \\ &= \Pi(x) + g(x) && \text{mod} \quad K \end{aligned} \tag{5.24}$$

where $g(x) = \text{mod}(f_1 + f_2 + 2\bullet f_2 \bullet x), K)$ can be calculated recursively

$$g(x+1) = \quad g(x) + 2 f_2 \quad \text{mod} \quad K \tag{5.25}$$

Since both $\Pi(x)$ and $g(x)$ are smaller than K, the above modulo operation degenerates to a comparison operation. Figure 5.12 shows an example of a pattern of QPP interleaver.

TABLE 5.1 Values of f_1 and f_2 for Different Sizes (K) of Information Block for LTE QPP Interleaver

K	f_1	f_2	:	K	f_1	f_2
40	3	10	:	**3264**	443	204
:	:	:	:	:	:	:
408	155	102	:	**6144**	263	480

K values are the block sizes which are the results of QPP interleaver.

QPP interleaver provides big room for design optimization of turbo codes, not only in terms of performance improvement but also from the aspect of improving the parallel processing capability. During the detailed design of the QPP interleaver, the coefficients f_1 and f_2 can be optimized (as seen in Table 5.1) so that LTE turbo codes have the following characteristics:

- Performance is better or equal to that of the turbo codes for Release 6 UMTS (WCDMA).
- Very regular structure for the interleaver. The minimum unit contains 8 bits (1 byte). When K is between 40 and 512, the step size is 1 byte. When K is between 512 and 1024, the step size is 2 bytes. When K is between 1204 and 2048, the step size is 4 bytes. When K is over 2048, the step size is 8 bytes.
- Small padding overhead, not to exceed 63 bits over an entire code block.
- Flexible size of structures to support a wide range of block lengths. Easy for parallel processing.

5.2.3 Link-Level Performance

Figure 5.13 shows the required SNRs for 1% block error rate (BLER) of LTE turbo codes with different block lengths under the AWGN channel [22]. Here the channel estimation is assumed ideal. Quadrature phase-shift keying (QPSK) modulation and code rate of 1/3 are assumed, without hybrid automatic retransmission request (HARQ). Hence the corresponding spectral efficiency is 2/3=0.667 bps/Hz. The block lengths include all the supported sizes of physical shared channels in LTE. There are multiple curves in Figure 5.13, corresponding to the simulation results by different companies in 3GPP. It is observed that by using a longer block length, the maximum gain is about 1.4 dB, compared to very short code blocks. This

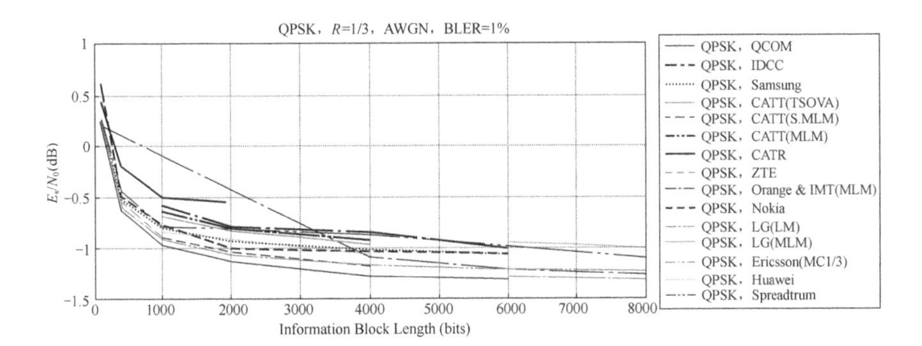

FIGURE 5.13 Performance of LTE turbo codes as a function of block length [22].

is called turbo interleaver gain. When the block length is increased to 1000 bits, most of the interleaving gain would be reaped.

It is also observed from Figure 5.13 that the performance of LTE turbo codes changes rather smoothly with the length of code block, e.g., BLER monotonically decreases as the block length increases, without the fluctuations observed in the turbo codes for UMTS. This allows the resource scheduler to carry out more precise link adaptation, and thus improve the overall system performance.

Figure 5.14 shows the curves of spectral efficiency vs. SNR for the AWGN channel, assuming ideal channel estimation, target BLER = 10% with single transmission (e.g., no HARQ retransmissions). There are three curves. The highest corresponds to the constrained channel capacity for 64-QAM. The middle one is derived from channel quality indicator (CQI). The lowest one is from the simulation where various overheads are considered, including control channels, cyclic prefix, reference signal, highest code rate = 0.93, and 2MHz guard band for 20 MHz. The gap between the simulation results (with overhead) and the theoretical performance bound is about 2–5 dB.

5.2.4 Decoding Complexity Analysis

The core computation inside each constituent decoder of Turbo is the BCJR algorithm [15] which is basically a serial processing. In order to improve the throughput, the sliding-window soft-input soft-output method is carried out when BCJR is implemented. In such decoding architecture, a code block is segmented into multiple sub-blocks, each equipped with an independent BCJR processing unit. Each sub-block is further divided into a number of windows. Within each window, the BCJR processing is serial. Metrics of the forward states at the end of m-th window would be used

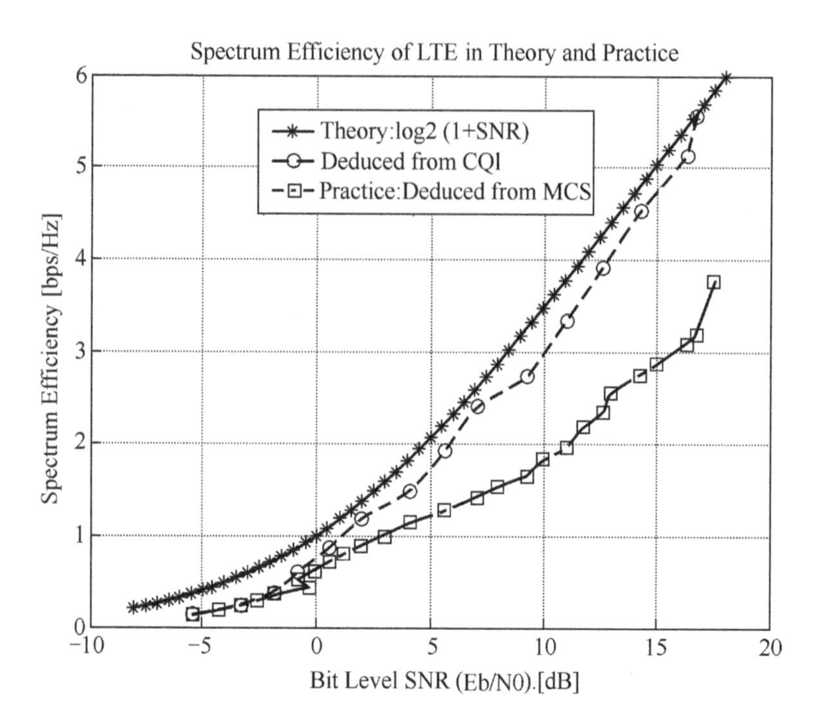

FIGURE 5.14 Performance of modulation and coding for LTE.

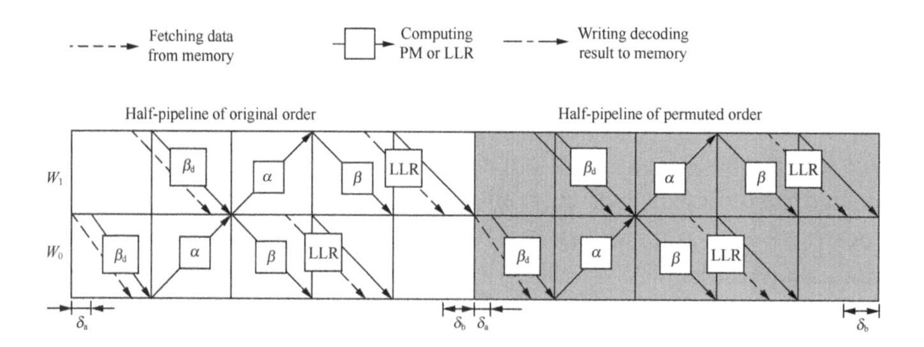

FIGURE 5.15 ACQ decoder architecture for turbo codes.

as the metrics of the forward states at the beginning of (m+1)-th window. The initialization of metrics of the backward states is a little more complicated and there are several different ways. One of the classic ways is called acquisition runs (ACQ) that can initialize the boundary condition of the windows and the sub-blocks. As Figure 5.15 shows, within each window, the dumb metric of backward state (β_d), the metric of forward state (α), the metric of backward state (β) and LLR need to be calculated.

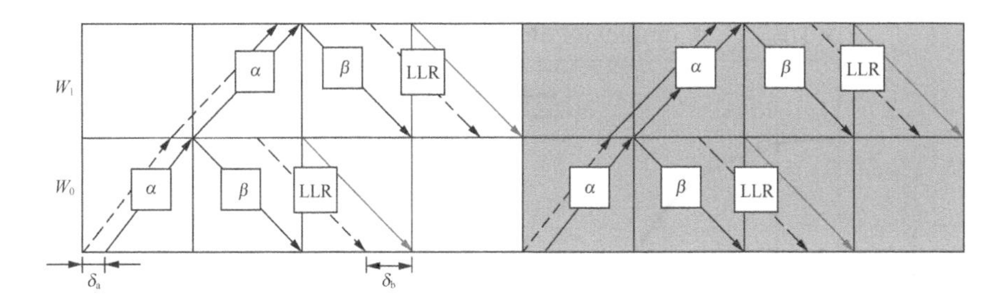

FIGURE 5.16 SMP decoding architecture for turbo codes.

TABLE 5.2 Decoding Complexity Comparison between Three Algorithms [23], Information Block Length = 4032 Bits

Algorithm	Log-MAP	Max-Log-MAP	TSOVA
No. of additions	774144	516096	262080
No. of max operation	258048	258048	68544
Table lookup	154828	0	0
Total	2580480	774144	330624

Another alternative way for the initialization of metric of backward state is called state-metric propagation (SMP). As Figure 5.16 shows, SMP is more suitable for very large-scale integrated (VLSI) circuits. It does not need the computation of dumb metric of backward states. In SMP, "β" and "α" of the previous iteration would be utilized.

The decoding complexities of turbo codes are listed and illustrated in Table 5.2 [23] and Figure 5.17 [24], respectively. In general, Log-MAP is the most complicated, followed by Max-Log-MAP and then Turbo with soft output Viterbi algorithm (TSOVA). In addition, the decoding complexity decreases as SNR increases. [24]. Overall speaking, turbo codes have good performance, but their decoding complexity is also relatively high.

5.3 TURBO CODES 2.0

In August 2016, a number of companies and research institutes proposed "Turbo codes 2.0" as the enhancement, including tail-biting turbo, a new puncturing method, new interleavers, etc. [6,7].

5.3.1 Longer Block Length

Turbo codes are suitable for long block lengths. The longer the block, the closer turbo codes can approach Shannon's limit. However, as the block

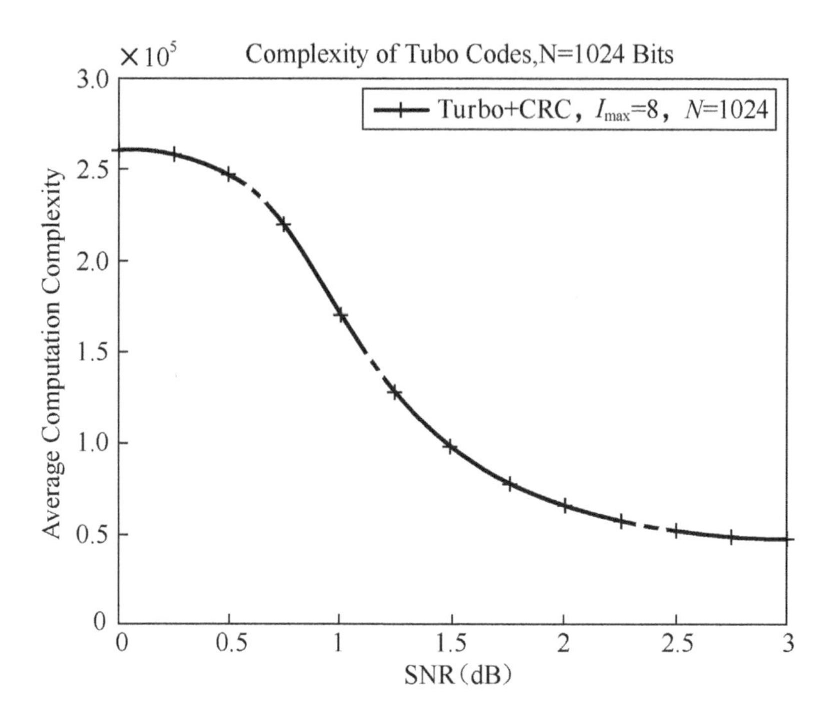

FIGURE 5.17 Decoding complexity of turbo codes [24].

length increases, the decoding complexity and latency also grow. This would limit its practical user. Certainly, a judicious increase of block length is still preferable in order to get some amount of performance gain without adding significant complexity. In LTE, the maximum block length is 6144 bits. Any longer block would be segmented into multiple sub-blocks, each being smaller or equal to 6144 bits. In [7], it was proposed to extend the block length to 8192 bits, with the purpose of supporting higher data rate and large blocks in 5G NR.

5.3.2 Even Lower Code Rate

In LTE, the mother code rate of turbo codes is 1/3. Various code rates can be obtained by puncturing or repetitions. By introducing more branches of parity bits, as illustrated in Figure 5.18, lower mother code rates of Turbo can be supported. Comparing the same repetition, a lower mother code rate setting would have better performance. As shown in Figure 5.19, with QPSK modulation and considering BLER=10%, compared to the repetition to get 1/5 code rate from 1/3 mother rate, the mother rate of 1/5 has about 0.5 dB performance gain [25].

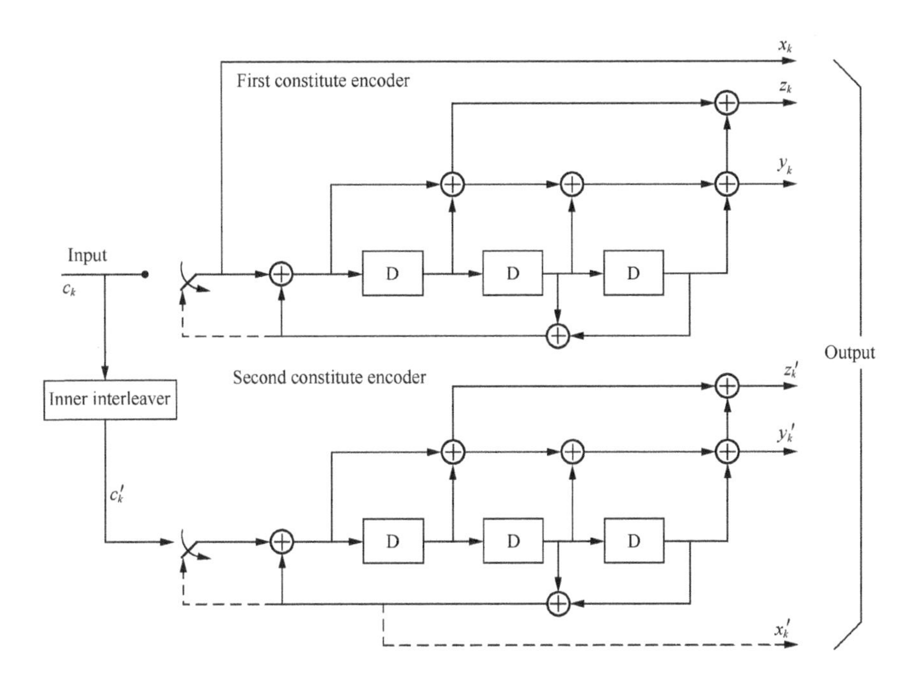

FIGURE 5.18 Turbo codes with low mother code rate (R = 1/5) [7].

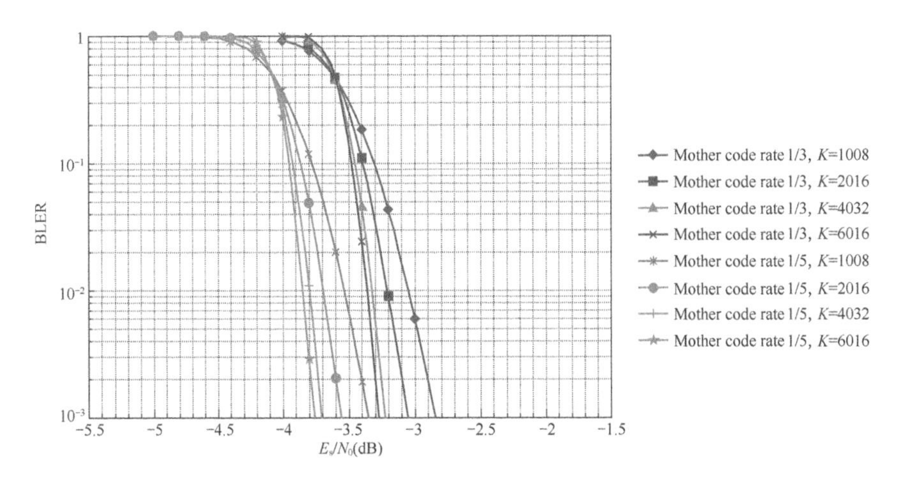

FIGURE 5.19 Performance gain with a mother code rate of 1/5 vs. repetition from 1/3 to 1/5 [25].

5.3.3 Tail-Biting Turbo Codes

In LTE Turbo, in order to terminate the status of the encoder back to 0 (to facilitate the decoding, e.g., no need to guess the end state), three tail bits should be appended to the end of the original information block. In a real implementation, these three bits can come from the last three bits of the encoded sequence (as seen in Figure 5.18, tapped from the output of the second shift register and the output of the third shift register, e.g., input switched to the lower branch). When the length of the information block is short (e.g., 40 bits), the tail bits would constitute noticeable overhead (3*3 bits, 9/(40*3+9)=6.9%). This would affect the competitiveness of turbo codes for short block length.

Tail-biting turbo codes [6,26] were proposed to reduce the overhead of tail bits. In tail-biting convolutional codes (TBCCs), since the convolutional code is non-recursive (e.g., feed-forward), it is easier to ensure the same state at the beginning and at the end, e.g., treat the last m information bits as the initial values for the shift registers (where m is the number of shift registers in the convolutional encoder). For tail-biting, since its constituent code is recursive convolutional, it is more complicated to align the beginning state and the end state. The basic method is to perform the cyclic encoding inside the two constituent encoders [6] (e.g., encoding twice), as shown in Figures 5.20 and 5.21.

Step 1. starting from all-zero states, perform convolutional encoding for the information bits (e.g., "01101101" as shown in Figure 5.20). Here, the output would be discarded. Only the end state S_k at time instant k needs to be stored. In the example in Figure 5.20, the initial state is $S_0=0$, and the end state is $S_k=5$.

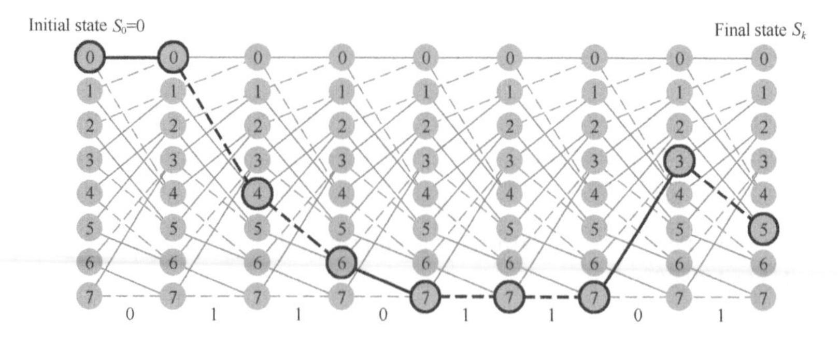

FIGURE 5.20 State transitions for tail-biting turbo codes in the first step of encoding.

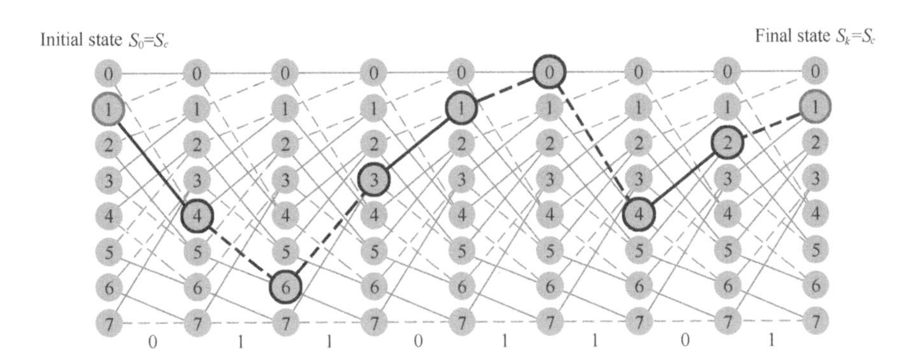

FIGURE 5.21 State transitions for tail-biting turbo codes in the second step of encoding.

TABLE 5.3 Lookup Table for Tail-Biting Turbo Codes

	***k* mod 7**					
S_k	**1**	**2**	**3**	**4**	**5**	**6**
0	0	0	0	0	0	0
1	6	4	3	2	5	7
2	3	5	4	6	7	1
3	5	1	7	4	2	6
4	7	2	1	5	6	3
5	1	6	2	7	3	4
6	4	7	5	3	1	2
7	2	3	6	1	4	5

Step 2. look up the value of the end state S_k from Table 5.3 (in the example of Figure 5.20, $S_k = 5$), the corresponding starting state S_c. According to Table 5.3 (second column), $S_c = 1$.

Step 3. starting from the state S_c (in this example, $S_c = 1$), perform the actual encoding per constituent code (in this example, the information bits are "01101101"). As seen in Figure 5.21, the starting status is $S_c = 1$. The end state is $S_k = 1$. That is, $S_k = S_c = 1$.

It should be pointed out that for different generator polynomials, Table 5.3 may be different. Also, according to [6], when the block length is a multiple of 7, a tail-biting code does not exist for recursive convolutional codes; thus, there are no tail-biting turbo codes.

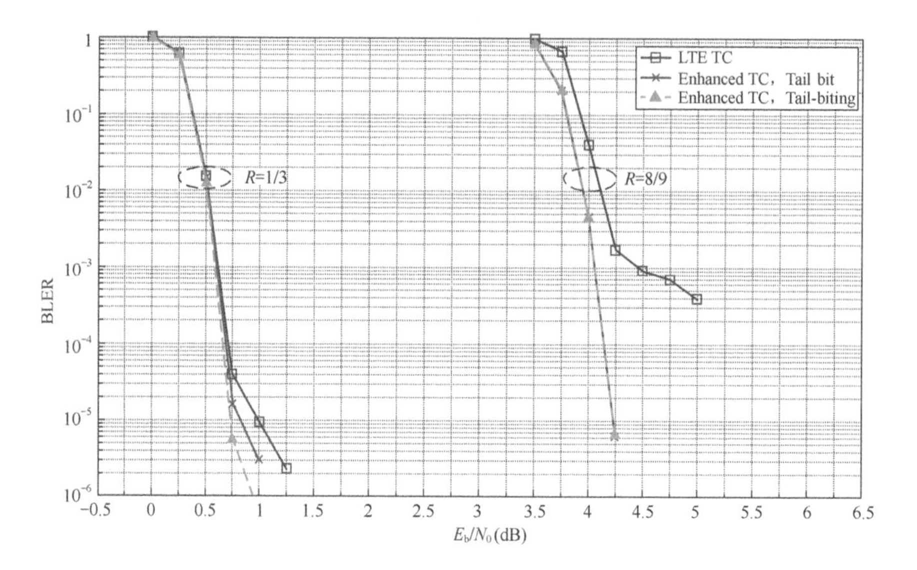

FIGURE 5.22 Performance of tail-biting turbo codes (K=6000 bits) [6].

As seen in Figure 5.22, a lower error floor is observed for tail-biting turbo codes [6]. This is due to the probability of having low-weight codewords generated by the regular tail bits. In addition, at a high code rate (R=8/9), 6000-bit information block length, and QPSK modulation, tail-biting turbo codes show 0.1 dB gain at BLER=10%. In [27], at R=1/2, BPSK and 40-bit information block length, 0.2 dB gain at BER=10^{-5} are observed for tail-biting turbo codes compared to the conventional turbo codes.

5.3.4 New Puncturing Method

In the new puncturing method [6], Hamming distance between the two constituent encoders after puncturing is optimized, as well as the extrinsic information. In addition, different puncturing patterns are introduced for systematic bits and parity bits (but the same between the two constituent encoders), respectively. This can further increase the Hamming distance after the encoding. In LTE, puncturing and interleaver are separately designed, which has no problems for long block length, but would lead to smaller Hamming distance for short block length.

5.3.5 New Interleaver

QPP interleaver in LTE is fixed and does not change with the code rate. In [6], the new interleaver can change with the code rate, in order to more

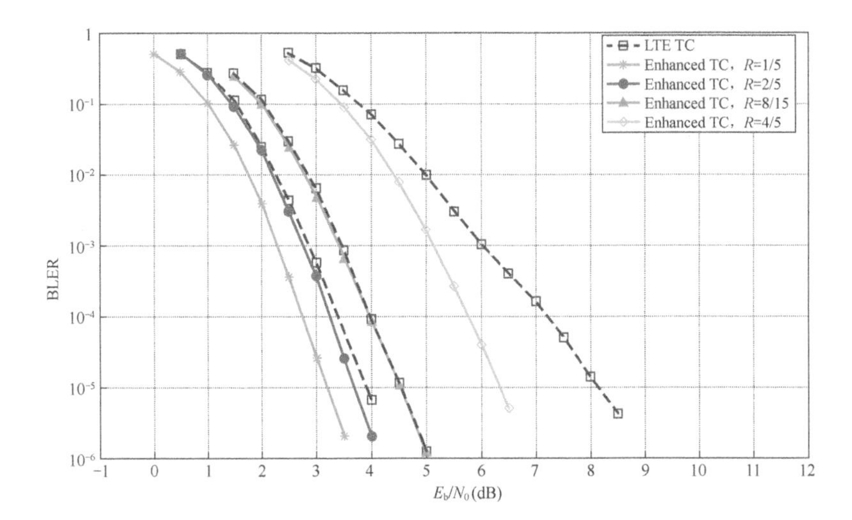

FIGURE 5.23 Performance of enhanced Turbo with new puncturing method and interleaver [6].

effectively reshuffle the information block and be more robust to the bursty interference. Its operation is

$$j = \Pi(i) = (P_i + S_{i \bmod Q}) \bmod K \tag{5.26}$$

where P and K are not divisible to each other. Q is a factor of K. S_i is a congruence class of Q and its value depends on K and code rate. By using this new interleaver, the conflict between read and write can be avoided in practical implementation, thus reducing the latency. Bursty errors can be better handled by different shuffling patterns for different code rates. As shown in Figure 5.23, by using the new puncturing method and new interleaver, the enhanced turbo codes can deliver 0.3 dB performance gain at a high code rate (R=4/5), information bit length=96 bits, QPSK modulation, and BLER=10%.

REFERENCES

[1] C. Berrou, et al., "Near Shannon limit error-correcting coding and decoding: Turbo Codes," *Proc. IEEE Intl. Conf. Communication (ICC 93)*, May 1993, pp. 1064–1070.

[2] 3GPP, TS25.212 V5.10.0-Multiplexing and channel coding (FDD) (Release 5), June 2005.

[3] 3GPP, TS36.212 V14.0.0-Multiplexing and channel coding (Release 14), Sept. 2016.

[4] 3GPP, R1-163662, Way Forward on channel coding scheme for 5G New Radio, Samsung, RAN1#84bis, April 2016.
[5] 3GPP, Draft Report of RAN1#84bis, April 2016.
[6] 3GPP, R1-167413, Enhanced turbo codes for NR: implementation details, Orange and Institut Mines-Telecom, RAN1#86, August 2016.
[7] 3GPP, R1-164361, Turbo code enhancements, Ericsson, RAN1#85, May 2016.
[8] R. G. Gallager, "Low-density parity-check codes," *IRE Trans. Inf. Theory*, vol. 8, Jan. 1962, pp. 21–28.
[9] R. G. Gallager, *Low density parity-check codes*, MIT, Cambridge, MA, USA, 1963.
[10] S. Y. Chung, "On the design of low-density parity-check codes within 0.0045 dB of the Shannon limit," *IEEE Commun. Lett.*, vol. 5, no. 2, 2001, pp. 58–60.
[11] S. Lin, *Error correction codes*, 2nd Edition (translated by J. Yan to Chinese), Mechanical Industry Press, Beijing, China, June 2007.
[12] P. C. Massey, "New developments in asymmetric Turbo Codes," *Proc. 2nd International Symposium on Turbo Codes*, September. 2000, pp. 93–100.
[13] O.Y. Takeshita, "Asymmetric turbo-codes," *IEEE International Symposium on Information Theory*, 1998, vol. 3, no. 3, p. 179.
[14] J. Deng, *Study on several key issues of Turbo codes*, Ph. D Thesis, Shanghai University, September 1999.
[15] L. Bahl, "Optimal decoding of linear codes for minimizing symbol error rate," *IEEE Trans. Inf. Theory*, vol. 20, no. 3, March 1974, pp. 284–287.
[16] S. ten Brink, "Convergence of iterative decoding," *Electron. Lett.*, vol. 35, no. 10, May 1999, pp. 1117–1119.
[17] S. ten Brink, "Convergence behavior of iteratively decoded parallel concatenated codes," *IEEE Trans. Commun.*, vol. 49, no. 10, Oct 2001, pp. 1727–1737.
[18] S. Benedetto, "Design of parallel concatenated convolutional codes," *IEEE Trans. on Comm.*, 1996, vol. 44, no. 5, pp. 591–600.
[19] S. Benedetto, "Design guidelines of parallel concatenated convolutional codes," *Proc. of IEEE Global Telecommunications Conference*, 1995, vol. 3, pp. 2273–2277.
[20] T. Richardson, "The geometry of turbo-decoding dynamics," *IEEE Trans. Inf. Theory*, 2000, vol. 46, no. 1, pp. 9–23.
[21] J. Sun, "Interleavers for turbo codes using permutation polynomials over integer rings," *IEEE Trans. Inf. Theory*, vol. 51, no. 1, Jan. 2005, pp. 101–119.
[22] 3GPP, R1-1610423, Summary of channel coding simulation data sharing, InterDigital, RAN1#86bis, October 2016.
[23] 3GPP, R1-1608768, Performance of Turbo codes with high speed decoding algorithm for NR, CATT, RAN1#86bis, October 2016.
[24] K. Chen, *Polar code theory and practical designs*, Ph. D Thesis, Beijing University Post & Telecommunications, March 2014.
[25] 3GPP, R1-166897, Turbo code enhancement and performance evaluation, LG Electronics, RAN1#86, August 2016.
[26] 3GPP, R1-061050, EUTRA FEC Enhancement, Motorola, RAN1#44bis, March 2006.
[27] F. Yang, *Turbo codes design and performance analysis for short block length*, Master Thesis, Xidian University, January 2013.

CHAPTER 6

Outer Codes

Liguang Li and Jun Xu

An outer code is usually implemented on top of a major channel coding. With respect to the outer code, the major channel coding is called inner code, as illustrated in Figure 6.1. For instance, polar code discussed in Chapter 3 is the inner code. When appended with cyclic redundancy check (CRC), the CRC would be considered as the outer code. Apparently, adding outer code would add more overhead, e.g., the need to increase the code rate of the inner code to keep the overall code rate the same. Then why to introduce outer codes?

6.1 CHANNEL CHARACTERISTICS AND OUTER CODES

In mobile communications, because of the multipath, Doppler, blocking, etc., the received signal quality would vary significantly, causing information errors during transmissions [1]. The errors can be random (sporadic) or bursty (localized with a big amount). Also, some coding schemes may lead to error propagation, resulting in large scale of errors.

In 5G NR, URLLC traffic of a high-priority user may puncture the resources of eMBB traffic of a low-priority user [2,3], as shown in Figure 6.2. In this case, the bursty error is likely for eMBB traffic.

For extremely short block lengths, e.g., Reed–Muller (RM) codes, no outer code is needed. Otherwise, the overhead is huge. Some codes such as low-density parity check (LDPC) have self-checking capability [4] and may not need outer codes or use small outer codes with mild effects on the code rate (e.g., CRC). From the protocol layers point of view, the outer code and inner code can be specified at different layers. For instance, inner codes such as turbo codes usually operate in the physical layer, whereas

DOI: 10.1201/9781003336174-6

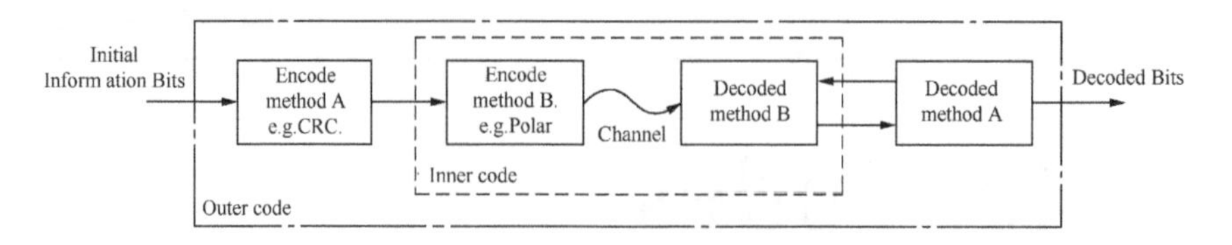

FIGURE 6.1 Relations between outer code and inner code.

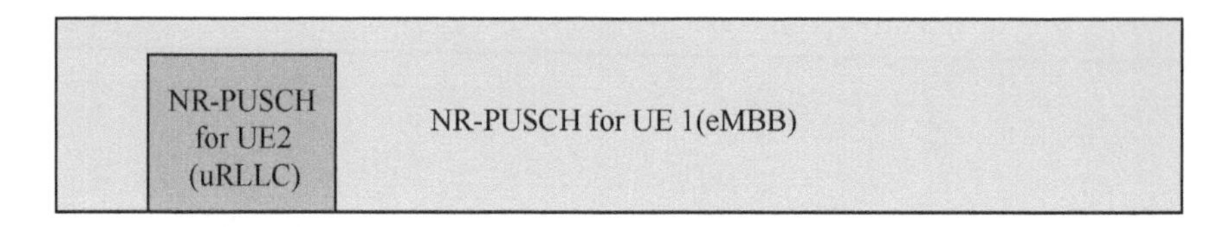

FIGURE 6.2 Low-priority eMBB traffic punctured by high-priority URLLC [2].

outer codes like Reed-Solomon (RS) code normally operate at the media access control (MAC) layer [5,6]. An inner code can provide soft information to an outer code, via soft detection/decoding algorithms.

In summary, the major function of outer codes is to correct the possible errors during the inner code decoding and reduce the error floor. Whether to use outer codes depends on the scenarios. Based on the collaboration between outer codes and inner codes, outer codes can have two types: explicit outer codes and implicit outer codes. Explicit outer codes are more common and will be discussed first.

6.2 EXPLICIT OUTER CODES

Explicit outer codes include more common outer codes and less common packet coding [1,2,7,8].

6.2.1 Common Outer Codes

Common outer codes include CRC, RS codes, and BCH codes.

- **CRC**. It is very compact and widely used for error detection (of course, CRC can also correct a small number of errors if used independently). CRC is called "genie" by Tal and Vardy [9], which demonstrates the powerfulness of CRC. As shown in Figure 6.3 [10], when 24-bit CRC is used, the polar code with mother block length $N=1024$, 1/2 code rate, and list size $L=8$, can outperform the polar code without CRC by about 0.4 dB at block error rate (BLER) $=10^{-3}$.

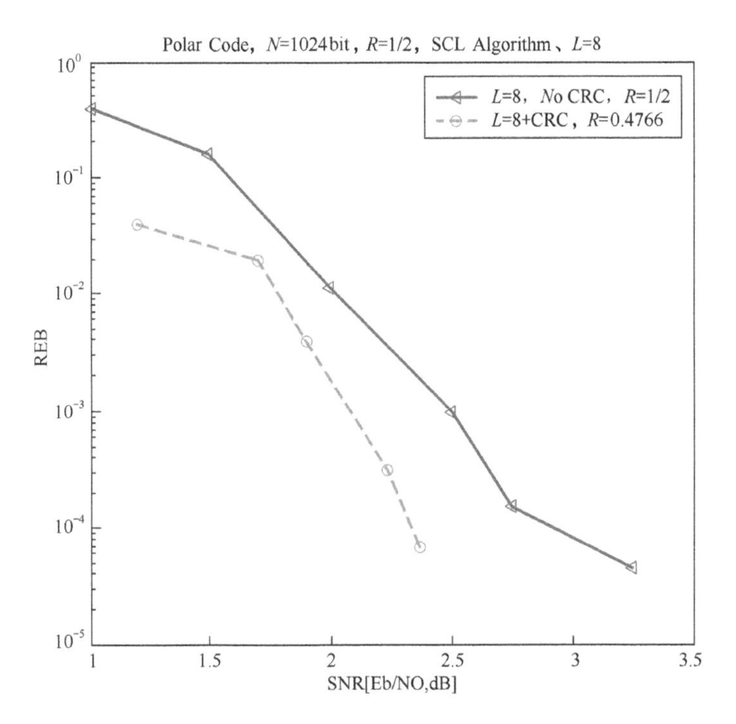

FIGURE 6.3 Performance gain by using outer CRC for polar codes [10].

- **Parity check (PC).** Its function is similar to CRC, except that the parity check bits are quite short (one bit may generate multiple PC bits based on the shift register structure [11])

- **RS codes [12,13].** RS codes were invented by Reed and Solomon in 1960 [12]. They have very powerful in correcting bursty errors. A (q–1, q–2t–1, 2t+1) RS code can correct t bursty errors. As seen in Figure 6.4 [10], when (255, 247, 9) RS is used, 0.25 dB performance gain is observed at BLER $=10^{-3}$ for polar code of $N=1024$, R = 1/2, $L=8$, compared to without RS code case. It is also observed that in the low BLER region, using RS code can ensure that BLER vs. signal-to-noise ratio (SNR) curves quickly enter the waterfall region.

- **BCH codes [13–15].** BCH codes were invented by Bose, Chaudhuri, and Hocquenghem in 1960 [14,15]. They are usually used for correcting random errors. As illustrated in Figure 6.4 [10], when (63, 57) BCH code is used, 0.1 dB performance gain is observed at BLER $=10^{-3}$ for polar code of $N=1024$, R $=1/2$, $L=8$, compared to without RS code case.

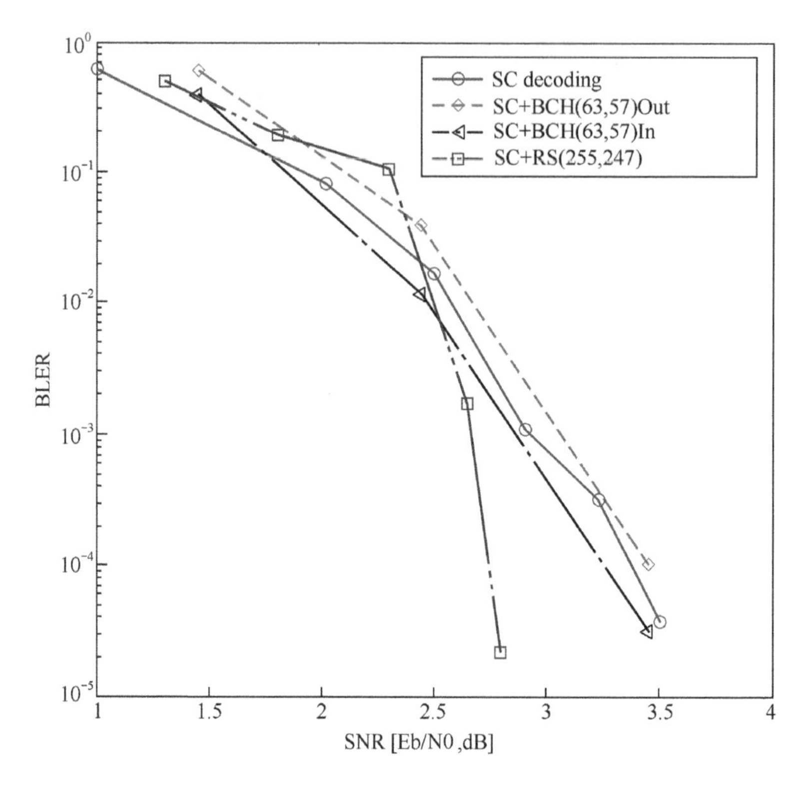

FIGURE 6.4 Performance benefit from RS code and BCH code [10].

6.2.2 Packet Coding

The basic idea of physical layer packet coding [1,2,7,8] is to add a sub-block to the traditional code blocks, or more specifically, to add a PC packet to all the code blocks. The motivation is to create a PC relationship between code blocks, in order to improve the reliability during the decoding.

In traditional packet transmissions, there is no relationship between code blocks, even when inter-block interleavers are used. It should be emphasized that certain puncturing is needed for each code block so that the sum of bits in the traditional code blocks and in the PC block would be the same as that of the original code blocks.

6.2.2.1 Solutions of Packet Coding

Key procedures of packet coding are shown in Figure 6.5: packet segmentation → appending CRC → channel encoding → generating parity bits for packet coding → bit selections → final packets to send. Compared to the traditional steps, packet coding and bit selections are added.

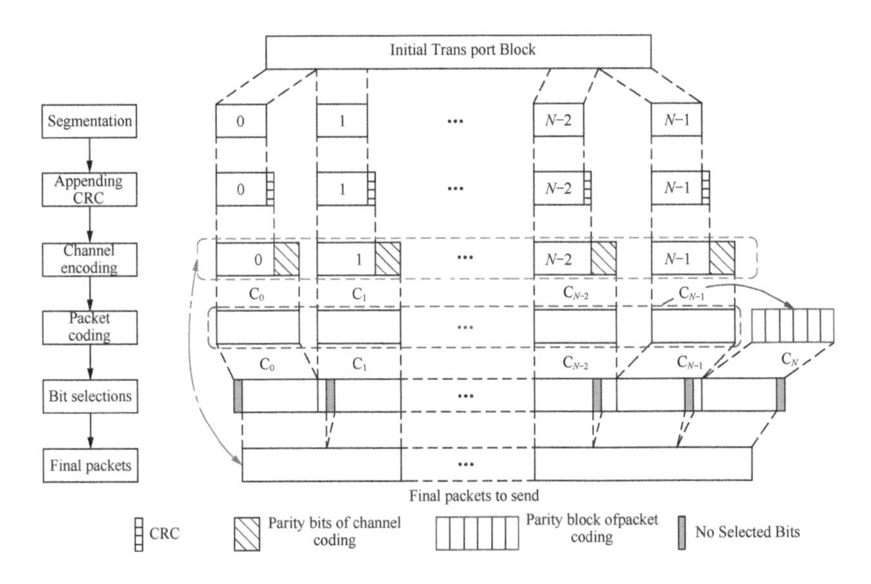

FIGURE 6.5 Illustration of packet coding.

In packet coding, a parity check packet is generated by an "exclusive-or" (XOR) operation over the corresponding bits overall all the code blocks. For instance, for N encoded blocks, i-th bits in all N code blocks form an information sequence S. To perform packet coding (e.g., XOR) on S, we get the PC packet C_N. Besides XOR, other forms of PC operations can be considered, such as multi-fold PC, Hamming codes, etc. To keep the overall code rate unchanged (when an extra PC packet is generated), a bit selection module is needed to puncture some bits.

Bit selection should ensure that the size of the final packet (including code blocks and packet coding block) is the same as the size of the original code blocks. The basic principle is to uniformly puncture the bits (same amount) over all code blocks. Here the inner code is assumed LDPC. Let us look at an example: code block length=672 bits, lift size=42, the size of the base matrix is M_b*N_b where N_b is fixed to be 16. The number of bits to be punctured can be calculated as:

- If the number of code blocks is no greater than 15, then 42 bits would be punctured in each LDPC code block. The rest (672-42*N) would be punctured by the packet coding block.
- If the number of code blocks is greater than 15, the punctured bits should be more evenly shared between the code blocks and the packet coding block.

42 42

0	0	0	0	0	0	0	0	0	0	0	0	0	−1	−1	−1
8	16	40	34	32	12	22	36	18	13	19	0	−1	0	−1	−1
30	20	18	22	38	2	6	28	32	37	26	21	31	−1	0	−1
40	24	12	20	10	14	2	30	16	19	34	18	−1	13	5	0

FIGURE 6.6 An example of punctured bits in an LDPC code block.

The rationale for such puncturing is as follows.

1. If the number of punctured bits is more than the lifting size, as Figure 6.6 shows when the first 84 bits are all punctured (an extreme case), all the PC equations in the min-sum LDPC decoder would become useless. Then no extrinsic information update is possible, leading to quite poor performance.
2. To keep as many as possible the bits in the packet coding block (e.g., less puncturing in the packet coding block, in order to fully utilize the PC relations).
3. To maintain even performance across different code blocks.

The bit selection process can be represented in a quantitative manner. Assuming there are N LDPC blocks where the numbers of bits to be punctured in each of the blocks are $e_0, e_1, e_2, \ldots\ldots, e_{N-1}$. The number of bits to be punctured in the packet coding block (e.g., parity packet) is f_0. When $N \leq 15$, their values should be

$$e_i = 42, \qquad i = 0,1,2,\ldots\ldots, N-1 \tag{6.1}$$

$$f_0 = 672 - \sum_{i=0}^{N-1} e_i \tag{6.2}$$

When $N > 15$, the values of $e_0, e_1, e_2, \ldots\ldots, e_{N-1}$ are as follows (the value of f_0 is the same as (Eq. 6.2)):

$$e_i = 1 + floor(672/(N+1)), \qquad i = 0,1,2,\ldots\ldots, G-1 \tag{6.3}$$

$$e_i = floor(672/(N+1)), \qquad i = G, G+1, G+2, \ldots\ldots, N-1 \tag{6.4}$$

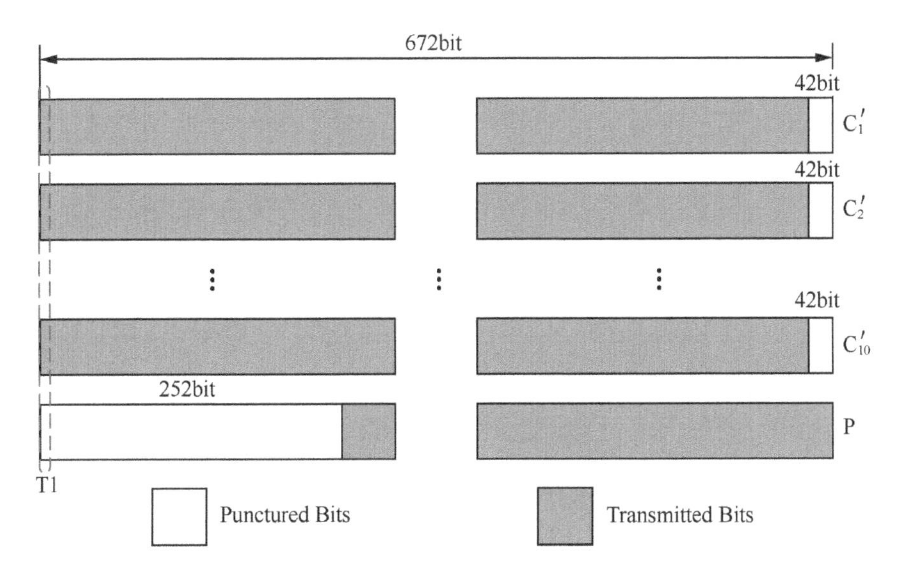

FIGURE 6.7 Illustration of Method 1 of bit selection.

where

$$G = 672 - (N+1) \bullet floor(672/(N+1)) \tag{6.5}$$

For example, assuming that $N = 10$, then $e_i = 42, \quad i = 0,1,2,......,9$, and $f_0 = 252$.

If $N = 100$, then $e_i = \begin{cases} 7, & for \ i = 0,1,2,\cdots,65 \\ 8, & for \ i = 66,67,\cdots,99 \end{cases}$ and $f_0 = 6$ 。

6.2.2.2 More Details of Bit Selections

The place of the puncturing within an LDPC code block would also affect the decoders' performance over the entire packet. Such optimization is mainly based on the simulation results. Let us still use the number of code blocks $N = 10$ as the example where 42 bits are punctured in each LDPC code block and 252 bits are punctured in the PC block.

Method 1: in each LDPC code block, the bits are punctured from the end. In the PC block, the bits are punctured from the beginning, as illustrated in Figure 6.7.

Simulation results of Method 1 are shown in Figures 6.8 and 6.9 for code rates of 1/2 and 13/16, respectively. The number of code blocks can be 2, 10, 50, and 100. It is seen that at a low code rate, performance gains can be observed in packet coding across a various number of code blocks. However, at a high code rate, the gain becomes zero or even negative in the

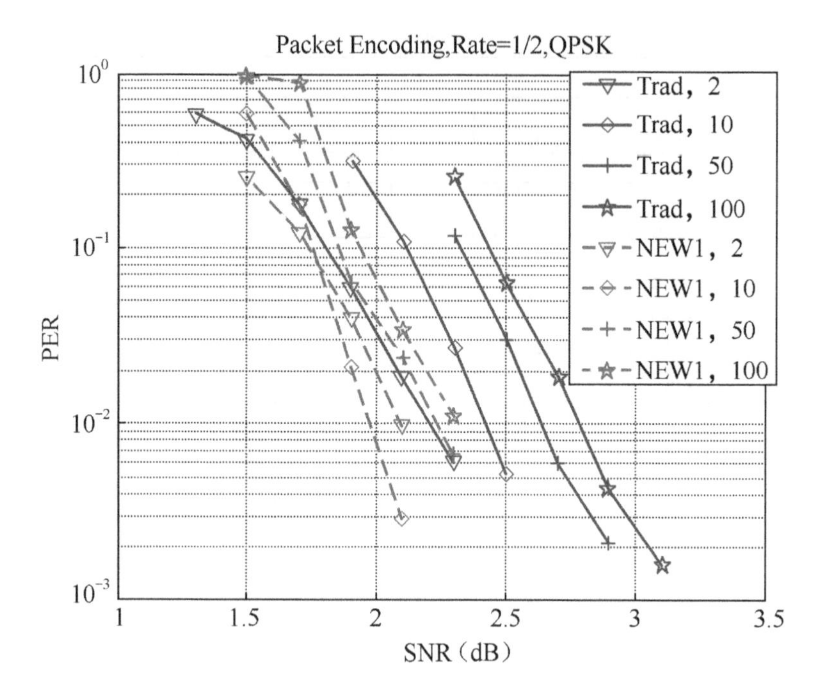

FIGURE 6.8 Simulation results of Method 1 for code rate of 1/2.

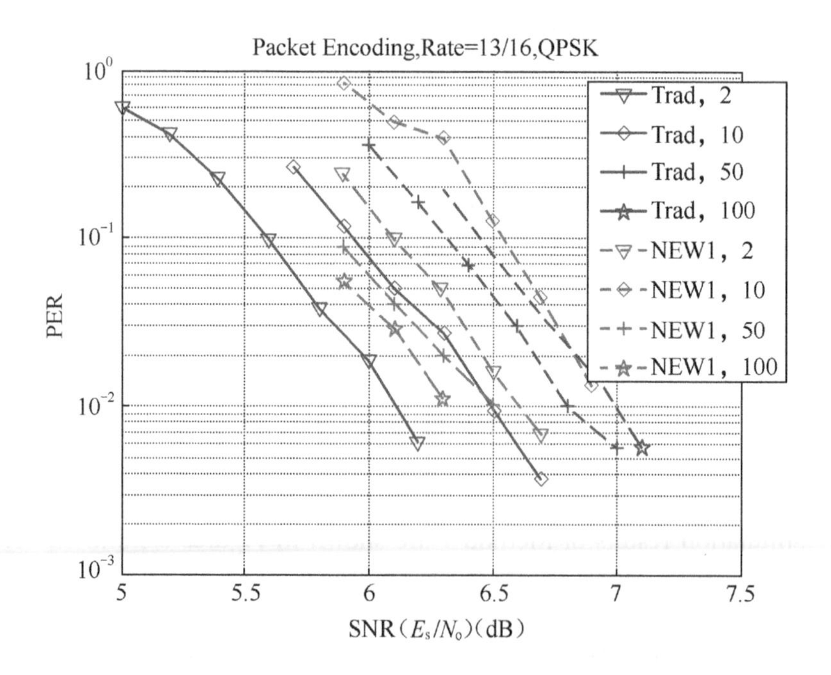

FIGURE 6.9 Simulation results of Method 1 for code rate = 13/16.

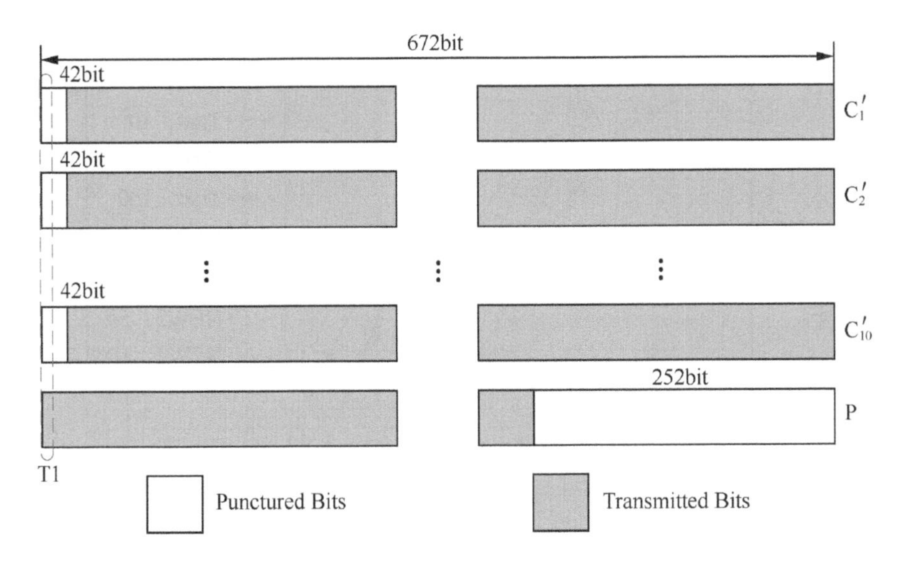

FIGURE 6.10 Illustration of Method 2 of bit selection.

case of 2 or 10 code blocks. The rule of thumb seems to be that the gain is more pronounced when there are a lot of code blocks (e.g., enough parity bits for correction).

Method 2: in each LDPC code block, the bits are punctured from the beginning. In the PC block, the bits are punctured from the end, as shown in Figure 6.10.

Simulation results are shown in Figures 6.11 and 6.12 for code rates of 1/2 and 13/16, respectively. The number of code blocks can be 2, 10, 50, and 100. It is seen that when the code rate is low, the performance gain is observed for various numbers of code blocks, except for the low gain in the 2-block case. When the code rate is 13/16, no gain can be observed for the two-block case, whereas gains can be observed for all other numbers of blocks.

Method 3: in each LDPC code block, the bits are punctured from the end, sequentially block by block. In the PC block, the bits are punctured from the beginning. There is no overlap of indices of the punctured bits, as illustrated in Figure 6.13.

Simulation results are shown in Figures 6.14 and 6.15 for code rates of 1/2 and 13/16, respectively. The number of code blocks can be 2, 10, 50, and 100. Performance gain is observed for various numbers of blocks and code rates.

Method 4: in each LDPC code block, the bits are punctured from the beginning, and sequentially block by block. In the PC block, the bits are

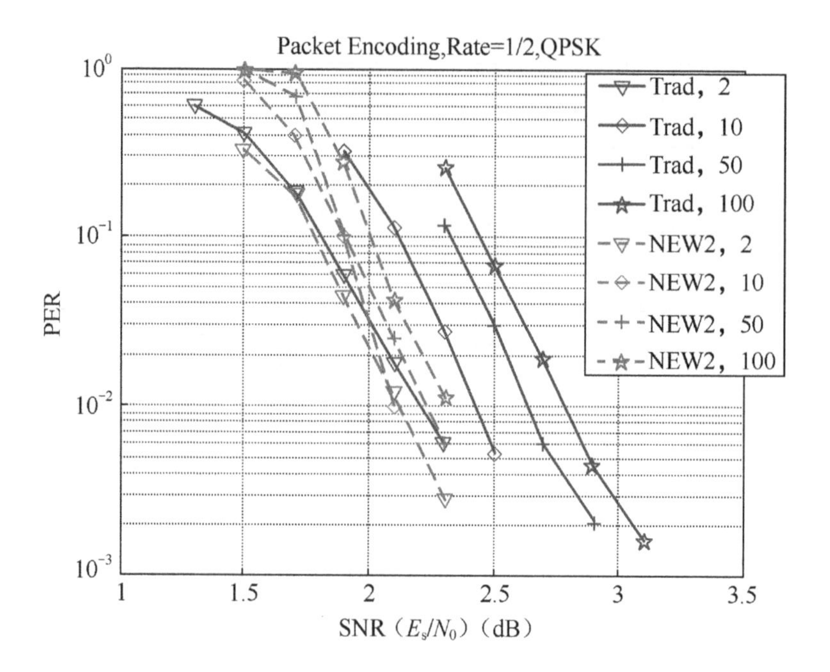

FIGURE 6.11 Simulation results of Method 2 for code rate = 1/2.

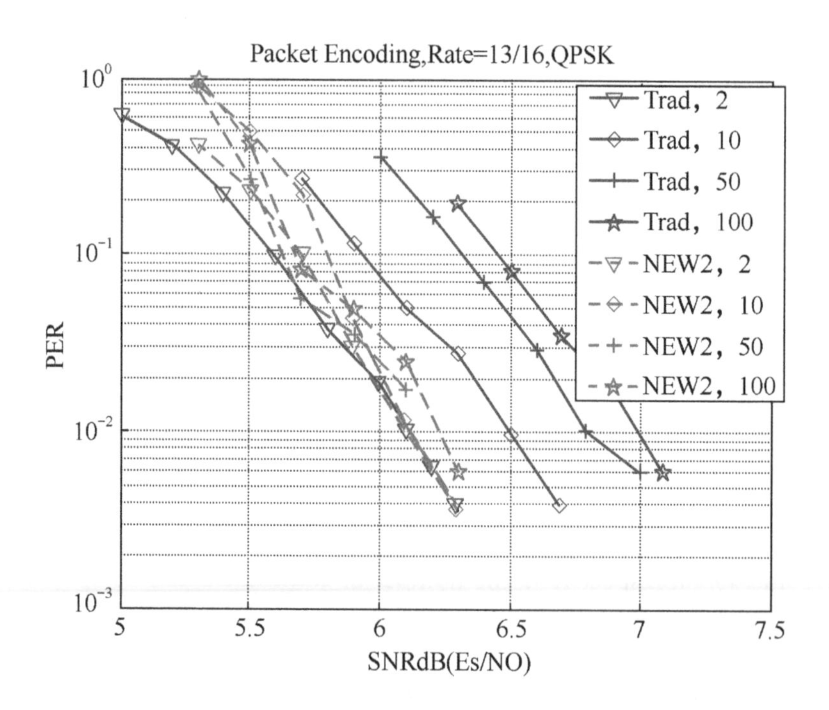

FIGURE 6.12 Simulation results of Method 2 for code rate = 13/16.

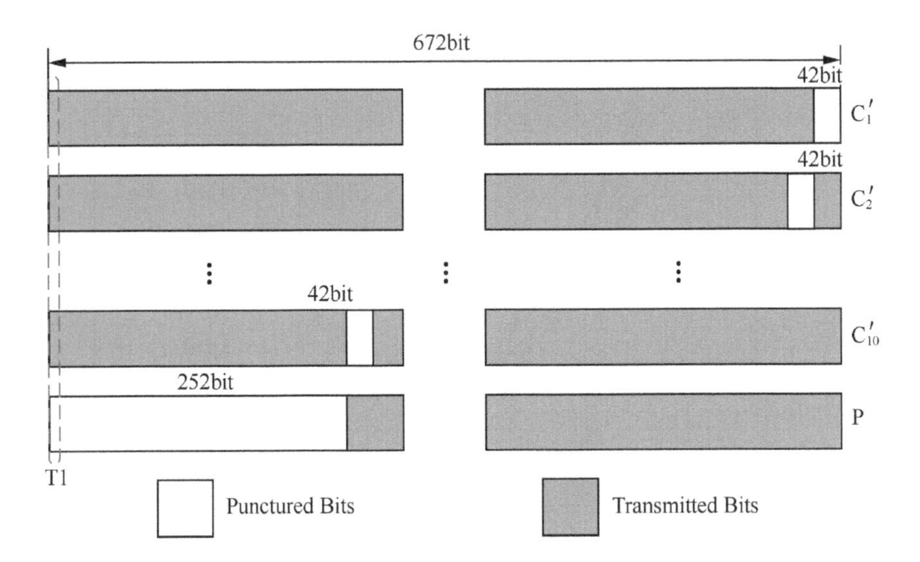

FIGURE 6.13 Illustration of Method 3 of bit selection.

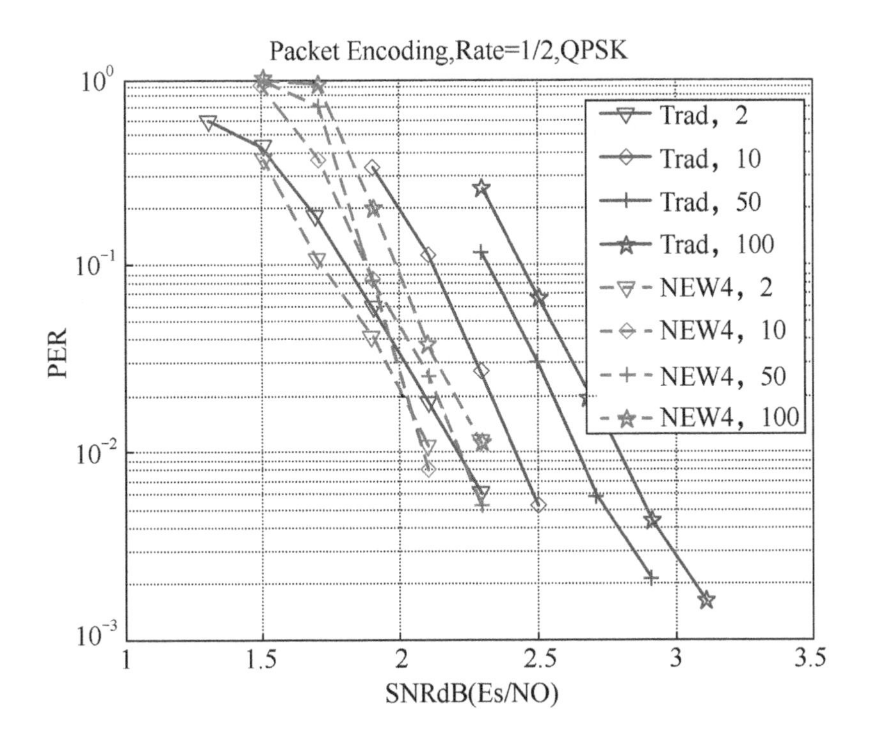

FIGURE 6.14 Simulation results of Method 3 for code rate = 1/2.

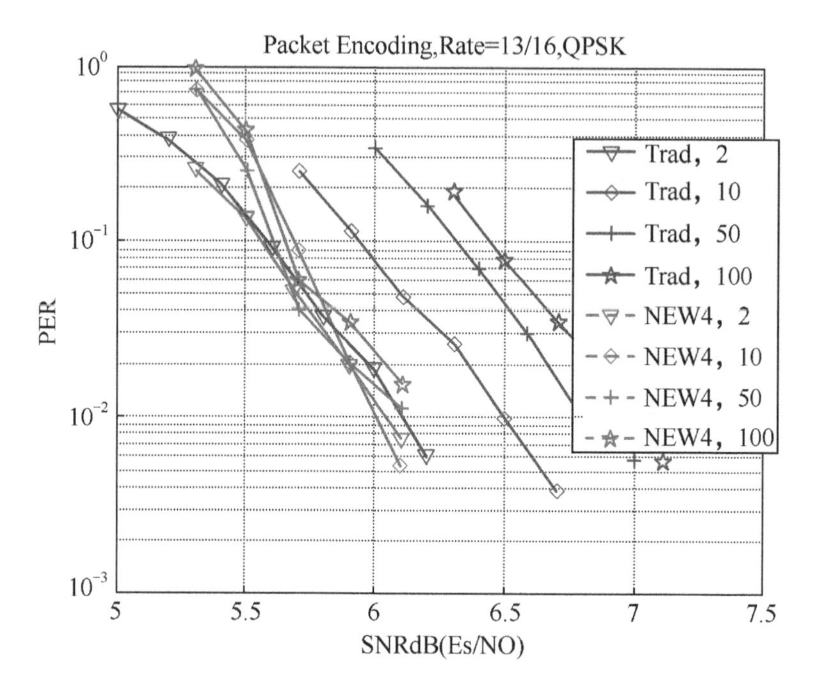

FIGURE 6.15 Simulation results of Method 3 for code rate=13/16.

punctured from the end. There is no overlap of indices of the punctured bits, as illustrated in Figure 6.16.

Simulation results are shown in Figures 6.17 and 6.18 for code rates of 1/2 and 13/16, respectively. The number of code blocks can be 2, 10, 50, and 100. It is seen that performance gain is observed in various code rates and numbers of code blocks.

From the above simulation results, it is seen that Methods 3 and 4 are slightly better than Methods 1 and 2.

In addition, according to the simulation results in [2], as illustrated in Figure 6.19, by using the outer code similar to the packet coding, the channel erasure-induced errors (due to the puncturing of eMBB bits by uRLLC traffic) can be effectively reduced so that the throughput can be increased noticeably ("o" curve in Figure 6.19). By contrast, when there is an outer code for protection, the throughput decreases quickly as the erasure probability increases ("*" curve in Figure 6.19).

6.2.2.3 *Decoding Algorithms*

From the encoding of packet coding, we can see that packet coding is a form of serially concatenated code (inner code+outer code): first to carry

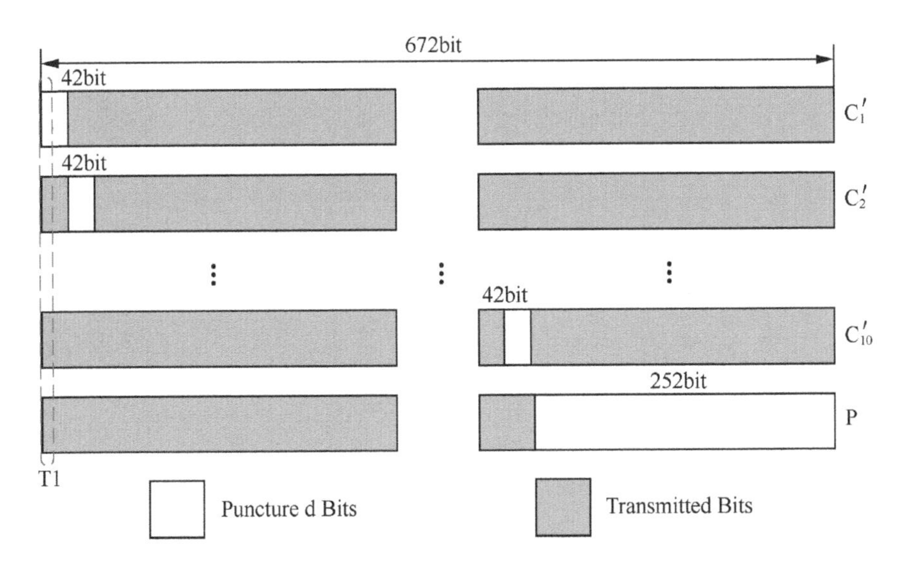

FIGURE 6.16 Illustration of Method 4 of bit selection.

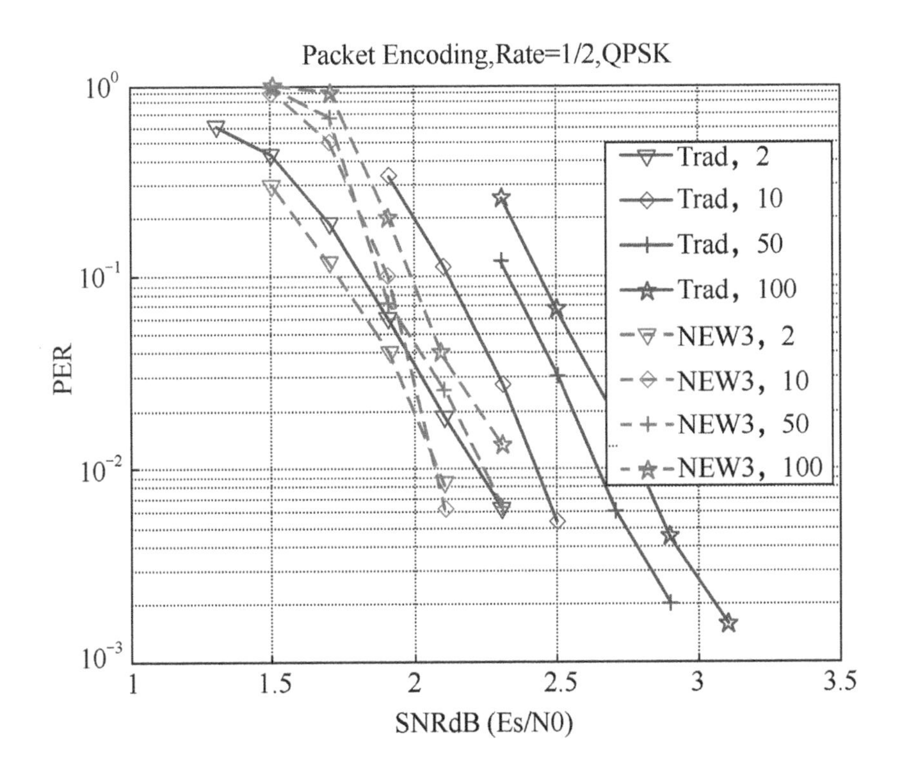

FIGURE 6.17 Simulation results of Method 4 for code rate = 1/2.

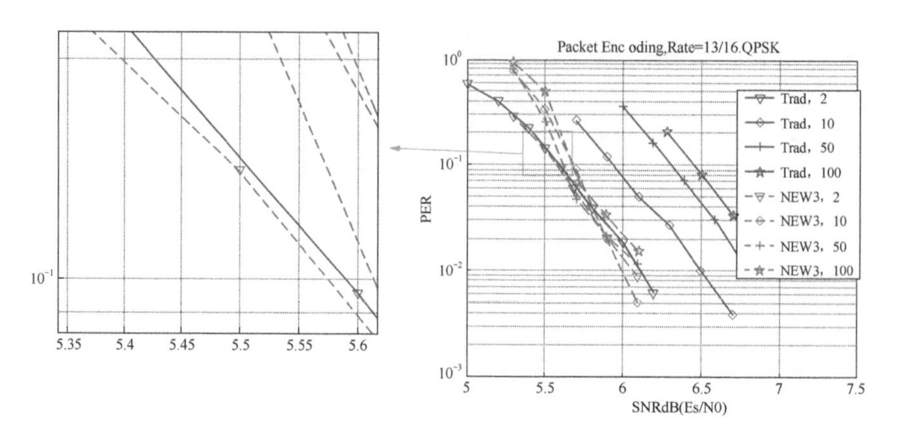

FIGURE 6.18 Simulation results of Method 4 for code rate = 13/16.

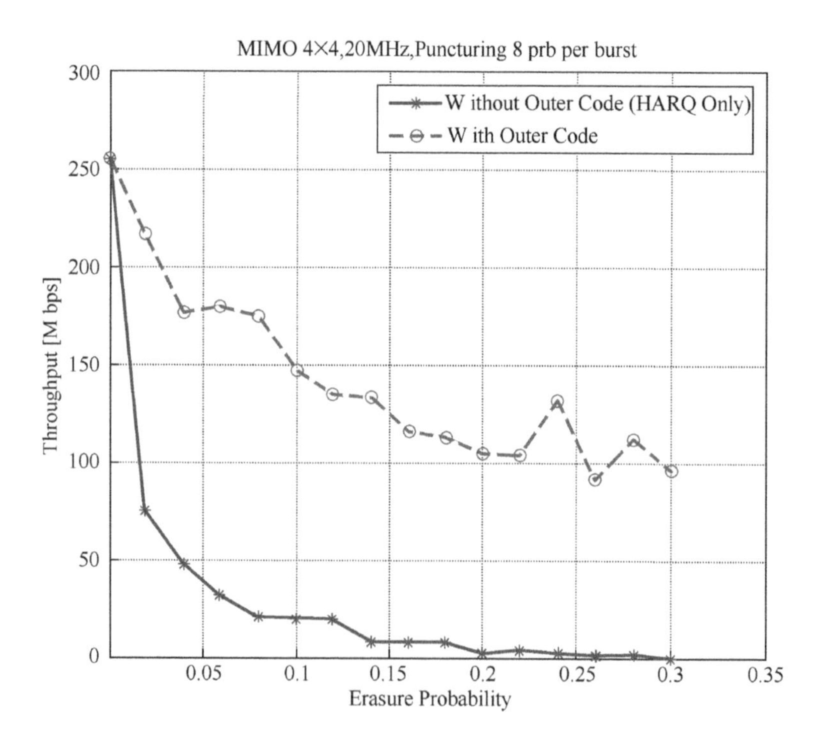

FIGURE 6.19 Link throughput between with and without outer code when the eMBB transmission is erased by bursty uRLLC traffic [2].

out LDPC encoding, followed by PC encoding of the entire codeword. Packet coding can also be considered as a form of block coding: encoding along the row using LDPC codes, encoding along the column using PC codes. Hence, there are multiple ways of its decoding.

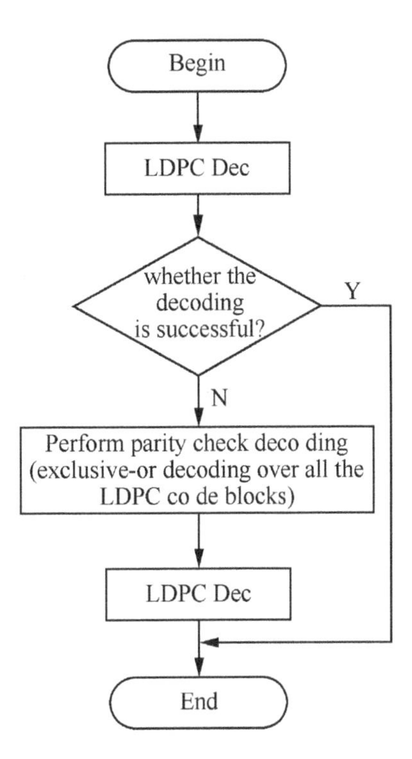

FIGURE 6.20 Flow chart of decoding Method 1 (first for LDPC decoding) for packet coding.

Method 1: conventional way: LDPC decoding → exclusive-or decoding → LDPC decoding

Treating packet coding as a form of serially concatenated code, the decoding algorithm would be: first to perform LDPC decoding, and to find whether the decoding is successful (by using CRC, or the PC H*C=0). If yes, then output the results directly; If not, then perform PC decoding (to sum LLR over all the LDPC code blocks), and then perform LDPC decoding, and output the decoding results. The above procedure is shown in Figure 6.20.

Method 2: PC decoding → LLR combining → LDPC decoding

As Figure 6.21 shows, PC decoding (exclusive-or decoding) should be performed first where the LLRs for combining are all from the receiver observations (prone to noise). When calculating the extrinsic information, a coefficient of 0.4 should be multiplied.

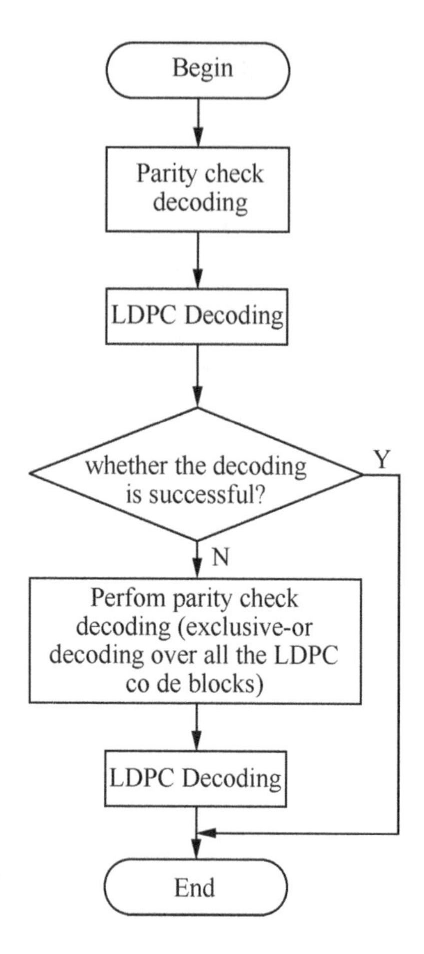

FIGURE 6.21 Flow chart of decoding Method 2 (first for PC decoding) for packet coding.

The min-sum algorithm is used for LDPC decoding. This method has an issue: decoding can only be started when all LDPC code blocks are received, which causes a long delay and a large memory size.

Method 3: multi-iteration.

During the decoding, the entire data packet is treated as a big block code: row encoding is through LDPC, and column coding is through PC. Then iterative decoding can be carried out in between so that different codewords can fully exchange the soft information. The flow diagram is shown in Figure 6.22. The

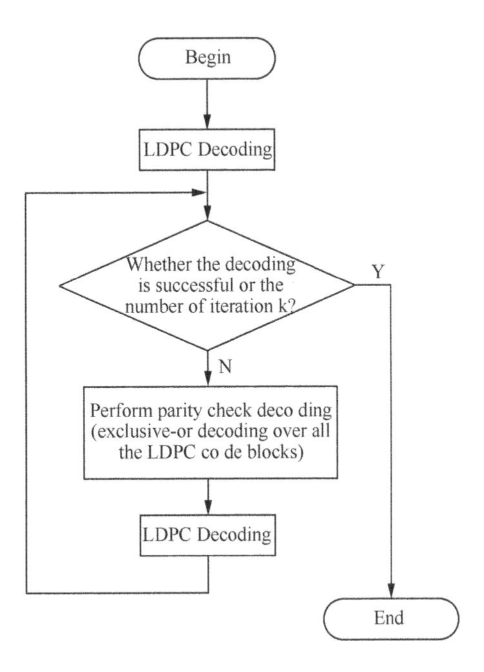

FIGURE 6.22 Flow chart of Method 3 (multi-iterations) for packet coding.

maximum number of iterations can be set to 7. During each iteration, LLRs inside the PC decoding (exclusive-or decoding) are from the demodulator, whereas other LLRs are from the LDPC decoder (need to be scaled by 0.5). The drawback of this method is multiple iterations required which leads to long processing delays.

Performances of LDPC using packet coding (as outer code) in the additive white Gaussian noise (AWGN) channel are shown in Figures 6.23 and 6.24 for code rates of 1/2 and 13/16, respectively. The number of LDPC code blocks is 20. "Trad" stands for traditional packet without outer code. "Method 1" refers to the conventional method which has been used in the earlier simulations.

It is observed that Method 1 and Method 2 perform almost the same. Their performance is better than traditional packets without an outer code by about 0.3 dB and 0.5 dB for code rates of 1/2 and 13/16, respectively. Method 3 (assuming max=7 iterations) can outperform the LDPC without outer code by about 0.7 and 1.0 dB for code rates of 1/2 and 13/16, respectively.

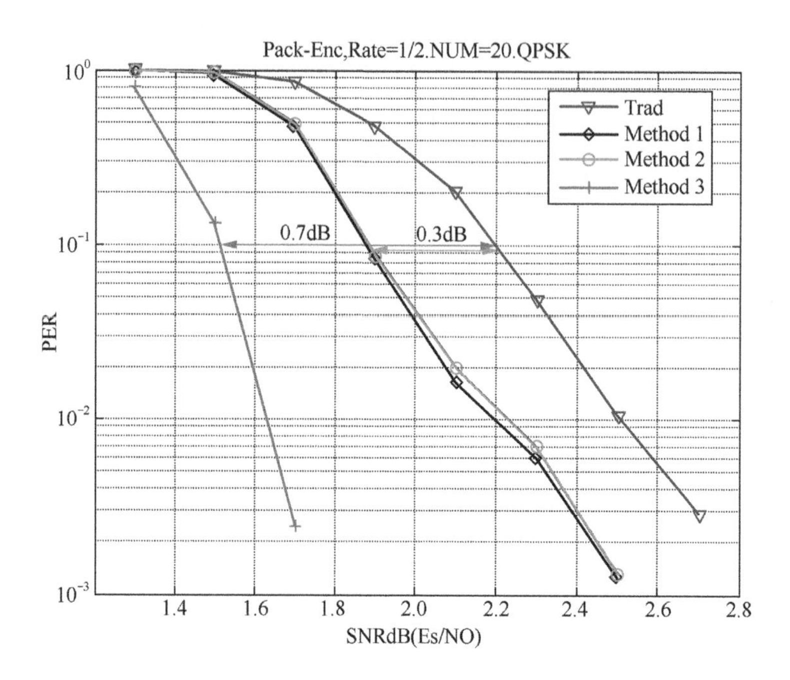

FIGURE 6.23 Performance comparison between different decoding algorithms for packet coding, R=1/2.

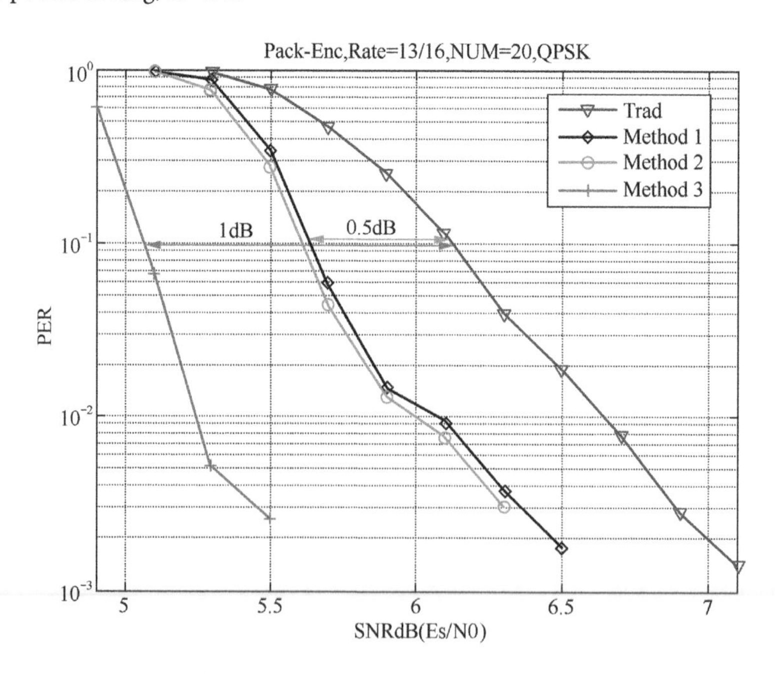

FIGURE 6.24 Performance comparison between different decoding algorithms for packet coding, R=13/16.

6.2.2.4 Performance of Packet Coding in Fading Channels

- **Fading channel 1.** classic Rayleigh fading which can be modeled as R=H×S+N, where R is the received signal, H is the complex coefficient of the channel, randomly generated with the variance=1, mean value=0. S is the modulation symbol transmitted. N is AWGN. Simulation results are shown in Figure 6.25 where "Trad" stands for traditional packet without outer code. "NEW1" refers to the conventional way of decoding for packet coding (without iteration). "NEW2" corresponds to multi-iteration decoding (max. #iterations=6). As seen in Figure 6.25, significant gains are observed in all the simulation settings.
- **Fading channel 2.** Extended Pedestrian A model (EPA). Simulation results are shown in Figures 6.26 and 6.27 for without inner interleaving inside an orthogonal frequency division multiplexing (OFDM) symbol, and with inner interleaving inside an OFDM symbol. It is seen that when there is no inner interleaving, a little gain is observed

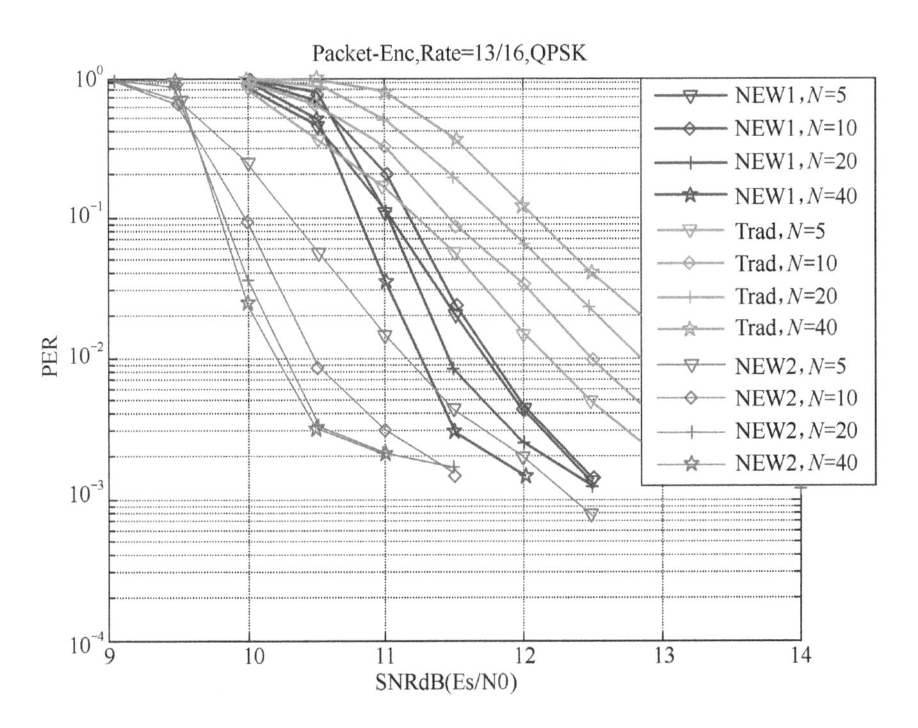

FIGURE 6.25 Performance of packet coding in a classic Rayleigh fading channel, code rate=13/16.

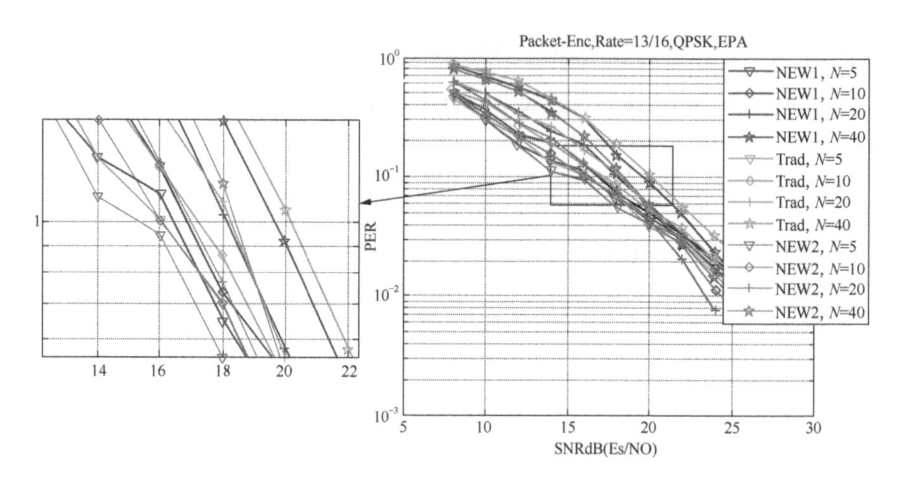

FIGURE 6.26 Performance of packet coding in EPA channel, code rate=13/16, without inner interleaving.

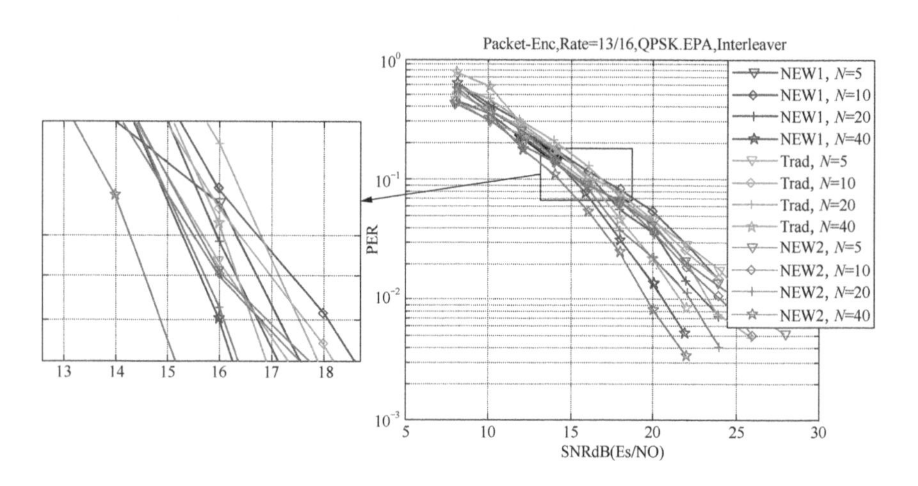

FIGURE 6.27 Performance of packet coding in EPA channel, code rate=13/16, with inner interleaving.

when the number of code blocks is small. The gain becomes more noticeable as the number of code blocks increases. Iterations can improve the performance.

- **Fading channel 3.** Extended Typical Urban model (ETU). Simulation results are shown in Figures 6.28 and 6.29 for without inner interleaving inside an OFDM symbol, and with inner interleaving inside an OFDM symbol. It is seen that when there is no inner interleaving,

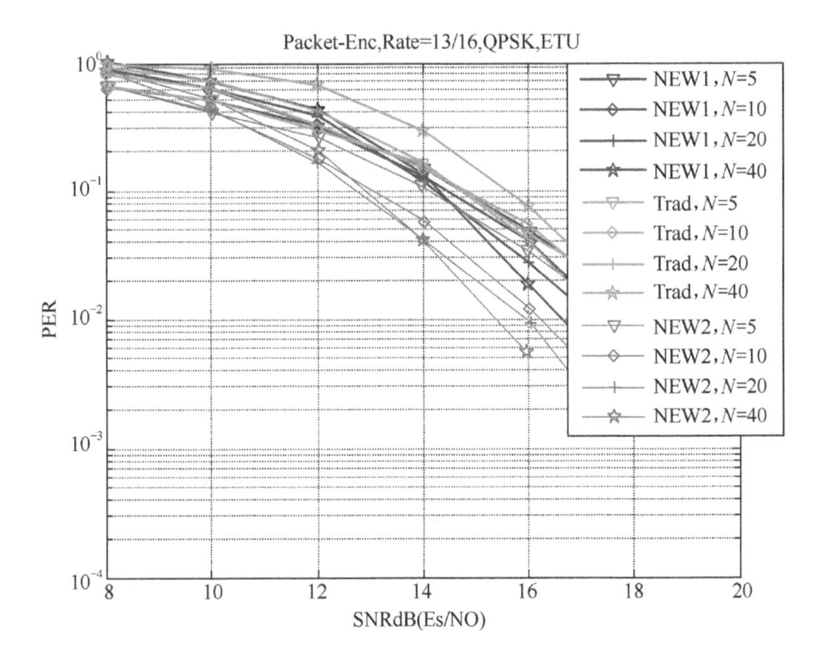

FIGURE 6.28 Performance of packet coding in ETU channel, code rate = 13/16, without inner interleaving.

a little gain is observed when the number of code blocks is small. The gain becomes more noticeable as the number of code blocks increases. Increasing iterations can improve the performance.

6.3 IMPLICIT OUTER CODES

Implicit outer codes refer to the integration of inner code and outer code so that it is hard to differentiate the exact coding schemes. During the decoding, we can either first decode the implicit outer code and then pass the information to the inner code, or first decode the inner code and then pass the information to the implicit outer code, or perform two decoding processes back and forth to pass the information to each other iteratively.

The 2D-polar code [16] discussed in Chapter 3 can be considered as implicit outer code + inner code, where the outer code is a block code and the inner code is a polar code. In turbo codes discussed in Chapter 5, one constituent code can be considered as the implicit outer code of the other constituent code.

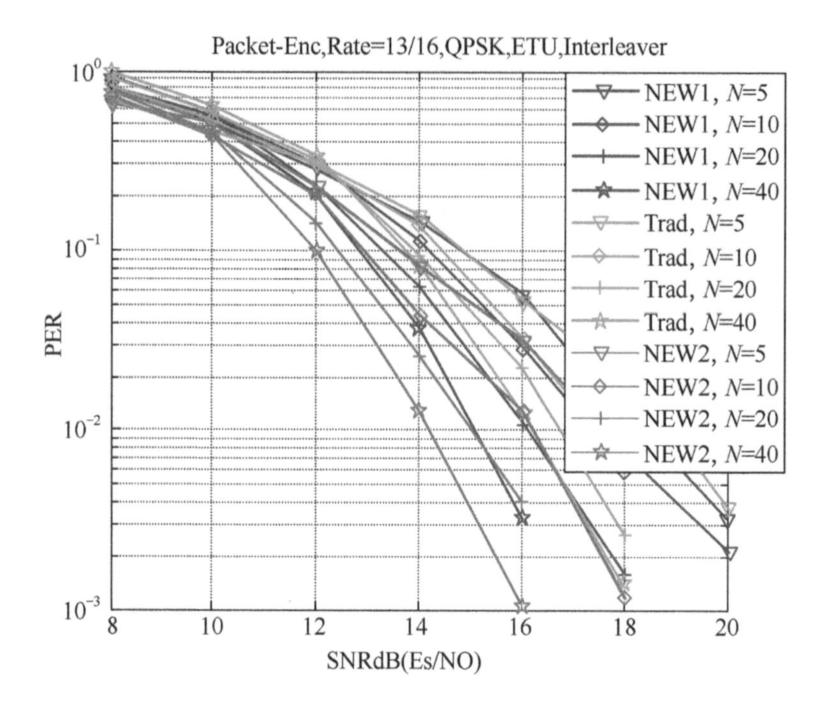

FIGURE 6.29 Performance of packet coding in ETU channel, code rate=13/16, with inner interleaving.

More interestingly, the LDPC code itself has PC capability. According to [4], LDPC code not only can provide regular coding capability but also would serve as an 8-bit CRC. In another word, the LDPC code includes eight implicit CRCs as the outer code.

6.4 SUMMARY

In this chapter, outer codes (especially the explicit outer codes) are discussed. When the block length is small, outer code is likely to be used. However, when the block length is large, outer codes can be used to improve the decoder's performance.

REFERENCES

1 3GPP, R1-1608976, Consideration on Outer Codes for NR, ZTE, RAN1#86bis, October 2016.
2 3GPP, R1-164703, Outer erasure code use cases and evaluation assumptions, Qualcomm, RAN1 #85, May 2016.
3 3GPP, R1-164667, Outer erasure code for efficient multiplexing, InterDigital Communications, RAN1 #85, May 2016.

4 3GPP, R1-1705864, Channel coding for PBCH, Nokia, RAN1#88bis, April 2017.
5 3GPP2, C.S0054-0 Version 1.0-CDMA2000 High Rate Broadcast-Multicast Packet Data Air Interface Specification, February 2004.
6 P. Agashe, "CDMA2000 high rate broadcast packet data air interface design," *IEEE Commun. Mag.*, vol. 42, no. 2, February 2004, pp. 83–89.
7 J. Xu, "Physical layer packet coding: inter-block cooperative coding for 5G," *IEEE Vehicular Technology Conference*, 15–18 May 2016, pp. 1–5.
8 ETSI, A Novel Hybrid ARQ Scheme Using Packet Coding, ETSI Workshop on future radio technologies – Air interfaces, January 2016, Sophia Antipolis, France.
9 I. Tal, A. Vardy, "List decoding of polar codes. Information Theory Proceedings (ISIT)," *2011 IEEE International Symposium.*
10 J. Wang. *Study on Polar encoding and decoding algorithms*, Haerbin Industrial University, Haerbin, China, June 2013.
11 3GPP, R1-1700088, Summary of polar code design for control channels, Huawei, RAN1 Ad-Hoc Meeting, USA, January 2017.
12 I. Reed and G. Solomon, "Polynomial codes over certain finite fields," *J. Soc. Ind. Appl. Math.*, vol. 8, no. 2, 1960, pp. 300–304.
13 S. Lin, *Error correction codes* (translated to Chinese by J. Yan), Mechanical Industry Press, June 2007.
14 A. Hocquenghem. Codes correcteurs d'erreurs. *Chiffres*, vol. 2, no.2, 1959, pp. 147–156.
15 R.C. Bose, DK Ray-Chaudhuri, "On a class of error correcting binary group codes," *Inf. Control*, vol. 3, no.1, 1959, pp. 68–79.
16 E. Arikan, "Two-dimensional polar coding," *Tenth International Symposium on Coding Theory and Applications (ISCTA'09)*, July, 2009, Ambleside, UK.

CHAPTER 7

Other Advanced Coding Schemes

Yifei Yuan, Mengzhu Chen and Focai Peng

Low-density parity check (LDPC) codes, polar codes, convolutional codes, and turbo codes discussed in previous chapters are all defined in binary domain GF(2). In fact, these codes can operate in a non-binary domain, e.g., GF(4) or higher. In this chapter, we will discuss non-binary LDPC codes [1,2], non-binary Repeat-Accumulate (RA) codes [3], lattice codes [4], and fountain codes [5] for link adaptation.

7.1 NON-BINARY LDPC CODES

7.1.1 Basic Idea

Non-binary LDPC codes [1,2] were first proposed by Davey and MacKay in 1998. To simplify the notions, binary LDPC is shortened to BLDPC. Non-binary LDPC is shortened to QLDPC which is defined in GF(q), where q is equal to an integer power of 2 and there are q elements in GF(q).

The encoding process of QLDPC is similar to that of BLDPC. Here we focus on the construction of parity check matrix **H** which should satisfy the sparsity requirement. Different from BLDPC, the elements in H are all defined in GF(q). This would make the parity check matrices of QLDPC more complicated than those of BLDPC. Similar to BLDPC, there are also two construction methods for the parity check matrix of QLDPC.

Decoding algorithms of BLDPC can also be considered for QLDPC. However, the decoding complexity of QLDPC is significantly higher.

DOI: 10.1201/9781003336174-7

Possible algorithms include belief propagation (BP), enhanced BP, extended Min-Sum (EMS) [6], etc. Among them, the enhanced BP is the most popular. BP algorithm requires a lot of multiplications and high complexity of the hardware. To reduce the complexity, decoding can be done in the log domain. FFT operation can be used in the enhanced BP to improve the throughput. Hence there are a number of algorithms derived from BP: Log-BP, FT-BP, Log-FFT-BP, etc.

QLDPC has the capability of reducing or eliminating short cycles (especially length-4 cycles), thus improving the performance. In data storage systems, and deep-space communications, the errors are often bursty. Since multiple bursty bit errors can be combined into less number of non-binary symbol errors, QLDPC is more robust to the bursty errors compared to BLDPC. As the speed of data communications gets higher and higher, high-order modulations are more frequently used. Non-binary codes are based on limited domain processing of high-order. Hence, it may be easier to combine QLDPC and high-order modulation.

7.1.2 Non-Binary LDPC Design for Bit-Interleaved Coded Modulation (BICM)

Figure 7.1 shows an example of non-binary LDPC of BICM.

The procedure of QLDPC with BICM is as follows:

1. Information bits randomly distributed are put to a QLDPC of the code rate R.
2. Encoded bits are modulated and then go through the AWGN channel.
3. Detector performs maximum likelihood (ML) detector on the received signal and outputs the log-likelihood ratio (LLR) of each symbol.

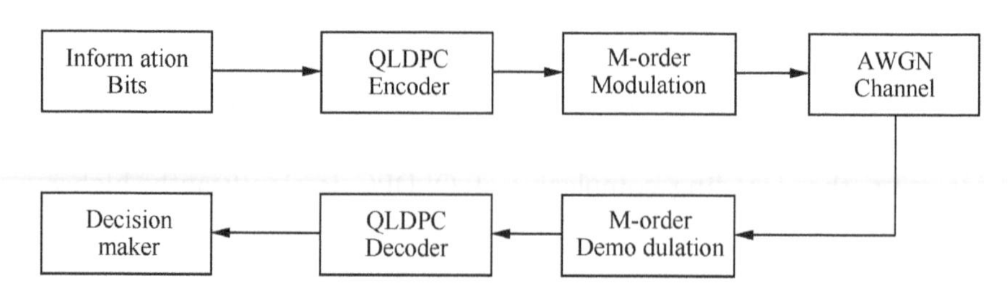

FIGURE 7.1 Non-binary LDPC with BICM.

4. LLRs are input to QLDPC decoder with iterative decoding algorithm to output the posterior probability of each bit.
5. Decision maker to output the decision for each bit. If the output of the decision maker is a valid codeword, the decoding is terminated. Otherwise, go back to Step (4) or until the maximum number of iterations is reached.

Figure 7.2 shows the simulation results of GF(16) (2, 4) regular LDPC and binary LDPC under the AWGN channel when the code rate is 1/2 with QPSK modulation. In GF(16) (2, 4) regular LDPC, "2" denotes column weight and "4" denotes row weight. It is observed that GF(16) (2, 4) regular LDPC is about 0.35dB better than GF(2) (3, 6) regular LDPC at bit error rate (BER)=6E-6 with QPSK modulation. In [7] it was shown that under the AWGN channel with BPSK, non-binary LDPC outperforms binary LDPC by about 0.2 dB at BLER=1%.

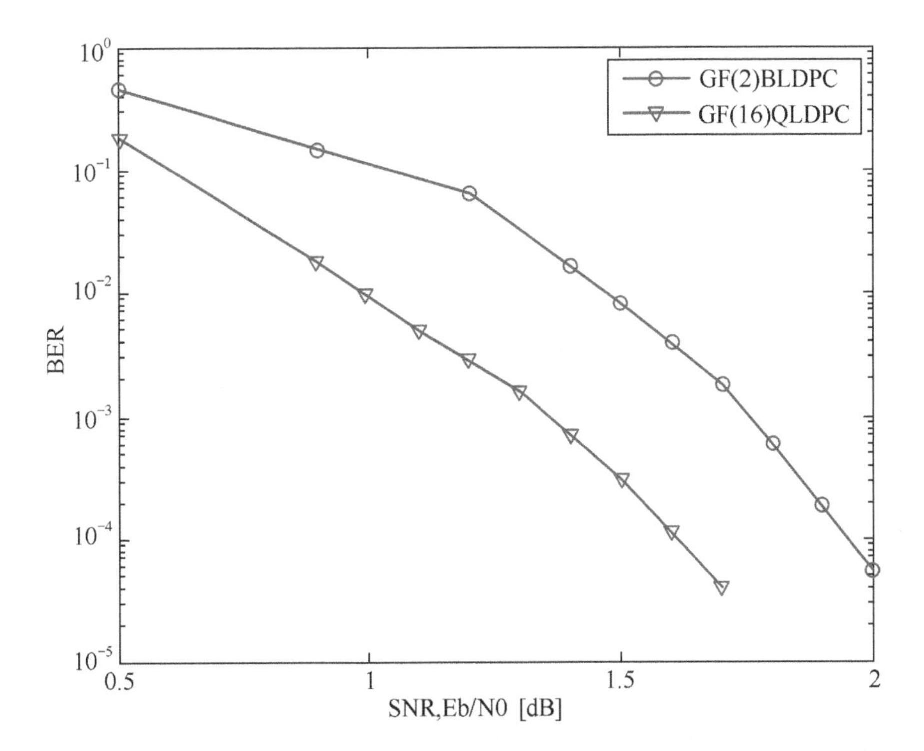

FIGURE 7.2 Performance comparison between GF(16) LDPC and binary LDPC codes, $R=1/2$.

7.1.3 Modulations for Non-Binary Codes

For high-order modulation like QAM, the reliability of each bit in the constellation points are quite different. Higher reliability means that the error probability of the bit is low. Lower reliability means that the error probability of the bit is high. For instance, in the constellation of LTE 64-QAM, a constellation contains 6 bits where the first two bits have the highest reliability. The middle two bits have moderate reliability. The last two bits have the lowest reliability. If the bits corresponding to a non-binary code are mapped to the bit positions with different reliability and across multiple modulation symbols, the diversity of constellation can be fully utilized. For fading channels, multiple bits corresponding to one element of GF(q) can be mapped to multiple modulation symbols, so that each element would experience different fades to maximize diversity.

Method 1: Interleaved Mapping. To apply different cyclic shifts for the same bit positions corresponding to each GF(q) element, and then mapped to modulation symbol column by column, as shown in Figure 7.3.

Method 2: IQ Mapping (IQ2). To sequentially map the binary bits corresponding to GF(q) codewords to I and Q of the modulation symbol. Different from conventional Gray mapping, here the set of bits

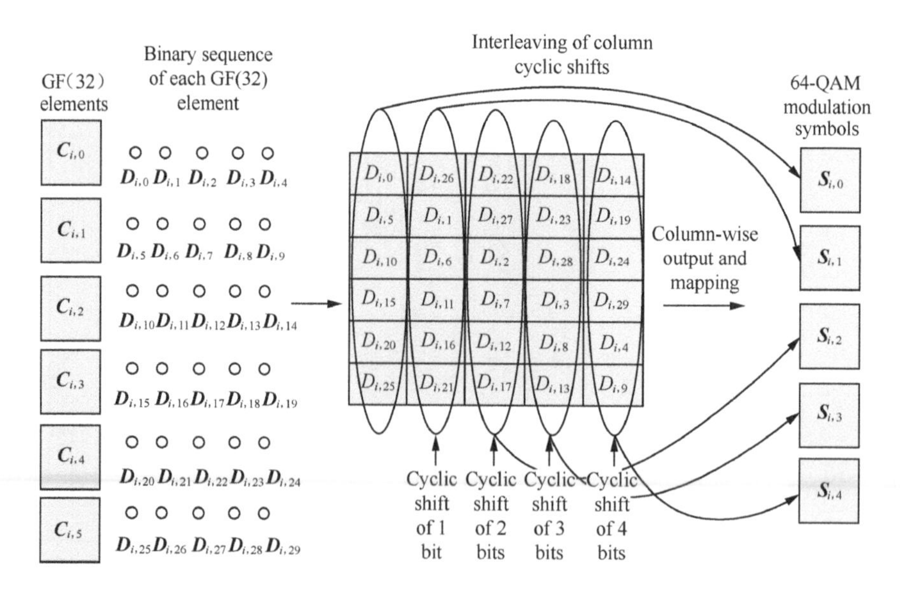

FIGURE 7.3 IM for non-binary codes.

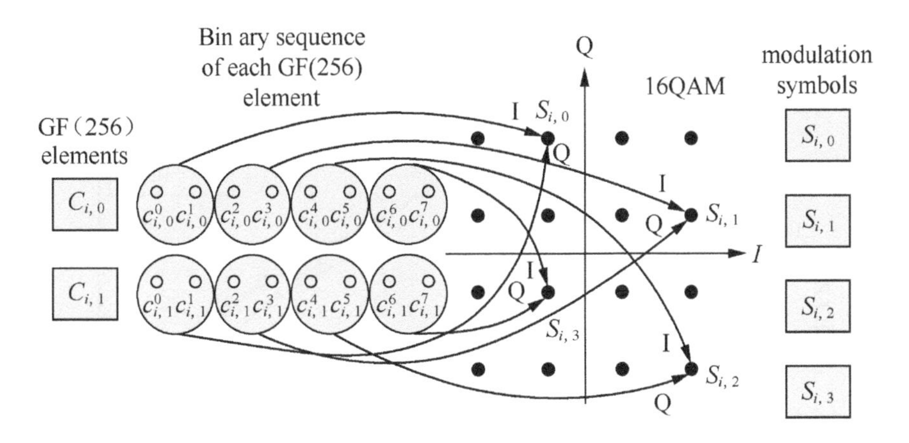

FIGURE 7.4 IQ2 mapping for non-binary codes.

corresponding to the first GF(q) codeword are mapped to I component, whereas the set of bits corresponding to the second GF(q) codeword are mapped to Q component, as illustrated in Figure 7.4.

Simulations are carried out to compare the performance between the above two mapping methods, the conventional Gray mapping, and binary turbo codes. The first simulation setting is ETU fading channel model, mobile velocity of 30 km/h, GF(16) LDPC, information block length = 2160 bits, code rate R = 3/4, 16 QAM, 10 MHz bandwidth, 1024-point FFT. The result is shown in Figure 7.5. It is seen that at BLER = 1%, interleaved mapping (IM) and I/Q mapping (IQ2) outperform binary turbo codes by about 0.45 and 0.15 dB, respectively.

In the second simulation, the assumptions are EPA fading channel, mobile velocity = 3 km/h, GF(16) LDPC, information block length = 2160 bits, code rate = 3/4, 16QAM modulation, 10 MHz bandwidth, 1024-point FFT. The result is shown in Figure 7.6. It is found that at BLER = 6%, the performance of Gray mapping is similar to that of turbo codes. IQ2 mapping and IM outperform turbo codes by about 0.4 and 0.65 dB, respectively.

In the third simulation, the assumptions are EPA fading channel, mobile velocity = 3 km/h, GF(64) LDPC, information block length = 2880 bits, code rate = 2/3, 64QAM modulation, 10 MHz bandwidth, and 1024-point FFT. The result is shown in Figure 7.7. It is found that at BLER = 6%, the performance of Gray mapping and IQ2 mapping is better than turbo codes by about 0.6 dB. IM outperforms turbo codes by about 0.9 dB.

In the fourth simulation, the assumptions are ETU fading channel, mobile velocity = 30 km/h, GF(64) LDPC, information block length = 2880

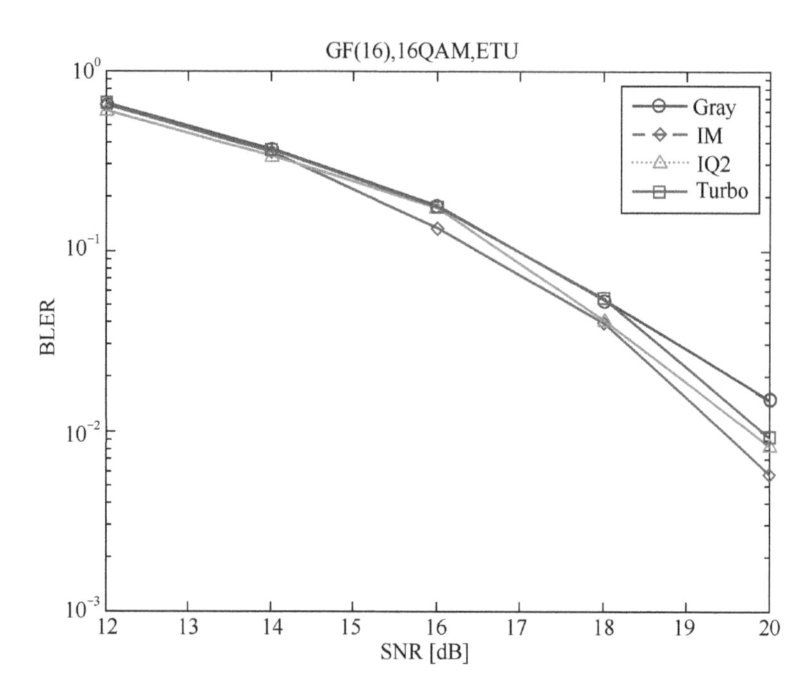

FIGURE 7.5 Performance of GF(16) LDPC for the ETU fading channel.

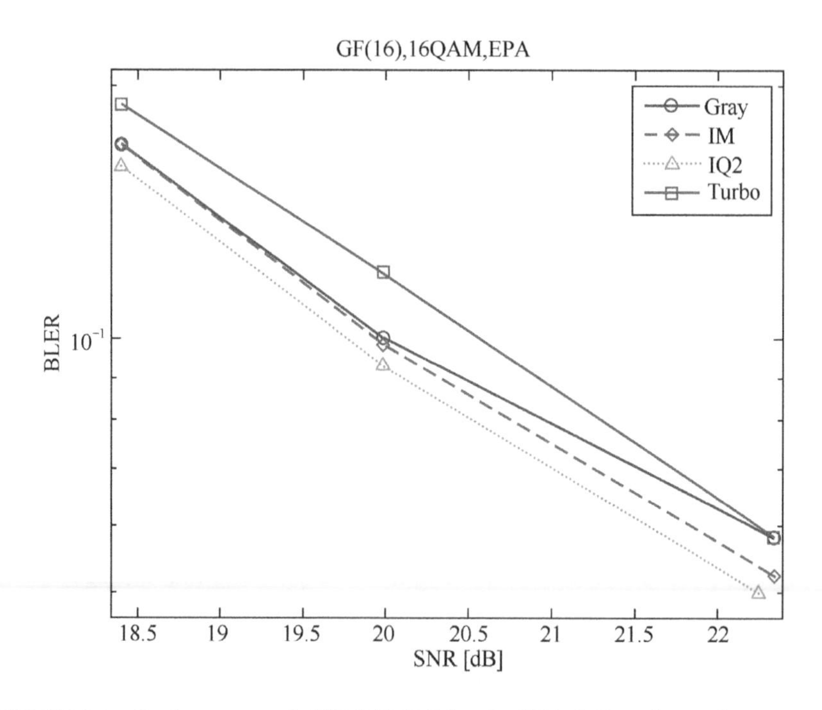

FIGURE 7.6 Performance of GF(16) LDPC for the EPA fading channel.

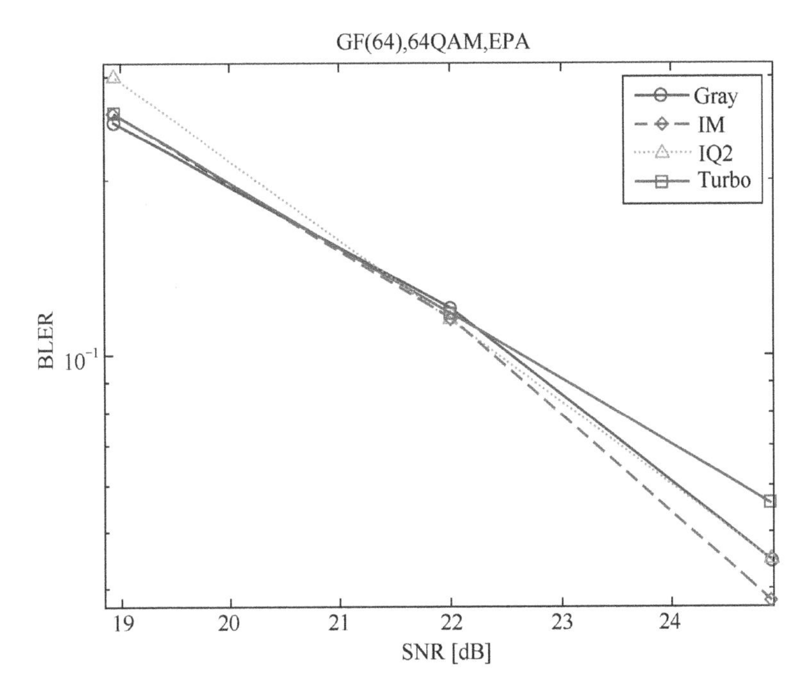

FIGURE 7.7 Performance of GF(64) LDPC for the EPA fading channel.

bits, code rate=2/3, 64QAM modulation, 10 MHz bandwidth, and 1024-point FFT. The result is shown in Figure 7.8. It is found that at BLER=3%, the performance of Gray mapping is better than turbo codes by about 0.15 dB. IM and IQ2 outperform turbo codes by about 0.4 dB.

From the above simulations, it is seen that under ETU and EPA fading channels, the optimized mapping schemes for non-binary LDPC (IM and IQ2 mapping) can provide 0.1~0.9 dB performance gain over turbo codes.

7.2 NON-BINARY RA CODES

RA codes can be considered as a type of turbo codes, or a type of LDPC codes, because the merits of turbo codes and LDPC codes are combined in RA codes: the simple encoding for turbo, and the parallel decoding for LDPC. In addition, non-binary RA codes have more freedom in selecting non-zero elements in a finite domain and can more easily avoid short cycles. Compare to binary turbo codes and binary RA codes, non-binary RA has stronger error correction capability, especially for high-order modulations.

Non-binary RA codes [3,7] can be obtained via weighted non-binary RA (WNRA) which includes repeater, interleaver, weighting module,

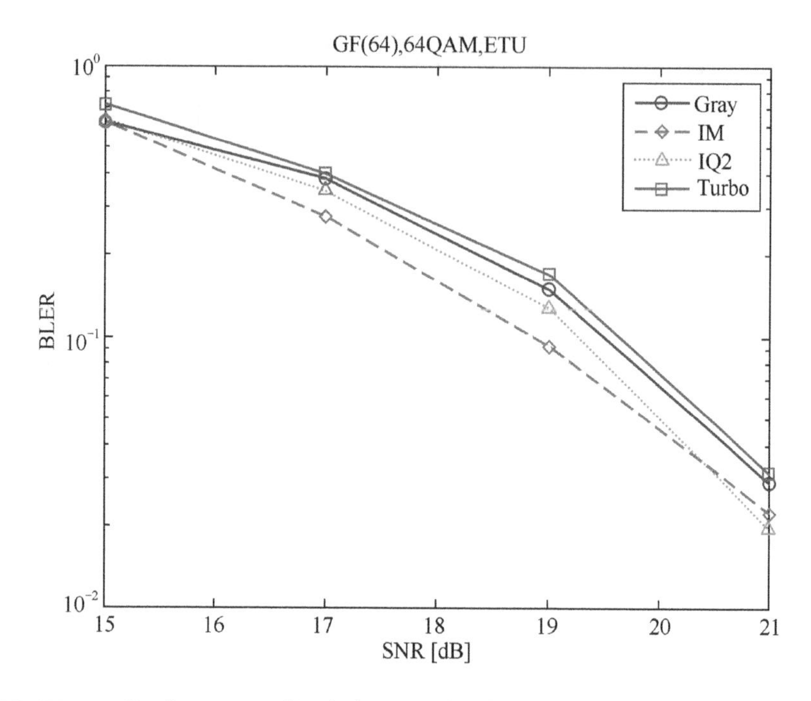

FIGURE 7.8 Performance of GF(64) LDPC for the ETU fading channel.

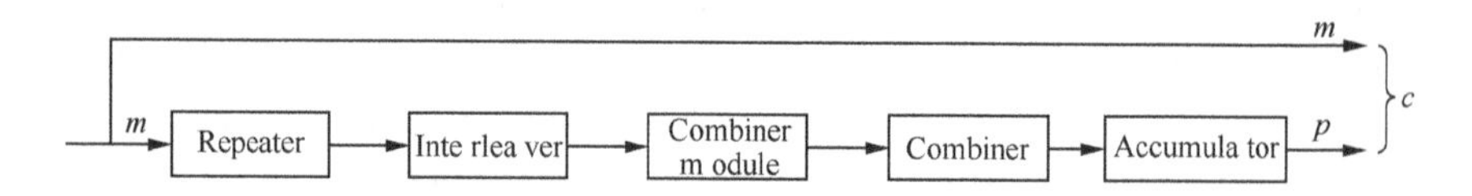

FIGURE 7.9 Encoder block diagram of the non-binary systematic RA code.

combiner, and accumulator, as shown in Figure 7.9. For a length k information bit stream $m=[m_1,m_2,m_3,......,m_k]$, after p repetitions, we get

$$m^{(1)}=m^{(2)}=......=m^{(p)}=[m_1,m_2,m_3,......,m_k] \quad (7.1)$$

In the interleaver of Figure 7.10, use p inner sequences $\Pi^{(1)},\Pi^{(2)},......,\Pi^{(p)}$ to interleave $m^{(1)},m^{(2)},......,m^{(p)}$, respectively, where $\Pi^{(1)}=[\pi_1^{(i)},\pi_2^{(i)},......,\pi_k^{(i)}]$, $i=1,2,......,p$. The output sequences are $B^{(1)},B^{(2)},......,B^{(p)}$, where $B^{(i)}$ is the output after $m^{(i)}$ is interleaved by $\Pi^{(i)}$. The interleaving processing can be represented as

$$B^{(1)}=[b_1^{(i)},b_2^{(i)},......,b_k^{(i)}]=[m_{\pi_1^{(i)}},m_{\pi 2_1^{(i)}},......,m_{\pi_k^{(i)}}], \quad i=1,2,......,p \quad (7.2)$$

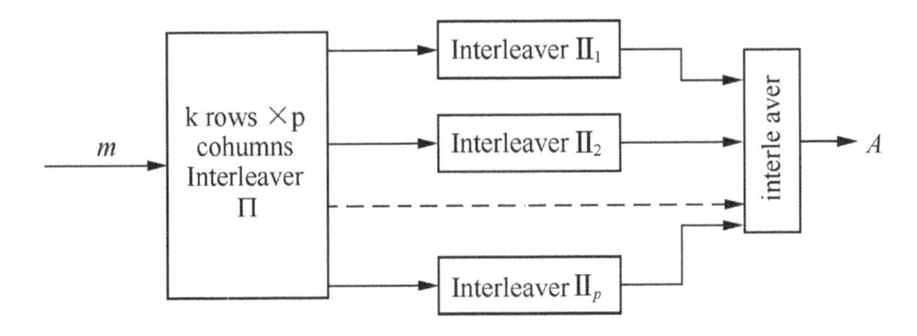

FIGURE 7.10 Interleaver structure for non-binary RA codes.

To concatenate the p output sequences (after being interleaved) to get

$$\begin{aligned} A &= [\theta_1, \theta_2, \ldots\ldots, \theta_{k \bullet p}] = [B^{(1)}, B^{(2)}, \ldots\ldots, B^{(p)}] \\ &= [b_1^{(1)}, \ldots\ldots, b_k^{(1)}, b_1^{(2)}, \ldots\ldots, b_k^{(2)}, \ldots\ldots, b_1^{(p)}, \ldots\ldots, b_k^{(p)}] \end{aligned} \tag{7.3}$$

Put the concatenated sequence A into the weighting module and the combiner. Assuming the weight vector is $W = [w_1, w_2, w_3, \ldots\ldots, w_{k \times p}]$, where $w_i \in GF(q)$, $i = 1, 2, \ldots\ldots, k \bullet p$. The parameter of the combiner is a. The output is

$$\begin{aligned} r_i &= \theta_{(i-1) \bullet a+1} \times w_{(i-1) \bullet a+1} + \ldots\ldots + \theta_{(i-1) \bullet a+a} \times w_{(i-1) \bullet a+a} \\ &= \sum_{m=1}^{a} a_{(i-1) \bullet a+m} \times w_{(i-1) \bullet a+m}, \quad i = 1, 2, \ldots\ldots, k \bullet p / a \end{aligned} \tag{7.4}$$

After going through the accumulator, to get $k \bullet p / a$ number of parity checks P. Assuming the accumulating factor $\alpha \in GF(q)$, $\beta \in GF(q)$, the accumulation operation can be represented as $1/(\alpha + \beta \bullet D)$. Then,

$$p_1 = r_1 / \alpha, \quad p_i = ((p_{i-1} \bullet \beta) + r_i) / \alpha, \quad i = 2, 3, \ldots\ldots, k \bullet p / a \tag{7.5}$$

The final output codeword is $c = [m_1, m_2, \ldots\ldots, m_k, p_1, p_2, \ldots\ldots, p_{k \bullet p/a}]$ with the length $n = k + k \bullet p / a$ and the code rate $R = k / n = k / (k + k \bullet p / a) = a / (a + p)$. When the number of repetitions p and the parameter of the combiner a are given, we can obtain the regular RA (p, a) codes. RA codes of different block lengths and code rates can be designed by changing the parameters k, p, and a.

Corresponding to the above encoding procedure, the parity check matrix of this RA code H can be partitioned into two parts $H=[H_1 \;\; H_2]$. H_1 is a sparse matrix of column weight p and row weight a. The distribution of non-zero elements is dependent on the parameter c of interleaver. The values of non-zero elements are determined by the weight vector of the weighting module. H_2 is a dual-diagonal matrix determined by the accumulator.

$$H_2 = \begin{bmatrix} \alpha & & & & \\ \beta & \alpha & & & \\ & \beta & \alpha \ldots & & \\ & & \vdots & \beta \;\; \alpha & \\ & & & \beta & \alpha \end{bmatrix} \tag{7.6}$$

Figure 7.11 is a bipartite graph of systematic RA code. When constructing non-binary RA codes, the design of the interleaver and weighting module is very important. Especially, when BP decoding is used, the existence of short cycles should be avoided, which is related to the structure of the parity check matrix H.

7.2.1 Interleaver

The interleaver consists of an outer row and column interleaver Π, p inner column interleavers $\Pi_1, \Pi_2, \ldots\ldots, \Pi_p$, and a multiplexer. The actual interleaving process is as follows:

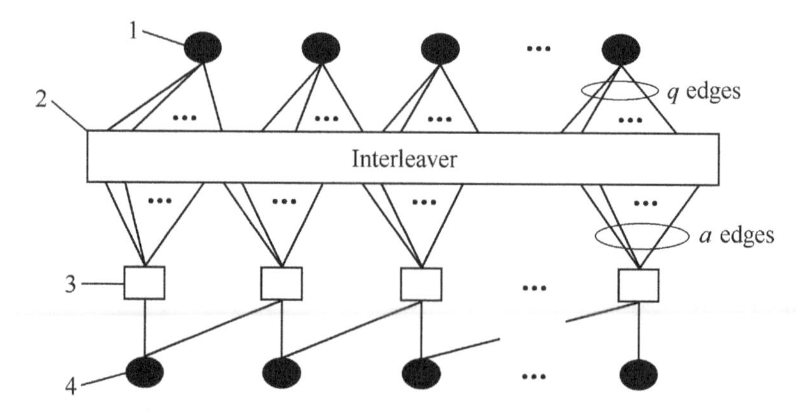

FIGURE 7.11 Bi-partite graph of systematic RA codes.

1. First put the information b (after the repeater) into the outer interleaver Π. Π is a k-row p-column interleaver where each column is the transposed information sequence $m=[m_1,m_2,m_3,\ldots\ldots,m_k]$.
2. Perform column interleave with each column. For instance, put the first column in Π to the interleaver Π_1 and get b'_p. Put the second column in Π to the interleaver Π_2 to get b'_2, and so on, till the p-th column in Π to the interleaver Π_p to get b'_p.
3. Multiplex the output sequences $b'_1,b'_2,\ldots\ldots,b'_p$ from Step (2) to get the entire output sequence b'.

The i-th inner column interleaver $\prod_i$ can be represented as follows:

$$\pi_i(x)=\begin{cases} c_i & x=1 \\ (\pi_i(x-1)+c_i)\bmod k & x\neq 1 \end{cases} \tag{7.7}$$

where $x\in\{1,2,\cdots,k\}$ denotes the positions before the interleaving, $\pi_i(x)$ denotes the positions after the interleaver $\prod_i$. c_i is a constant which does not change within a $\prod_i$. For any to two inner columns interleavers $\prod_i$ and $\prod_j$, if $i\neq j$, then $c_i\neq c_j$. R_i is used to represent the row set in the matrix H_1 that corresponds to the interleaver Π_i as represented in Inline Figure 1. d_i represents the distance between any two non-zero elements R_i in each row. It can be seen from Eq. (7.7) that for all $x\in\{1,2,\cdots,k\}$, when $\pi_i(x)>\pi_i(x-1)$, then $d_i\in\{c_i,2c_i,\ldots,(a-1)c_i\}$; when $\pi_i(x)<\pi_i(x-1)$, then $d_i\in\{k-c_i,k-2c_i,\ldots,k-(a-1)c_i\}$.

Inline Figure 7.1
YTR

Figure 7.12 shows the two common types of short cycles in parity check matrices. The short cycles within H_1 are of the first type, and the short cycles across H_1 and H_2 are of the second type. Because of the simple structure of H_2, short cycles are more likely in H_2.

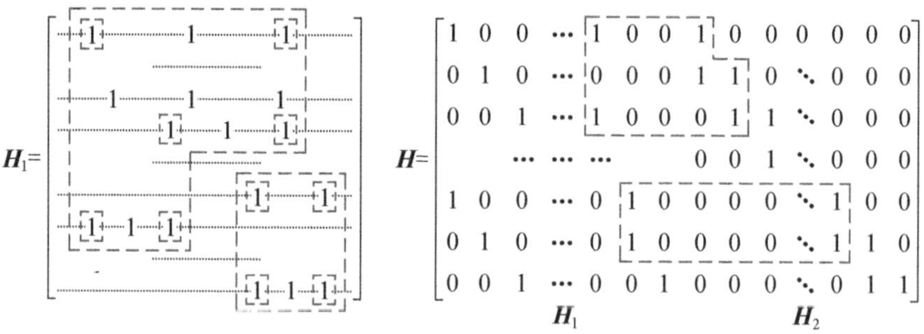

First type cycle of length-4 & 6

Second type cycle of length-4 & 6

FIGURE 7.12 Short cycles in parity check matrices.

In order to construct a cyclic shift type of parity check matrix H that does not have length-4 cycles, parameters k, a and c should satisfy the following conditions:

1. k is divisible by a;
2. $c_i < k/a$;
3. c_i is a prime number, and not mutually divisible by each of $t_1,\cdots,t_s$, respectively;
4. $d_i \neq d_j$ where $i, j \in \{1,2,\cdots,p\}, i \neq j$, $x \in \{1,2,\cdots,k\}$. After further study, it is found that within the parameter set of the interleaver, to select those parameters of RA codes that can construct the parity check matrices that do not have length-6 cycles of the first type and length-6 cycles of the second type. In order to construct the parity check matrix H that is free from the length-6 cycle, parameters should satisfy the two additional conditions;
5. $d_i + d_j \neq d_l$ where, $i, j, l \in \{1,2,\ldots,p\}$, $i \neq j, i \neq l, j \neq l$, and $d_i \in \{c_i, 2c_i, \ldots, (a-1)c_i, k-c_i, k-2c_i, \ldots, k-(a-1)c_i\}$, $i \in \{1,2,\cdots,p\}$;
6. $d_j \neq 2 \times tap$, $j = 1,2,\ldots,k$, where d_j' denotes the distance between any two non-zero elements in j-th column in H_1. tap denotes the spacing between two parallel diagonals in H_2 (for instance, tap$=1$ in H_2 of Figure 7.12).

It is seen from Figure 7.12 that Condition (5) ensures that no length-6 cycle would exist within three different rows in H_1, thus avoiding the length-6 cycles of the first type. In addition, search through a *tap* that satisfies Condition (6) and then determine the structure according to the value of *tap*, to avoid the length-6 cycles of the second type. If the block length is not very short, it would not be difficult to find the parameters *c* and *tap* for the interleaver, and then construct parity check matrices that do not have length-4 and length-6 cycles.

7.2.2 Weighting Module

The input signal sequence M of length k, after p repetitions gives the sequence b, after the permutation of the interleaver to get the sequence b'. The weight vector is $W=(w_1,w_2,\cdots,w_{k\times p})$, where $w_i \in GF(q), i=1,2,...,kp$. The values of weight vector determine the values of non-zero elements in H_1. When the parameter of the combiner is a, $w_1,w_2,\cdots,w_a$ would be the non-zero elements in the first row of H_1. $w_{i\times a+1},w_{i\times a+2},\cdots,w_{(i+1)a}$ would be the non-zero elements in the i-th row. To ensure the quasi-cyclic property of H_1 as seen in Inline Figure 1, the weight vector here should have the format $W=\left[\underbrace{w_1...w_1}_{i_1}\underbrace{w_2...w_2}_{i_2}...\underbrace{w_p...w_p}_{i_p}\right]$, and satisfy the following three conditions:

1. $w_j \in GF(q), j=1,2,...,p$;
2. $k=t_1^{n_1}\times t_2^{n_2}\times\cdots t_s^{n_s}$ ($t_1,\cdots,t_s$ are prime numbers that differ from each other);
3. $i_1,i_2,\cdots,i_p$ are zeros or an integer number of k, and satisfying $i_1+i_2+...+i_p=p^*k$.

Based on the design principle of interleaver sequence, to reduce the possibility of short cycles in the quasi-cyclic parity check matrices, normally we can set $i_1 = i_2 = ... = i_p = k$.

7.2.3 Combiner and Accumulator

For every a number of elements in the weight vector, summation is conducted in $GF(2^s)$ domain to get a sequence $R=[r_1,r_2,...,r_{k^*p/a}]$ of length $k\times p/a$. In the finite domain, the accumulator can be represented as $1/(\alpha+\beta D)$ where $\alpha,\beta \in GF(2^s)$. The corresponding H_2 is seen in the right part of Figure 7.12. In addition, a generalized accumulator structure like

$1/\left(\alpha-\beta D^{t}\right)$ or $1/\left(\alpha-\beta D^{t}-\gamma \mathrm{D}^{s}\right)$ can be used for design optimization, where t and s are all positive integers.

7.2.4 Decoding

Decoding algorithms for non-binary RA codes can be the BP, Log-BP or enhanced Min-Sum algorithms [6] for non-binary LDPC which are all based on factor graphs.

7.3 LATTICE CODE

In the Euclidean space R^{m} of the m-dimensional real domain, a set can be formed by n linearly independent vectors. When the coefficients of the linear combination are integers, the set would be called an n-dimensional lattice [8] in m-dimensional R^{m} space. The n-dimensional lattice Λ can be represented via an $n \times n$ generation matrix $\mathbf{G}$ as

$$\mathbf{G}=[\mathbf{g}_1,\mathbf{g}_2,\ldots\ldots\mathbf{g}_n]=\begin{bmatrix} g_{11}, g_{12},\ldots\ldots, g_{1n} \\ g_{21}, g_{22},\ldots\ldots, g_{2n} \\ \vdots \quad \ldots\ldots \quad \vdots \\ g_{n1}, g_{n2},\ldots\ldots, g_{nn} \end{bmatrix} \tag{7.8}$$

where $\mathbf{g}_1,\mathbf{g}_2,\ldots\ldots,\mathbf{g}_n \in R^n$ are all real numbers.

Any point in the Lattice Λ can be represented by the generation matrix $\mathbf{G}$ and the integer coefficient vector $\mathbf{u}$ as

$$\Lambda=\mathbf{u}\bullet\mathbf{G} \tag{7.9}$$

where $\mathbf{u}=[u_1,u_2,\ldots\ldots,u_n]$ and $u_1,u_2,\ldots\ldots,u_n \in Z^n$ are all integers.

Lattice codes were first proposed by Forney from Codex Company of the USA in 1988 [9]. The signal codes proposed by Shalvi in 2003 belong also to lattice codes. In 2007, N. Sommer from Tel Aviv University of Israel proposed a lattice code based on the LDPC which is called low-density lattice codes (LDLC) [4]. LDLC is a practical code and can approach the channel capacity for the AWGN channel [11,12]. Also, its decoding complexity grows just linearly with the block length [10].

The codeword x of LDLC can be obtained by an n-dimensional linear transformation of the integer-value information vector b, e.g., $x=Gb$. G is called the generation matrix of LDLC, made up of n linearly independent column vectors (e.g., G is non-singular). Its determinant should satisfy $|\det(G)| = 1$. Define the parity check matrix $H = G^{-1}$ which should be sparse.

Similar to the case of LDPC, the Tanner graph and parity check matrix can also be used to represent LDLC. Together with the sparsity, we can use the BP type of iterative decoding for LDLC. However, the difference from LDPC is that the processing of encoding and the channel are operated in the real domain, which significantly simplifies the convergence analysis. It is proved in [11,12] that by using iterative decoding, the performance of LDLC can approach Shannon's limit for AWGN. Also, compare with other channel coding schemes, LDLC codes can work without redundancy bits, e.g., R = 1. Hence LDLC can approach channel capacity by channel coding gain and shaping gain, without increasing the signal bandwidth [13]. The properties of LDLC can be summarized as follows:

- The parity check matrix can be constructed based on permutation matrix or cyclic shift. The non-zero elements in the matrix should be sparse and can take real numbers.
- The degrees of all the columns are equal to a constant d.
- Code rate = 1.
- Encoding can be efficiently carried out iteratively by the Jacobi algorithm. The low complexity of encoding.
- BP algorithm is to pass the probability density function over $(-\infty, +\infty)$, rather than the LLR. The low density of LDLC allows iterative decoding whose complexity is linear with the block length. The convergence rate is fast.
- Approaching Shannon's limit within 0.5 dB [4].

An example of using LDLC for AWGN is illustrated in Figure 7.13. First, a parity check matrix **H** of good performance needs to be constructed. Then to perform iterative encoding using the parity check matrix, followed by the shaping operation to limit the transmit power of the codewords. After receiving the signal corrupted by AWGN of finite power, BP iterative decoding is carried out. The output of the decoder is put through the modulo which outputs the recovered information bits.

Let us consider a special LDLC, called Latin Square LDLC which has the following unique property.

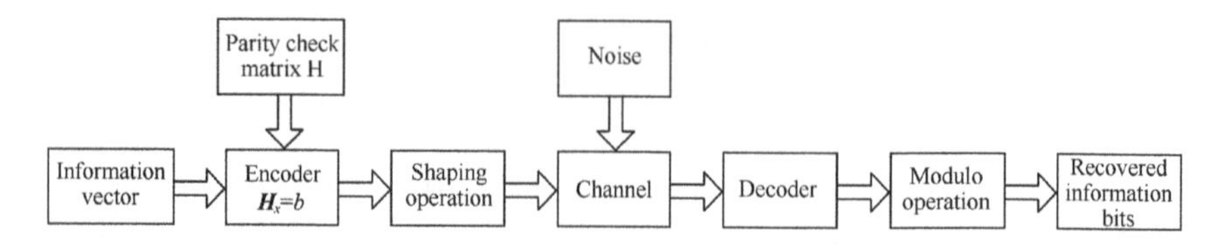

FIGURE 7.13 Block diagram of transmitter and receiver side processing for LDLC in AWGN channel.

- For each column and each row of the parity check matrix H, there are d non-zero elements. Their positions, order, and values can be different. Define the d non-zero elements in descending order as the generation sequence. If $h_1 > h_2 > h_3 ... > h_d$, then the generation sequence of LDLC is $\{h_1, h_2, h_3, ... h_d\}$.

Different from the parity check matrix **H**, the generation matrix of LDLC $G = H^{-1}$ does not have sparsity property. The computation complexity and memory requirement would be in the order of $O(n^2)$, e.g., increasing with the square of the block length, if we directly calculate the matrix inverse. By exploiting the low sparsity of the parity check matrix H, the Jacobi iterative algorithm for the solution of a set of linear equations $H \cdot x = b$ can be used for highly efficient encoding. However, in order to use the Jacobi iterative algorithm, the main diagonal elements should be non-zero. If there are zero elements in the diagonal, a certain transformation is needed beforehand. The encoding operation of LDLC can be represented as

$$x^{(t)} = \tilde{b} - \tilde{H} \cdot x^{(t-1)} \tag{7.10}$$

Set the initial codeword vector as $x^{(0)} = 0$. t is the index of the iterations. The matrix $\tilde{H}$ is obtained by elementary transformation of the parity check matrix **H**, and then divided by the largest element in each row, with the diagonal element cleared.

$$\tilde{H} = D^{-1} H_1 - I \tag{7.11}$$

where H_1 is the matrix after the elementary transformation of the parity check matrix. D^{-1} is the diagonal matrix formed by the reciprocals of the main diagonal elements of H_1. $\tilde{b} = D^{-1} \cdot b$. Since $\tilde{H}$ is sparse, the computation complexity and memory required are all in the order of $O(n)$. For the parity check matrix **H**, the following two conditions should be met so that the iterative algorithm would converge.

- Size of H should be large (e.g., $n \geq 100$) but with low degree ($d \leq 10$)
- The value of the largest element in the generation vector $h_1 = 1$ and also satisfying

$$\alpha = \sum_{i=2}^{d} {h^2}_i \Big/ h_1^2 < 1 \tag{7.12}$$

To better match the power-limited AWGN channel, shaping needs to be carried out after the encoding, with the purpose of limiting the power of the codeword to avoid the excessive power of the codeword. More specifically, the encoded codeword should be confined within a region. It is found by researchers that the optimal shaping would result in a spherical Voronoi region. However, such an optimal algorithm is too complex for practical engineering. The following super-cube-based shaping algorithm is a type of sub-optimal algorithm that can be practically implemented.

First to carry out matrix decomposition $H = TQ$ where $\boldsymbol{T}$ is a low triangle matrix and $\mathbf{Q}$ is an orthogonal matrix. Assuming $b' = b - Lk$, our purpose is to find k, so that $x' = H \cdot b'$ is restricted in a super-cube. Substitute in $H = TQ$, we get $T \cdot \tilde{x} = b'$, where $\tilde{x} = Qx'$. After getting $\tilde{x}$, the shaped codeword can be obtained via $x' = Q^T \tilde{x}$. The obtained codeword belongs to a rotated super-cube. While the shaped codeword is no longer uniformly distributed, significant shaping gain can be achieved.

Via simulations, it can be verified that the average power of the codewords is reduced after the shaping. The simulation parameters are as follows.

The codeword length is 500. The column or row weight is 5, and the generation vector is $h = \{1, \sqrt{7}, \sqrt{7}, \sqrt{7}, \sqrt{7}\}$. The information source can take three values $\{0,1,2\}$. The simulation results are shown in Figures 7.14 and 7.15. It is seen that after the shaping, the information source ranges mostly within $[-1,1]$. The values of ± 2 are quite rare. The amplitude fluctuation of the codeword before the shaping is relatively large. After the shaping, the amplitude is confined within $-L/2 \leq x \leq L/2$. Using the equation

$$P_{av} = \frac{1}{N} \sum_{i=1}^{N} \|x_i\|^2 \tag{7.13}$$

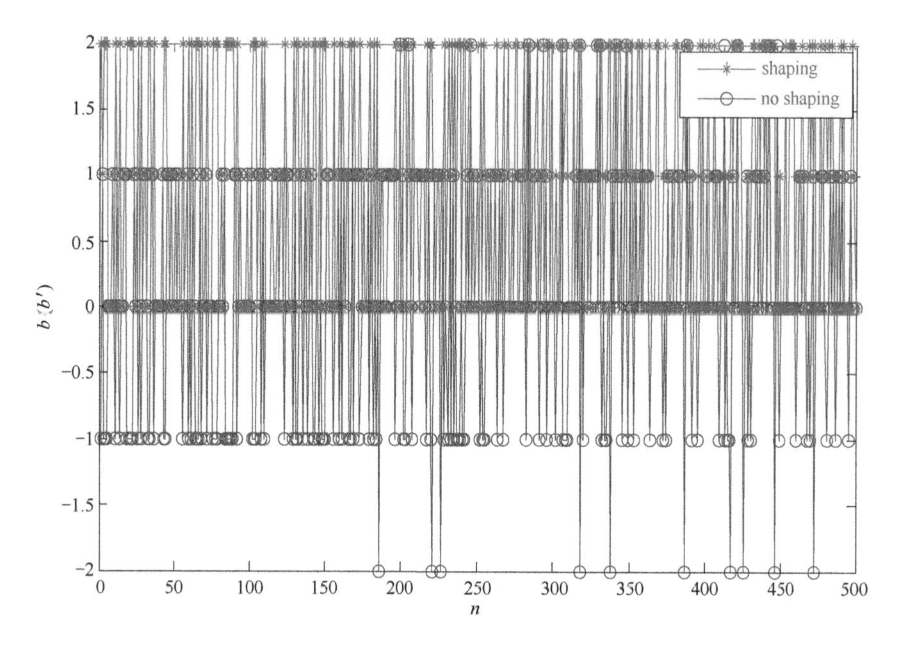

FIGURE 7.14 Values of information source before and after the shaping.

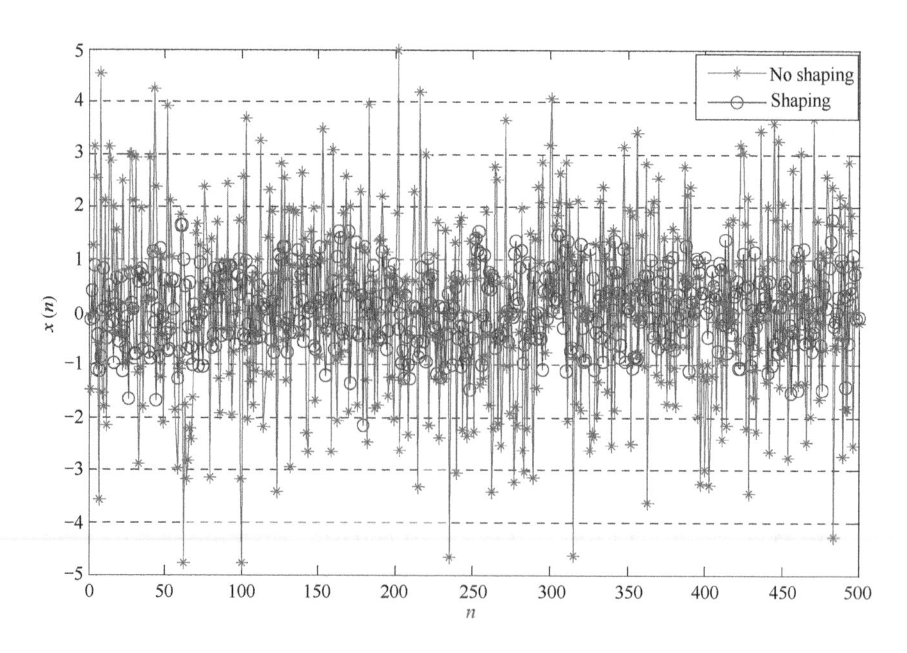

FIGURE 7.15 Values of the codeword before and after the shaping.

TABLE 7.1 Simulation Parameters of LDLC Codes

		Block Length of Information Bits	**LDLC GF(*q*)/Turbo Code Rate+QAM Order**	**Spectral Efficiency (Bit/Modulation Symbol)**
Group 1	LDLC	200/500/1200	GF(2)	2
	Turbo	200/512/1216	1/2 Turbo+16QAM	
Group 2	LDLC	200/500/1200	3	3.2
	Turbo	192/512/1152	4/5 Turbo+16QAM	
Group 3	LDLC	200/500/1200	4	4
	Turbo	200/512/1216	4/5 Turbo+32QAM	
Group 4	LDLC	200/500/1200	8	6
	Turbo	184/472/1152	7/8 Turbo+128QAM	

to calculate the average power of the codeword. Before the shaping, $P_{av}\_n=1.96\times10^3$. After the shaping, $P_{av}\_s=2.5\times10^2$.

Simulations are carried out for LDLC and turbo codes with high-order modulations, under different spectral efficiencies. Simulation parameters are listed in Table 7.1.

Simulation results and the analysis are as follows:

Group 1. Figure 7.16 shows the performance comparison between binary LDLC and turbo code of R=1/2 16QAM. The spectral efficiency is 2 bits/modulation symbol. It is observed that turbo+16QAM performs better. For the information block size of 1216 bits, 0.4 dB gain is seen for the turbo code at BER=10^{-5}, compared to 1200-dimensional LDLC. The BER curves of the turbo code are also sharper, e.g., decreasing faster as Eb/No increases.

Group 2. when the spectral efficiency is 3.2 bits/modulation symbol, performances of GF(3) LDLC and turbo+higher-order modulation are compared in Figure 7.17. It is observed that LDLC outperforms the turbo code. At BER=10^{-5}, 0.6 dB performance gain is seen for 1200-dimensional LDLC over 1216-bit Turbo+16QAM.

Group 3. when the spectral efficiency is 4 bits/modulation symbol, performances of GF(4) LDLC and turbo+higher-order modulation are compared in Figure 7.18. It is observed that LDLC outperforms the turbo code. At BER=10^{-5}, 0.8 dB performance gain is seen for 1200-dimensional LDLC over 1216-bit Turbo+32QAM.

Group 4. when the spectral efficiency is 6 bits/modulation symbol, performances of GF(8) LDLC and turbo+higher-order modulation

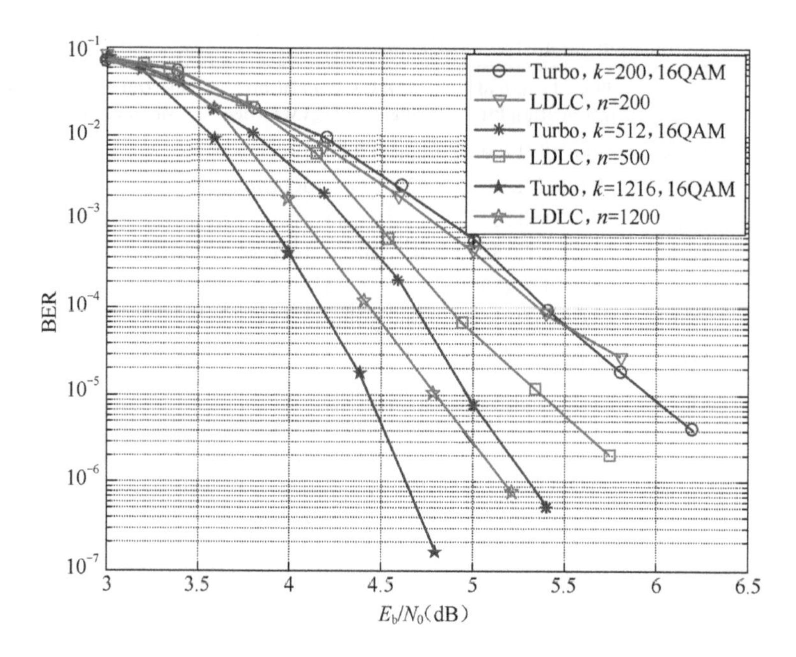

FIGURE 7.16 Performance comparison between binary LDLC and turbo+16QAM, spectral efficiency=2 bits/modulation symbol.

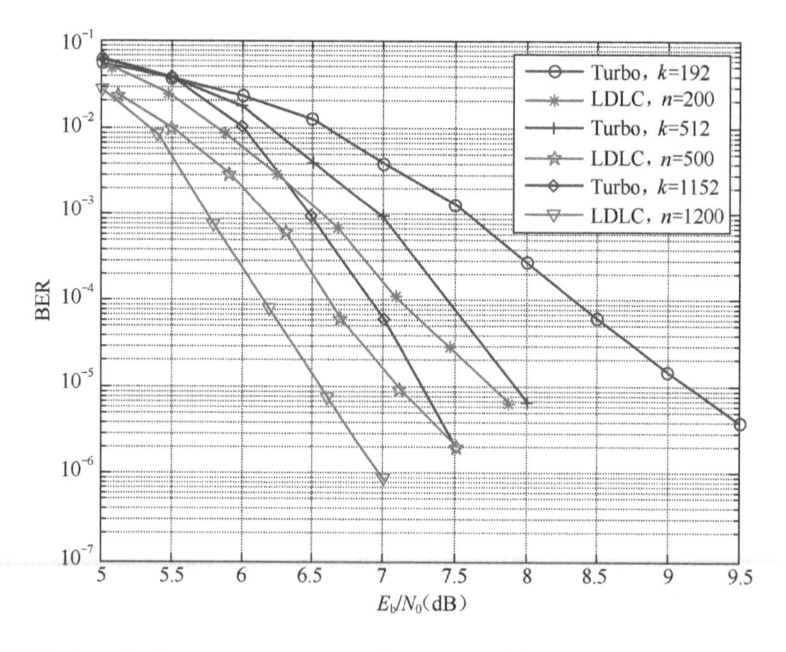

FIGURE 7.17 Performance comparison between GF(3) LDLC and turbo+16QAM, spectral efficiency=3.2 bits/modulation symbol.

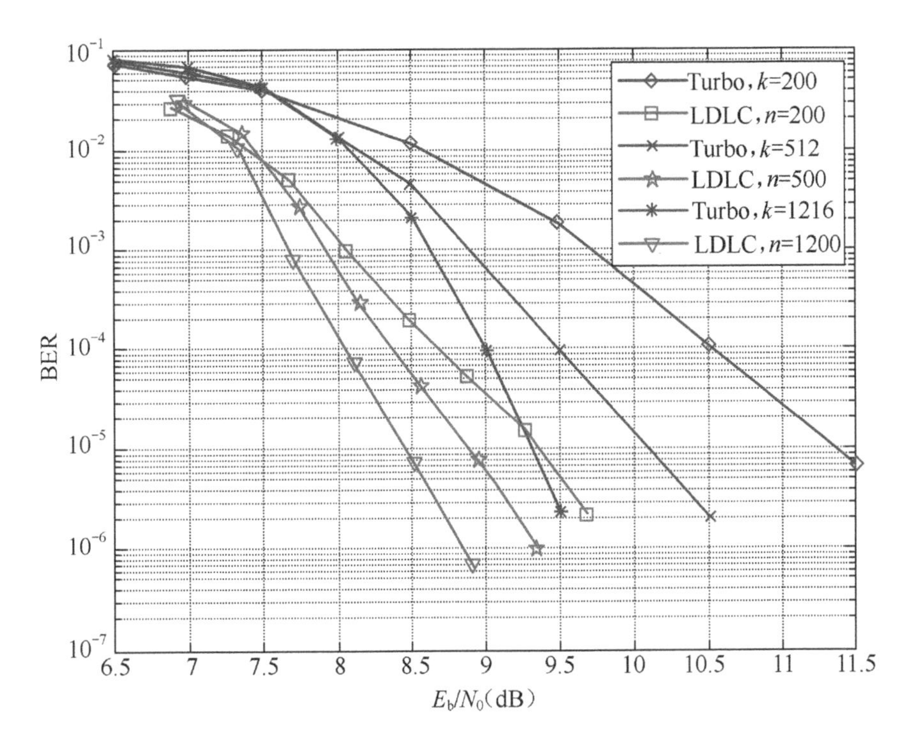

FIGURE 7.18 Performance comparison between GF(4) LDLC and turbo+32QAM, spectral efficiency=4 bits/modulation symbol.

are compared in Figure 7.19. It is observed that LDLC outperforms the turbo code. At BER=10^{-5}, 1.2 dB performance gain is seen for 1200-dimensional LDLC over 1216-bit turbo+128QAM.

The above simulations show that when the spectral efficiency is relatively low (e.g., 2 bits/modulation symbol), turbo+higher-order modulation performs better than binary LDLC. When the spectral efficiency is relatively high (e.g., 3.2 bits/modulation symbol), non-binary LDLC performs better than turbo+higher-order modulation. As the spectral efficiency increases, the performance benefit of non-binary LDLC becomes more significant.

7.4 ADAPTIVE CHANNEL CODING BASED ON RATE-LESS CODES

Spinal codes [5,14] are a type of rate-less code suitable for time-varying channels and can approach Shannon's limit. Its core idea is to continuously apply a Hash function to the input information bits, together with the Gaussian mapping function, to continuously generate pseudo-random

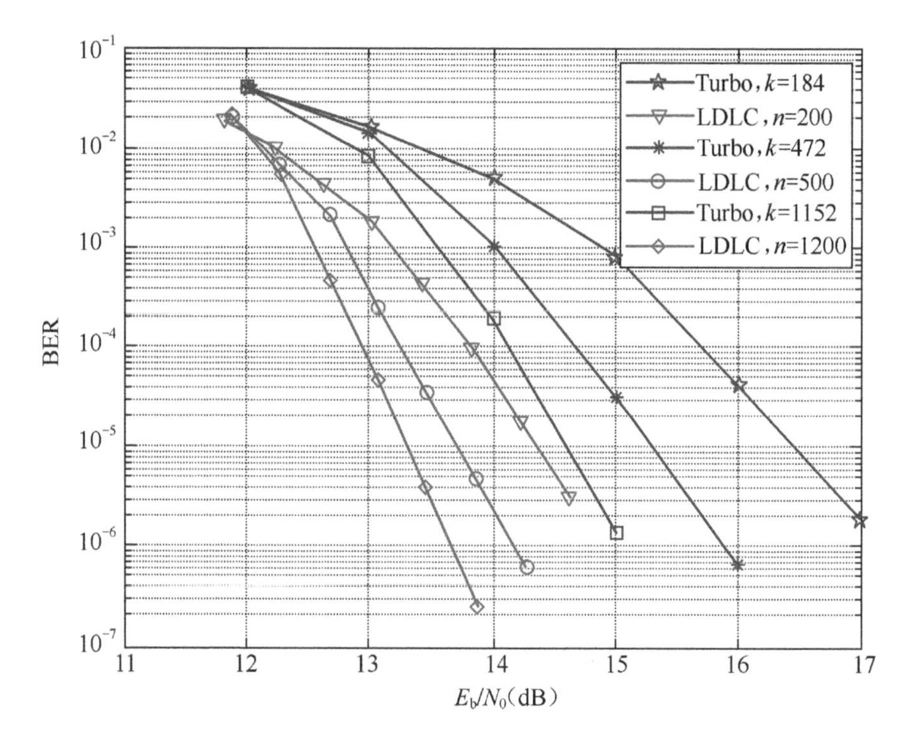

FIGURE 7.19 Performance comparison between GF(8) LDLC and turbo+128QAM, spectral efficiency=6 bits/modulation symbol.

truncated Gaussian signals. Compared to other channel coding schemes, spinal codes can approach Shannon's limit when the block length is short. The encoder of spinal codes is illustrated in Figure 7.20.

An n-bit information sequence is first segmented into n/k segments of equal length. Then starting from the initial state s_0 of the Hashing function $h:\{0,1\}^v \times \{0,1\}^k \rightarrow \{0,1\}^v$, to subsequently generate n/k v-bit Hash states. Then by using Gaussian mapping [5], e.g., $f:\{0,1\}^v \rightarrow \mathbb{X}^{v/c}$ to map every c bits into a Gaussian symbol, so that the binary Hash vector is mapped to the channel input.

The transmission of spinal codes is in a unit of passage (denoted as "pass" in Figure 7.20, circulated in the dashed block). The transmission would not stop till the decoding is successful. From the encoder structure, it is seen that spinal codes resemble convolutional codes, e.g., both of them are tree-like codes. However, since the number of states is huge, decoding algorithms based on the trellis structure (e.g., Viterbi algorithm, BCJR algorithm) cannot be used here. The first practical decoding algorithm for spinal codes is based on a greedy search along the code tree, which is called

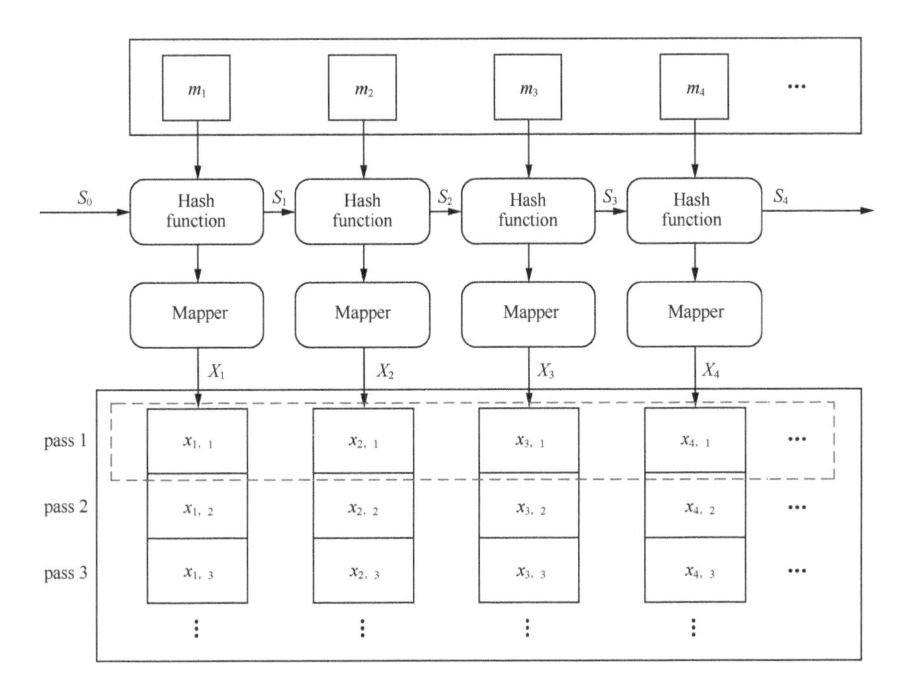

FIGURE 7.20 Illustration of the encoder of spinal codes.

a bubble decoder. In this algorithm, there would be $B2^k$ nodes to search in each layer where B is a parameter of bubble decoder. The complexity of the decoder is in the order of $O\left(nB2^k\left(\log B+k+v\right)\right)$ [5].

To reduce the decoder's complexity, forward stack decoding (FSD). The basic idea is to partition the code tree into $\left\lceil \frac{n}{kD} \right\rceil$ layers. Each layer, except the last layer, is constructed by a depth-D tree. (If n/k can be evenly divided by D, the depth of the last year is also D.) Then in each layer to carry out stack decoding in terms of ML metric and to find the optimal node of depth D. The initial decoding nodes of the next layer are composed of this optimal node together with sub-optimal depth-D nodes for other B-1 nodes of this layer.

If in a certain layer, no optimal depth-D node can be found, then put B number of D-depth nodes to the next layer. Figures 7.21 and 7.22 show the performance and complexity of spinal code using bubble decoding algorithm and FSD decoding in the AWGN channel. The length of the information is $n=48$, and the number of segments is $k=8$. The parameter of the two decoding algorithms **B** is 16. The maximum length of the stack is 6656. It is observed that at low SNR, the performance of the spinal code is closer to the Shannon's limit. However, for high SNR (e.g., higher order of

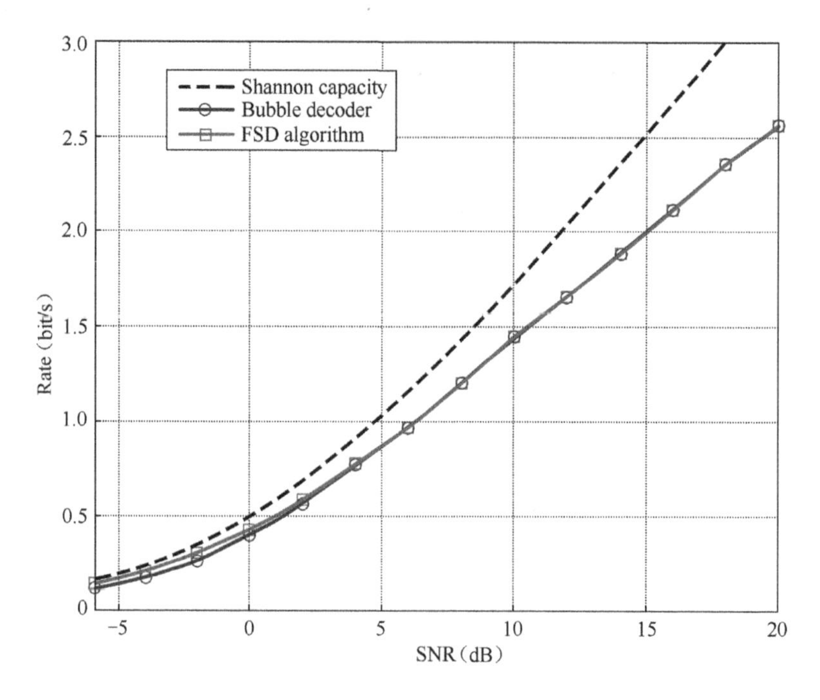

FIGURE 7.21 Performance of bubble decoding and FSD for spinal code.

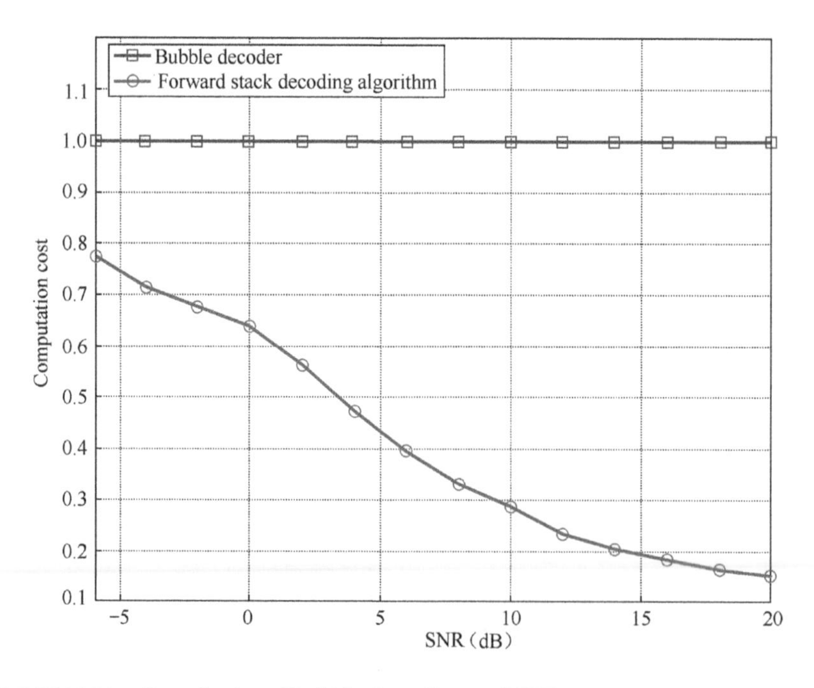

FIGURE 7.22 Complexity of bubble decoding and FSD.

modulation order), the gap to Shannon's limit widens. Similar simulation result is seen in [15].

Since the amount of computation for bubble decoding is a fixed number which is used as a normalization factor in Figure 7.22. It is found that the complexity of FSD decreases as SNR increases. At SNR=20 dB, the computation complexity of FSD is only about 15% of that of bubble decoding.

7.5 STAIRCASE CODES

Staircase codes were proposed by B.P. Smith in 2012 [16]. Since the arrangement of its codeword is similar to stairs, it is called stair codes. The motivation of staircase codes is as follows. First, for very high data rate communications such as 100 Gbps optical communications, the equipment should be of low cost. The processing has to be of low latency and high speed. Hence, iterative decoding with many iterations would not be desirable. Also, the error floor should be extremely low, e.g., 10^{-15}. Lastly, for optical communications, the code rate is extremely high, e.g., R=239/255, which requires the code to perform well at a high code rate.

7.5.1 Encoding

Considering the code blocks as shown in Figure 7.23: $B_0, B_1, B_2, ...B_i, ...$ Each code block B_i is an $m \times m$ matrix, where $i \in Z^+$ a positive integer. To simplify the discussion, the elements in B_i are assumed GF(2) (in fact, staircase codes can operate in non-binary domain). Normally, the first code block B_0 should be initialized to all 0. Each row of the two code blocks $[B_{2k}^T, B_{2k+1}]$, $k=0,1,2,3,......$ in horizontal direction is a constituent code, for example, Hamming code, BCH code, RS code, Golay code, etc. Each column of the two code blocks $[B_{2k+1}, B_{2k+2}^T]$, $k=0,1,2,3,......$ in vertical direction is a constituent code. It is noted that each code block on the left of the stair B_{2k} $k=0,1,2,3,......$ is the transposed version (e.g., $B_2^{'} = B_2^T$).

Assuming that the constituent code is of systematic form, e.g., Golay code, and considering the matrix transpose (e.g., B_2^T) as a special form of interleaving (only half of the systematic bits participate in the interleaving), staircase codes can be thought of as simplified turbo codes, or a convolutional block code (parity check bits would be continually passed down), as illustrated in Figure 7.24. Hence, staircase codes can be considered as the combination of convolutional codes and block codes.

In Figure 7.24, the input information bits are $[S_1, S_2]$. The output systematic bits are $[S_1^T, S_2]$ which are not different from the input bits, except the orders are changed. The parity bits P_1 are only related to the systematic

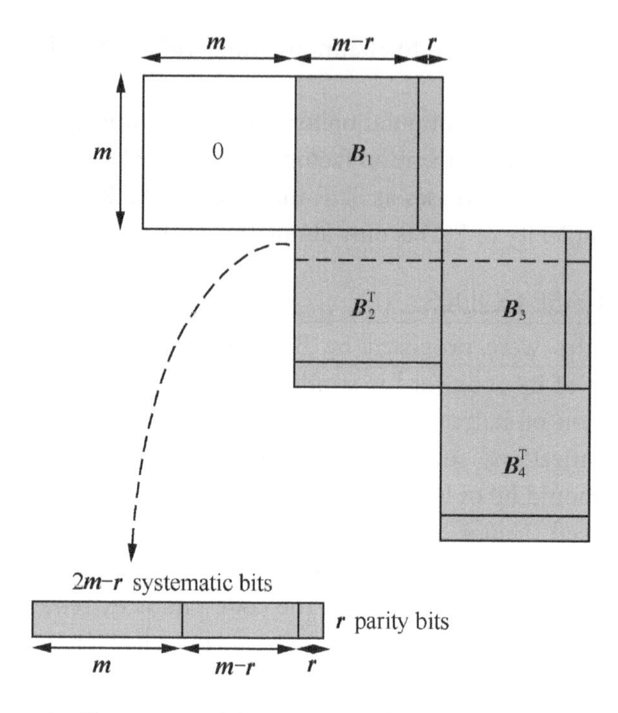

FIGURE 7.23 An illustration of the structure of staircase codes [17].

bits S_1 (as well as the information bits and parity bits of the previous stair-step). The parity bits P_2 are related to the systematic bits $S_1 \times S_2$ and the parity bits P_1. When linear block codes are used as the constituent code, the total complexity of encoding is on the order of $O(N)$, where N is the block length.

7.5.2 Decoding

Decoding of staircase codes can be based on the syndrome method which is more efficient [16] than the BP algorithm for LDPC. For the received code block pair $[B_{2k}^T, B_{2k+1}]$, $k = 0,1,2,3,\ldots\ldots$, first, carry out the decoding in the horizontal direction. By using the syndrome method, at most $(r\text{-}1)/2$ errors in this row can be corrected. Then, perform decoding in the vertical direction. Then update the syndromes in horizontal and vertical directions. Since the vertical direction parity bits of P_1 would interact with the horizontal parity bits of P_2, the horizontal parity bits of P_2 would interact with the vertical parity bits in P_1 of the next stair level, the decoding process and the update of the syndrome are a cyclic iterative process. This process will continue until it reaches the maximum number of iterations.

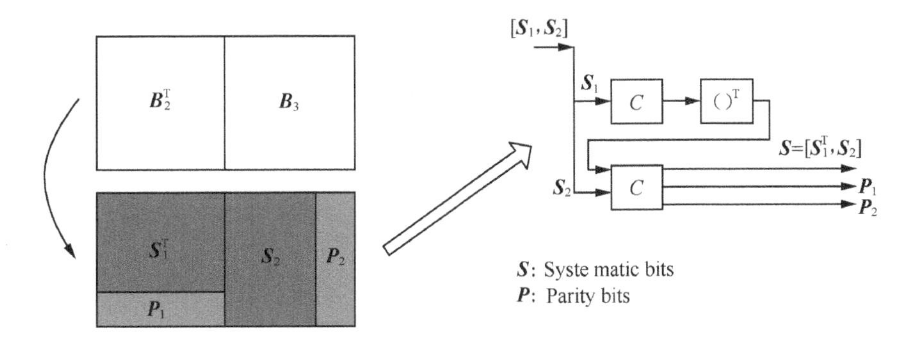

FIGURE 7.24 Staircase code resembling turbo codes.

When linear block codes are used for the constituent code, the total complexity of decoding is on the order of $O(I \bullet N + I \bullet N) = O(2 \bullet I \bullet N)$ where "2" refers to decoding and updating the syndrome. I denotes the number of iterations. N is the code block size. Considering that the decoding latency for high-speed optical communications should be small, the number of iterations is often set to ~15.

7.5.3 Performance

The BER performance of a staircase code is shown in Figure 7.25. It is seen that at the code rate R=239/255 and BER=10^{-15}, its performance can approach Shannon's limit within 0.56 dB and is better than RS code (255, 239) by about 3.5 dB. No error floor is observed at ER=10^{-15} for staircase code.

7.5.4 Future Direction

Staircase codes are a type of relatively simple channel coding scheme used for optical fiber communications with good performance. In [18], the application of staircase codes to high-speed wireless communications was discussed. The possible future directions are

- New arrangement (interleaving), for instance, each stair level consists of three code blocks (two are overlapped)
- Other constituent codes, e.g., extended RM, LDPC [19]
- New way of bit collection to further improve the code rate
- New decoding schemes, e.g., iterative decoding for each constituent code

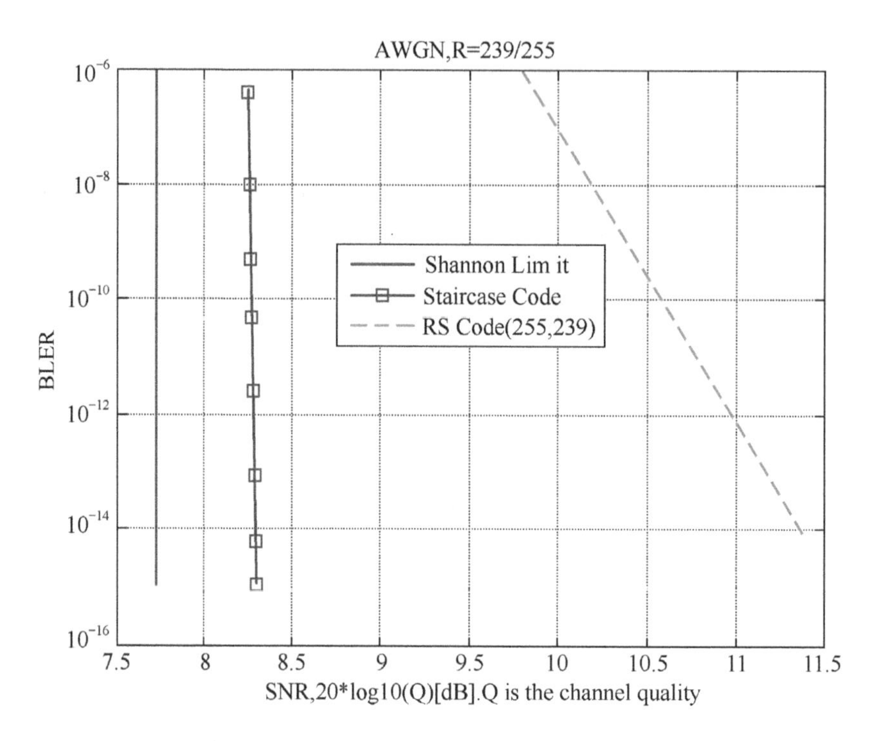

FIGURE 7.25 Performance of staircase code [16].

In summary, some advanced channel coding schemes (e.g., non-binary LDPC) have excellent performance. Certainly, their complexities are higher. As the hardware becomes more powerful, these codes may find more of their use in the future.

REFERENCES

[1] M.C. Davey, "Low-density parity-check codes over GF(q)," *IEEE Commun. Lett.*, vol. 2, no. 6, June 1998, pp. 165–167.

[2] D. J. C. Mackay, *Evaluation of Gallager codes of short block length and high rate applications*, Springer, New York, 2001, pp. 113–130.

[3] G. Tu, *Belief propagation decoding algorithm for repetition accumulation codes*, Master thesis, Xidian University, 2014.

[4] N. Sommer, "Low-density lattice codes," *IEEE Trans. Inf. Theory*, 2008, 54 (4), pp. 1561–1585.

[5] J. Perry, "Spinal codes," *ACM SIGCOMM Conference on Applications*, 2012, pp. 49–60.

[6] X. Ma, "Low complexity X-EMS algorithms for nonbinary LDPC codes," *IEEE Trans. Commun.*, vol. 60, no. 1, Jan 2012, pp. 9–13.

[7] W. Lin, *Non-binary LDPC—design, construction and decoding*, Ph. D thesis, Xidian University, April 2012.

[8] B. Qian, *Study on low-density lattice code (LDLC)*, Master thesis, Beijing Jiaotong University, March 2-13.
[9] D. J. Forney, "Coset codes. II. Binary lattices and related codes," *IEEE Trans. Inf. Theory*, vol. 34, no. 5, 1988, pp. 1152–1187.
[10] O. Shalvi, "Signal Codes," *Proc. Inf. Theory Workshop*, 2003, pp. 332–336.
[11] R. Urbanke, "Lattice codes can achieve capacity on the AWGN channel," *IEEE Trans. Inf. Theory*, vol. 44, no. 1, 1998, pp. 273–278.
[12] U. Erez, "Achieving 1/2 log (1+SNR) on the AWGN channel with lattice encoding and decoding," *IEEE Trans. Inf. Theory*, vol. 50, no. 10, 2004, pp. 2293–2314.
[13] X. Wang, *Key technologies of modulation and coding for next generation communications system*, Ph. D thesis, Northeast University (China), July 2009.
[14] J. Perry, "Rateless spinal codes," *Proc. of 10th ACM Workshop on Hot Topics in Networks - HotNets '11*, 2011, pp. 1–6.
[15] J. Li, *Decoding algorithm and complexity study for rate-less spinal codes*, Master thesis, Xidian University, November 2015.
[16] B. P. Smith, "Staircase codes—FEC for 100 Gb/s OTN," *IEEE/OSA J. Lightwave Technol.*, vol. 30, no. 1, Jan. 2012, pp. 110–117.
[17] A. Sheikh, "Probabilistically-Shaped Coded Modulation with Hard Decision Decoding and Staircase Codes," submitted to IEEE Trans. on Information Theory.
[18] P. Kukieattikool, "Staircase codes for high-rate wireless transmission on burst-error channels," *IEEE Wireless Commun. Lett.*, vol. 5, no. 2, April 2016, pp. 128–131.
[19] IETF, "Simple Low-Density Parity Check (LDPC) Staircase—Forward Error Correction (FEC) Scheme for FECFRAME," Internet Engineering Task Force, Request for Comments—6816, Category—Standards Track, ISSN—2070-1721.